SEMICONDUCTORS AND SEMIMETALS

VOLUME 21

Hydrogenated Amorphous Silicon

Part B

Optical Properties

Semiconductors and Semimetals

A Treatise

Edited by R. K. WILLARDSON

WILLARDSON CONSULTING
SPOKANE, WASHINGTON

ALBERT C. BEER

BATTELLE COLUMBUS LABORATORIES
COLUMBUS, OHIO

SEMICONDUCTORS
AND SEMIMETALS

VOLUME 21
Hydrogenated Amorphous Silicon
Part B
Optical Properties

Volume Editor
JACQUES I. PANKOVE

RCA/DAVID SARNOFF RESEARCH CENTER
PRINCETON, NEW JERSEY

1984

ACADEMIC PRESS, INC.
(Harcourt Brace Jovanovich, Publishers)

Orlando   San Diego   San Francisco   New York   London
Toronto   Montreal   Sydney   Tokyo   São Paulo

O 94022

ACADEMIC PRESS, INC.
Orlando, Florida 32887

United Kingdom Edition published by
ACADEMIC PRESS, INC. (LONDON) LTD.
24/28 Oval Road, London NW1 7DX

Library of Congress Cataloging in Publication Data
(Revised for volume 21B-21D)
Main entry under title:

Semiconductors and semimetals

Includes bibliographical references and indexes.
Contents: v. 1-2. Physics of III-V compounds --v. 3.
Optical properties of III-V compounds--   --v. 21,
pt. B, C, D. Hydrogenated amorphous silicon.
1. Semiconductors--Collected works.   2. Semimetals--
Collected works.   I. Willardson, Robert K.   II. Beer,
Albert C.
QC610.9.S47   537.6'22   65-26048
ISBN 0-12-752148-8 (21B)

Contents

Chapter 4   The Vibrational Spectra of a-Si : H

P. J. Zanzucchi

Chapter 5   Electroreflectance and Electroabsorption

Yoshihiro Hamakawa

Chapter 6   Raman Scattering of Amorphous Si, Ge, and Their Alloys

Jeffrey S. Lannin

Chapter 7   Luminescence in a-Si : H

R. A. Street

Chapter 8   Photoconductivity

Richard S. Crandall

Chapter 9   Time-Resolved Spectroscopy of Electronic Relaxation Processes

J. Tauc

Chapter 10   IR-Induced Quenching and Enhancement of Photoconductivity and Photoluminescence

P. E. Vanier

Chapter 11   **Irradiation-Induced Metastable Effects**

H. Schade

Chapter 12   **Photoelectron Emission Studies**

L. Ley

List of Contributors

Numbers in parentheses indicate the pages on which the authors' contributions begin.

NABIL M. AMER, *Lawrence Berkeley Laboratory, University of California, Berkeley, California 94720* (83)

G. D. CODY,* *Corporate Research Science Laboratories, Exxon Research and Engineering Company, Linden, New Jersey 07036* (11)

RICHARD S. CRANDALL, *RCA/David Sarnoff Research Center, Princeton, New Jersey 08540* (245)

YOSHIHIRO HAMAKAWA, *Faculty of Engineering Science, Osaka University, Toyonaka, Osaka, Japan* (141)

WARREN B. JACKSON, *Xerox Corporation, Palo Alto Research Center, Palo Alto, California 94304* (83)

JEFFREY S. LANNIN, *Department of Physics, Pennsylvania State University, University Park, Pennsylvania 16802* (159)

L. LEY, *Max-Planck-Institut für Festkörperforschung, Stuttgart, Federal Republic of Germany* (385)

JACQUES I. PANKOVE, *RCA/David Sarnoff Research Center, Princeton, New Jersey 08540* (1)

H. SCHADE, *RCA/David Sarnoff Research Center, Princeton, New Jersey 08540* (359)

R. A. STREET, *Xerox Corporation, Palo Alto Research Center, Palo Alto, California 94304* (197)

J. TAUC, *Division of Engineering and Department of Physics, Brown University, Providence, Rhode Island 02912* (299)

P. E. VANIER, *Metallurgy and Materials Science Division, Brookhaven National Laboratory, Upton, New York 11973* (329)

P. J. ZANZUCCHI, *RCA/David Sarnoff Research Center, Princeton, New Jersey 08540* (113)

* Present address: Corporate Research Science Laboratories, Exxon Research and Engineering Company, Annandale, New Jersey 08801.

Foreword

This book represents a departure from the usual format of "Semiconductors and Semimetals" because it is a part of a four-volume miniseries devoted entirely to hydrogenated amorphous silicon (a-Si : H). In addition, this group of books — Parts A – D of Volume 21 — has been organized by a guest editor, Dr. J. I. Pankove, an internationally recognized authority on this subject. He has assembled most of the who's who in this field as authors of the many chapters. It is especially fortunate that Dr. Pankove, who has made important original contributions to our understanding of a-Si : H, has been able to devote the time and effort necessary to produce this valuable addition to our series. In the past decade, a-Si : H has developed into an important family of semiconductors. In hydrogenated amorphous silicon alloys with germanium, the energy gap decreases with increasing germanium content, while in alloys with increasing carbon content the energy gap increases. Although many applications are still under development, efficient solar cells for calculators have been commercial for some time.

In Volume 21, Part A, the preparation of a-Si : H by rf and dc glow discharges, sputtering, ion-cluster beam, CVD, and homo-CVD techniques is discussed along with the characteristics of the silane plasma and the resultant atomic and electronic structure and characteristics.

The optical properties of this new family of semiconductors are the subject of Volume 21, Part B. Phenomena discussed include the absorption edge, defect states, vibrational spectra, electroreflectance and electroabsorption, Raman scattering, luminescence, photoconductivity, photoemission, relaxation processes, and metastable effects.

Volume 21, Part C, is concerned with electronic and transport properties, including investigative techniques employing field effect, capacitance and deep level transient spectroscopy, nuclear and optically detected magnetic resonance, and electron spin resonance. Parameters and phenomena considered include electron densities, carrier mobilities and diffusion lengths, densities of states, surface effects, and the Staebler – Wronski effect.

The last volume of this miniseries, 21, Part D, covers device applications, including solar cells, electrophotography, image pickup tubes, field effect transistors (FETs) and FET-addressed liquid crystal display panels, solid state image sensors, charge-coupled devices, optical recording, visible light emitting diodes, fast modulators and detectors, hybrid structures, and memory switching.

R. K. WILLARDSON
ALBERT C. BEER

Preface

Hydrogenated amorphous silicon, a new form of a common element, is a semiconductor that has come of age. Its scientific attractions include a continuously adjustable band gap, a usable carrier lifetime and diffusion length, efficient optical transitions, and the capability of employing either n- or p-type dopants.

Furthermore, it can be fabricated very easily as a thin film by a technology that not only inherently escapes the expense of crystal perfection but also requires significantly smaller amounts of raw materials.

The discovery of a new material endowed with wondrous possibilities for very economical practical applications naturally attracts many researchers who invariably provide new insights and further vision. Their mediation and experimentation build up rapidly and lead to a prolific information flow in journals and conference proceedings.

The initial cross-fertilization generates an overload of data; books are written that attempt to digest specialized aspects of the field with state-of-the-art knowledge that often becomes obsolete by the time the books are published a year or two later.

We have attempted to provide this book with a lasting quality by emphasizing tutorial aspects. The newcomer to this field will not only learn about the properties of hydrogenated amorphous silicon but also how and why they are measured.

In most chapters, a brief historical review depicts the evolution of relevant concepts. The state of the art emerges, and a bridge to future developments guides the reader toward what still needs to be done. The abundant references should be a valuable resource for the future specialist.

We hope that this tutorial approach by seasoned experts satisfies the needs of at least one generation of new researchers.

CHAPTER 1

Introduction

Jacques I. Pankove

RCA/DAVID SARNOFF RESEARCH CENTER
PRINCETON, NEW JERSEY

In this introductory chapter the contents of this volume are reviewed, and some new information that is relevant to the topic of optical properties of a-Si : H is introduced. The authors of this volume have been chosen for their eminent contributions in the various approaches that deepen our insights into this material. Each author was asked to provide a tutorial presentation of his technique, to review the current knowledge, and to suggest what more needs to be done.

The most readily interpretable properties of a semiconductor come from optical absorption measurements. The photon energy probes pairs of states that can be coupled optically and defines their energy separation, while the intensity of the measured interaction gives the product of the number of such states and their coupling coefficient (or transition probability). This is why the first few chapters are devoted to absorption in a-Si : H.

In Chapter 2 Cody shows how one can distinguish between transitions in crystalline and amorphous semiconductors. In the latter, there is no requirement of momentum conservation. Hence, if parabolic densities of states are assumed for valence and conduction bands, one would expect that for photon energy hv greater than that of the energy gap E_g the absorption coefficient would vary as $(1/hv)(hv - E_g)^2$, an empirical relation first derived by Tauc to determine the energy gap E_g. All amorphous semiconductors seem to obey this Tauc relation. The absorption coefficient data used to determine the Tauc gap are larger than 10^4 cm^{-1}. At lower values of absorption coefficient, one finds an exponential dependence, the Urbach edge, that has been associated with the perturbation of band edges. This perturbation is enhanced in a-Si : H by thermal vibrations; hence the slope of the absorption edge is temperature dependent. The slope also is affected by structural changes, such as those that occur when hydrogen is evolved. Cody uses the concept of "focus point," where all the exponential edges for a given material converge, establishing what might be called an unperturbed, zero-temperature band gap (different from Tauc's band gap).

Alloys of a-Si : H with various tetrahedrally bonding elements have energy

1

gaps that can be adjusted by the alloy composition. The absorption edge correspondingly shifts with composition. Thus, in a-Si_xC_{1-x}: H, the energy gap shifts to higher energies (Sussmann and Ogden, 1981), whereas in a-Si_xGe_{1-x}: H the opposite is found (Nakamura *et al.*, 1982). Recent studies of the optical absorption edge in periodic layered structures or "superlattices" consisting of alternating layers of a-Si: H and a-Si: N_x: H (Abeles and Tiedje, 1983) and in single-well a-$Si_{0.2}C_{0.8}$: H/a-Si: H/a-$Si_{0.2}C_{0.8}$: H heterostructures (Munekata and Kukimoto, 1983) show that the band gap E_g increases and the exponential band tail broadens with decreasing a-Si: H layer thickness. For instance, for a layer thickness of 8 Å the band gap is 2.2 eV compared to the bulk value of 1.7 eV. The experimental results have been interpreted as quantum size effects resulting from the two dimensional confinement of the electrons and holes in the a-Si: H layers.

Below the exponential absorption edge, there are transitions that involve deep levels. These deep-level-to-tail state transitions have such low absorption coefficients that their study requires very sensitive techniques such as photothermal deflection spectroscopy (PDS); this is discussed by Amer and Jackson in Chapter 3. In PDS, the heat generated by the absorbed radiation causes a gradient in the refractive index of the transparent medium contacting the sample. The resulting gradient in refractive index deflects a probing laser beam that grazes the sample. This fine technique detects absorption coefficients as low as 10^{-2} cm^{-1}. Using PDS, Amer and Jackson find that the absorption between 0.8 and 1.3 eV is proportional to the number of dangling bonds determined by electron spin resonance. Photothermal deflection spectroscopy also reveals that prolonged exposure to light causes an increase in this characteristic absorption that can be interpreted as an increase in the number of dangling bonds.

At still lower photon energies, one enters the important realm of photon – phonon interaction. This allows the probing of atomic vibrational modes, in particular, those involving hydrogen. In Chapter 4 Zanzucchi describes how one can distinguish among SiH, SiH_2, SiH_3, $(SiH_2)_n$, and also among SiF, SiO, etc. This spectral range employs Fourier transform spectroscopy as well as standard infrared transmission. The sensitivity can be greatly increased by the technique of multiple internal reflection spectroscopy.

Modulation spectroscopy is a derivative technique that emphasizes the most rapid changes in a spectrum, changes that signal the onset of an important transition. There are many methods of generating a modulation, e.g., fluctuation of the wavelength, temperature, or pressure. The method described by Hamakawa in Chapter 5, as a tool to study changes in the absorption edge, uses a modulation of the electric field at the surface of the a-Si: H film. The detected signal is a derivative of the reflectance spectrum or the absorption edge, revealing singularities or structure in the spectrum.

The resulting characteristic structure can be identified with the optical gap or with other critical parameters of a semiconductor. Thus, one can follow a widening of the energy gap as the hydrogen concentration increases, see changes that depend on the substrate temperature or power in the discharge during deposition, or see changes when other elements (such as carbon) are added to a-Si:H. Hamakawa introduces a new technique, back surface reflection electroabsorption spectroscopy, that utilizes a modulation of the electric field at the rear surface of the a-Si:H film while the light traverses the film twice. This yields information in the spectral region where the film is transparent. Therefore, electroreflectance and electroabsorption are complementary differential techniques, the first probing at high photon energies and the second at lower energies.

With reflectance spectroscopy one can probe a-Si:H over a broader range of photon energies, e.g., 1–10 eV, and apply a Kramers–Kronig analysis to the data to extract the imaginary part of the dielectric function—the part that depicts the absorption processes. Thus, Ewald *et al.* (1979) find that the ε_2 peak of a-Si:H is intense and occurs at 3.55 eV, whereas in a-Si it is weaker and occurs at 3.05 eV. These spectra do not show the fine structure of crystalline Si, which has a main peak at ~4.2 eV. The position and intensity of the ε_2 peak for a-Si:H depend on the deposition temperature and on subsequent heat treatment.

Reflected light is usually elliptically polarized. If one analyzes the two polarization components of the reflected light, one obtains a very sensitive probe of surface properties. Ellipsometry can detect and characterize the properties of films on surfaces with even less than monolayer coverage (Aspnes, 1982). This technique can be used to determine the film thickness and its refractive index, to study the oxidation of a surface, or to follow changes after heat treatment. It is a powerful tool for monitoring in situ the deposition of a-Si:H (Perrin and Brevillon, 1981–1982).

The structure of a solid can be explored conveniently by Raman scattering. In this technique, discussed by Lannin in Chapter 6, one measures the change in the spectrum of a monochromatic light after it is scattered by the sample. The spectrum of the scattered light bears the imprint of phonon modes allowed by the solid. Thus, it gives insight into the order or disorder of the material. Features in the Raman scattering spectrum are primarily a function of short-range order involving changes in bonding distances, bond angles, coordination number, and electron states. Usually, the Raman spectrum shows peaks corresponding to transverse acoustic (TA) and transverse optical (TO) phonon resonances. The linewidth of the features in the Raman scattering spectrum is a measure of disorder. An inverse correlation is found between the disorder and the Tauc band gap. The addition of hydrogen to a-Si increases the intensity of the peaks in the Raman scattering spectrum and reduces their linewidth. Hence, the incorporation of hydrogen appears to assist structural order.

Another technique for probing the phonon spectra of a-Si : H is the use of neutron scattering (Kostorz, 1979). Although there is no extensive discussion of neutron scattering in our volume, it is worth bearing in mind that two methods are available: (1) The effect of phonon scattering on the energy of neutrons can be measured as a change in propagation direction or (2) the effect of the phonon interaction on the time of flight of the neutrons can be measured. In the latter case, the difference in travel times for a pulse of neutrons with and without sample is related to the energy lost (or gained) by the neutrons in the scattering event. Since the neutron scattering cross section for hydrogen is very large, it would be interesting to follow this approach in greater detail, and this would be an opportunity to study the isotope shift in the presence of deuterium.

Luminescence, as a source of information about optical transitions, complements what is learned from absorption measurements. Whereas absorption couples occupied states at lower energies with empty states at higher energies, in luminescence the emission spectrum represents the convolution of filled upper states and empty lower states. The disorder of a-Si : H relaxes momentum conservation requirements and makes the transitions more probable, as evidenced by the efficient low-temperature emission peaking in the 1.2 – 1.4 eV range. From the efficiency of the process one can learn what the competing nonradiative transitions might be. In Chapter 7 Street explores in great depth the aspects of luminescence in a-Si : H: e.g., the trapped carriers recombining by radiative tunneling between localized states separated by an average distance on the order of 50 Å and the carriers thermalizing to traps that are distributed in energy. Street shows how one can distinguish between geminate and nongeminate recombination and how phonon interaction leads to a strong Stokes shift. The nonradiative processes may also involve tunneling; in some cases Auger processes may dominate the recombination; and of course surface recombination is always present and can be important in the thinnest films and in those films that have internal voids. Then, there is the remarkable temperature dependence of the luminescence efficiency on the temperature appearing in the numerator of the exponent (a linear so-called inverse Arrhenius plot) — a dependence that had been first observed for the threshold current of injection lasers. The luminescence efficiency also decreases in the presence of dangling bonds, thus showing a negative correlation with the number of characteristic spins. Another emission peak (usually weaker) appears at 0.8 to 0.9 eV and depends on defects, such as dangling bonds, following either dehydrogenation, doping, or exposure to radiation (see Chapter 11). Time-resolved spectroscopy (TRS) can follow the decay of luminescence over a broad distribution of decay times, from 10 nsec to 10 msec. Time-resolved spectroscopy demonstrates the tunneling-assisted transitions between electrons and holes located in distant traps or between donors and acceptors, sometimes revealing their Coulomb interaction. Street also reviews observations made in the presence of electric and magnetic fields.

Recent time-resolved photoluminescence measurements by Wilson *et al.* (1983) show that the recombination lifetime is about 8 nsec at 15°K in a great variety of a-Si:H samples having different quantum efficiencies. This early luminescence is attributed to a localized exciton of ~2 Å extent. At room temperature, the recombination lifetime drops by a factor of five and the efficiency by a factor of 10. The temperature dependence of the initial luminescence efficiency follows an inverse Arrhenius dependence with a characteristic temperature of 100°K (about four times larger than in the CW case).

The photoluminescence spectra of a-Si:H superlattices (Abeles and Tiedje, 1983) or single-well heterostructures (Munekata and Kukimoto, 1983) exhibit a relatively broad peak around 1.4 eV, characteristic of bulk a-Si:H films. The major difference from bulk a-Si:H, observed by Abeles and Tiedje, is that the exponential temperature dependence of the luminescence efficiency is much weaker. From the optical and photoluminescent results the authors conclude that the distribution of localized states broadens as the layer thickness decreases.

Brodsky (1980) shows that the observed characteristic energies of photoluminescence, optical absorption, and activated conductivity can be explained by the inherent inhomogeneity of the a-Si:H system. In his quantum well model the potential barriers around Si–H bonds are assumed to be repulsive to band-edge holes and electrons. Confinement of these carriers in Si-rich regions with at least one dimension as small as approximately 15 Å results in spatially overlapping quantum wells for electrons and holes. Typical well barriers of 0.55 (0.2) eV for holes (electrons) give quantized well states about 0.25 (0.1) eV removed from the 1.2 eV band gap of Si. The 2.0 eV barrier separation is identified as the mobility gap, the 1.7–1.8 eV local-to-extended threshold is the Tauc optical edge, and the 1.4 eV photoluminescence arises from recombination via quantum well states. The model assumes the strength and 4–5 Å range of the Si–H repulsions, the approximate retention of crystalline Si band structure in the Si-rich regions, and the role of Si–H bonds as the principal source of disorder for the breakdown of optical selection rules. These assumptions were examined and supported in subsequent papers (DiVincenzo *et al.*, 1981, 1983; Brodsky and DiVincenzo, 1983). Brodsky points out the analogy between the Anderson–Mott valence and conduction localized states caused by disorder, in this case random Si–H bonds in the Si host, and the spatially correlated quantum well states.

Photoconductivity involves two phenomena: one optical, photon absorption, and the other electronic, charge transport. In fact, the photocurrent can be used to monitor light absorption, a-Si : H being an excellent photodetector material. However, because the transport process is very complex, the interpretation of photoconductivity can be very difficult. In Chapter 8 Crandall treats this problem, pointing out the difference between primary and secondary photocurrents. In the primary photocurrent, one measures the motion of photogenerated carriers, because blocking contacts prevent the influx of external charges. In the secondary photocurrent, the contacts are ohmic (or forward biased) so that the photogenerated carriers modulate the conductivity of the photoconductor and, depending on the relative trapping time and transit time, considerable gain can be obtained. Thus, traps play a crucial role in secondary photocurrents. Recombination centers, especially dangling bonds, are also important, because they determine the survival of photogenerated carriers. Further treatment of electronic properties that affect transport will be found in Volume 21C.

Time-resolved photoinduced absorption spectroscopy is discussed by Tauc in Chapter 9. This technique monitors the relaxation processes a short time after optical excitation. Although fast detectors are available with response time close to 1 nsec, to monitor events in the time range of 0.5 psec to 1.5 nsec, sophisticated correlation techniques must be used. Part of the exciting pulse is delayed by a known time interval to probe the change in absorption induced by the carriers generated by the excitation pulse. The presence of hot carriers that can be excited to still higher energies causes an increase in absorption, whereas thermalized carriers that fill the bands (thereby reducing the number of available initial and final states) cause an increased transmission. Both trapping and geminate recombination effects can be seen in photoinduced absorption.

When several levels are involved in optical transitions, it is possible to perturb the kinetics of these transitions by changing the population of intermediate states. The independent direct address of these intermediate states can be achieved by optical pumping with photons of energy lower than that of those producing the primary process. In Chapter 10 Vanier discusses several experiments in which photoconductivity and photoluminescence are modulated by an auxiliary beam of lower energy infrared photons. In photoconductivity, the primary excitation is from valence to conduction band. The number of carriers available for photoconductive transport can either be increased or decreased by frustrating or enhancing the recombination process via optical pumping to or from recombination states. Similarly, in photoluminescence, although the optical transition is from states in the conduction-band tail to states in the valence-band tail, some recombination occurs via dangling bonds or other nonradiative recombination centers. Hence, here also changing the population of these competing centers will affect the luminescence. Vanier explores many effects, including the kinetics of various intermediate transitions, by varying the modulation frequency of the auxiliary infrared beam.

Metastable states can be induced in a-Si:H by irradiation with light or x rays, by bombardment with electrons or ions, or by injection of electron–hole pairs. These metastable states appear to result from electron–hole pair recombination, because, in the light-induced case, when an electric field is simultaneously applied to sweep away the photogenerated carriers, the metastable states do not form (Staebler *et al.*, 1981). Metastable states persist after the irradiation has been turned off and the resulting changes in electrical and optical properties can be monitored: the dark conductivity increases, the photoluminescence efficiency at ~1.3 eV decreases, a new luminescence peak appears at ~0.8 eV, and the spin resonance of dangling bonds increases. All these changes can be reversed by annealing at ~200°C, and some of the changes are removed by applying an electric field in the dark (Lang *et al.*, 1982; Swartz, 1984). In Chapter 11 Schade reviews most of these observations and discusses various hypotheses explaining these phenomena. Schade also reviews observed changes in the infrared vibrational spectra suggesting changes in hydrogen bonding after irradiation. Of all the models advanced to explain the metastable states, we are partial to the breaking of weak Si–Si bonds resulting from the energy released by an electron–hole recombination. High incident energy also can break weak bonds directly. The weak bonds are those between Si atoms that are stressed by distance or by gross angular deformation. The removal of an electron from a stressed bond leaves two dangling bonds that may reconstruct upon annealing. These metastable dangling bonds act as traps and nonradiative recombination centers that lower the dark conductivity and the luminescence efficiency and increase the characteristic spin resonance. Materials that have few defects, such as those made by the homogeneous chemical vapor deposition (homoCVD) technique (see Volume 21A, Chapter 7) or those prepared in the presence of fluorine, which etches weakly bonded Si during the deposition (Dalal *et al.*, 1981), exhibit no metastable effects (Madan *et al.*, 1979).

Han and Fritzsche (1983) found that light exposure produces two kinds of defects, one responsible for a drop in the mobility–lifetime product $\mu\tau$ and another one responsible for sub-band-gap absorption. The metastable states that reduce $\mu\tau$ can be created as efficiently at low temperature as at room temperature, whereas the metastable states that increase absorption at ~1 eV seem to be temperature activated. The two defects anneal at slightly different temperatures. Thus, by controlling the temperature during irradiation, it is possible to create and saturate the $\mu\tau$-killing defect first and then to produce, at a higher temperature, the 1 eV absorbing centers. Later on, it is possible to anneal the $\mu\tau$-affecting centers before annealing the 1 eV absorbing centers. The resulting plot of $\mu\tau$ versus absorption at 1 eV forms a roughly rectangular hysteresis loop demonstrating the separate accessibility of the two types of defects. A dual-beam photoconductivity experiment (infrared quenching or enhancement) supports the hypothesized independence of the two metastable states.

Zhang *et al.* (1983) studied light-induced changes in infrared absorption in sputtered a-Si:H. They found a consistent increase in the absorption for the three dominant vibrational modes at 600 cm^{-1}, 850 cm^{-1}, and 2000 cm^{-1}. There was no change in the spectrum, but the intensity increased after illumination and decreased back to the original value upon annealing. Their model is that there are many infrared-inactive hydrogen atoms possibly clustered in voids. When weak Si–Si bonds are broken, the extra H moves in to tie the broken bonds, thus increasing the IR activity. Upon annealing the same H moves back.

Photoluminescence and its fatigue have been observed also in siloxene, Si:H:O with the composition 6:6:3 (Hirabayashi *et al.,* 1983). This material is a disordered layered compound having an optical gap of 2.9 eV and emitting at about 2.5 eV. Luminescence fatigue is very large, causing a drop in luminescence of one order of magnitude under UV irradiation. The luminescence efficiency recovers after annealing at near room temperature. The spin density of dangling bonds increases during irradiation, maintains its increased value in the dark, then reverts to the original value upon annealing. Another interesting observation is that the luminescence efficiency of this material decreases with increasing temperature according to an inverse Arrhenius dependence, proportional to $\exp(-T/T_0)$, where $T_0 \sim 120°$K. There is no evidence of Stokes shift in this material.

High-energy photons, such as ultraviolet or x rays, excite electrons from the depths of the valence band and from core levels to levels high in the conduction band, far above the vacuum level. These hot electrons, and those resulting from the subsequent Auger effect (within a depth of less than ~ 100 Å from the surface), escape into vacuum, where their energy can be analyzed. In Chapter 12 Ley shows how this information is acquired, and how one can deduce valuable insights from photoelectric emission spectroscopy (varying the photon energy and measuring the energy distribution of emitted electrons). Thus, one can learn the distribution of states throughout the valence band and high into the conduction band, and one can watch the downward motion of the upper edge of the valence band as the hydrogen content is increased. One can identify collective oscillations called plasmons, and one can follow the shift of the Fermi level as a function of doping. Ley points out that it is possible to monitor small shifts of the core levels caused by changes in bonding energy when Si bonds to H, O, F, etc. Indeed, photoelectric emission spectroscopy appears to be one of the most powerful tools for studying solids and especially their surfaces.

Aoyagi *et al.* (1981) have developed an optical technique for determining simultaneously the lifetime and diffusion coefficient of carrier pairs in amorphous materials. In this technique, a pulsed laser beam is split into two beams that cross over the sample and produce by interference a gratinglike pattern of photogenerated carriers. A grating is in fact produced because the excited carriers induce a change in refractive index. The line spacing of this grating is adjusted by the angle between the two split beams. The pattern decays with the characteristic lifetime τ of the carriers and smears by carrier diffusion. A probing beam is deflected by this grating and then disappears with a single exponential decay with a time constant $\mathcal{T}$. Different values of $\mathcal{T}$ are obtained when the grating pitch Λ is changed. From the relationship $1/\mathcal{T} = 1/\tau + 4\pi^2 D/\Lambda$ between the decay time of the probe signal and the grating pitch Λ, one can determine the carrier diffusion coefficient D and the lifetime τ. A plot of $1/\mathcal{T}$ versus $4\pi^2/\Lambda^2$ yields an intercept $1/\tau$ at $1/\Lambda^2 = 0$ and the slope gives the value of D. By varying the wavelength of the pulsed laser one can generate the grating at various depths in the sample and thereby characterize the material as a function of depth. Thus, it was found that in a solar cell structure consisting of a-SiC:H on a-Si:H on a substrate, the values of D and τ decreased in the following order: bulk > a-Si:H surface > a-Si:H/substrate interface > a-SiC:H/a-Si:H interface (Komuro *et al.,* 1983b). When this technique was applied to the study of the Staebler–Wronski effect by Komuro *et al.* (1983a), it was found that the Staebler–Wronski effect is a bulk and not a surface effect and that the effect is attributable to a decrease in carrier lifetime by one order of magnitude instead of a change in diffusion coefficient or mobility.

REFERENCES

Abeles, B., and Tiedje, T. (1983). *Phys. Rev. Lett.* **51**, 2003.
Aoyagi, Y., Segawa, Y., Namba, S., Sahara, T., Nishihara, H., and Gamo, H. (1981). *Phys. Status Solidi B* **67**, 669.
Aspnes, D. E. (1982). *Thin Solid Films* **89**, 249.
Brodsky, M. H. (1980). *Solid State Commun.* **36**, 55.
Brodsky, M. H. (1981). *In* "Fundamental Physics of Amorphous Semiconductors" (F. Yonezawa, ed.), p. 56. Springer-Verlag, Berlin and New York.
Brodsky, M. H., and DiVincenzo, D. P. (1983). *Physica* **117B/118B**, 971.
Dalal, V. L., Fortmann, C. M., and Eser, E. (1981). *AIP Conf. Proc.* **73**, 15.
DiVincenzo, D. P., Bernholc, J., and Brodsky, M. H. (1981). *J. Phys. Colloq. Orsay, Fr.* **42**, C4-137.
DiVincenzo, D. P., Bernholc, J., and Brodsky, M. H. (1983). *Phys. Rev. B* **28**, 3246.
Ewald, E., Milleville, M., and Weiser, G. (1979). *Philos. Mag. B* **40**, 291.
Han, D., and Fritzsche, H. (1983). *J. Non-Cryst. Solids* **59 & 60**, 649.

Hirabayashi, I., Morigaki, K., and Yamanaka, S. (1983). *J. Phys. Soc. Jpn.* **52,** 671.

Komuro, S., Aoyagi, Y., Segawa, Y., Namba, S., Masuyama, A., Okamoto, H., and Hamakawa, Y. (1983a). *Appl. Phys. Lett.* **42,** 807.

Komuro, S., Aoyagi, Y., Segawa, Y., Namba, S., Masuyama, A., Okamoto, H., and Hamakawa, Y. (1983b). *J. Non-Cryst. Solids* **59 & 60,** 501.

Kostorz, G. (ed.) (1979). "Treatise on Materials Science and Technology," Vol. 15. Academic Press, New York.

Lang, D. V., Cohen, J. D., Harbison, J. P., and Sergent, A. M. (1982). *Appl. Phys. Lett.* **40,** 474.

Madan, A., Ovshinsky, S. R., and Benn, E. (1979). *Philos. Mag. B* **40,** 259.

Munekata, H., and Kukimoto, H. (1983). *Jpn. J. Appl. Phys.* **22,** L544.

Nakamura, G., Sato, K., and Yukimoto, Y. (1982). *Proc. Conf. IEEE Photovoltaic Specialists, 17th, 82 CH1821-8,* 1331.

Perrin, J., and Drevillon, B. (1981–1982). *Acta Electron.* **24,** 239.

Staebler, D. L., Crandall, R., and Williams, R. (1981). *Appl. Phys. Lett.* **39,** 733.

Sussmann, R. S., and Ogden, R. (1981). *Philos. Mag. B* **44,** 137.

Swartz, G. A. (1984). To be published.

Wilson, B. A., Hu, P., Harbison, J. P., and Jedju, T. M. (1983). *Phys. Rev. Lett.* **50,** 1490.

Zhang Pei-Xian, Tan Cui-Ling, Zhu Quiong-Rui, and Peng Shao-qi (1983). *J. Non-Cryst. Solids* **59 & 60,** 501.

CHAPTER 2

The Optical Absorption Edge of a-Si:H

G. D. Cody†

CORPORATE RESEARCH SCIENCE LABORATORIES
EXXON RESEARCH AND ENGINEERING COMPANY
LINDEN, NEW JERSEY

I. Introduction

1. THE OPTICAL ABSORPTION EDGE OF CRYSTALLINE SEMICONDUCTORS

The optical absorption edge of a crystalline semiconductor can be easily defined experimentally. The sharp rise in the optical absorption at a characteristic energy suggests the concept of an optical gap E_G, which, of course, emerges directly from fundamental calculations of the allowed energy levels

† Present address: Corporate Research Science Laboratories, Exxon Research and Engineering Company, Annandale, New Jersey.

11

of a semiconductor. Precise measurements of the absorption edge have led to the discovery of new phenomena, e.g., the spectra of excitons, as well as the coupling, in the optical absorption process, between the excited electrons and holes and the characteristic vibrations of the lattice. Indeed, the examination of the shape and position of the absorption edge as a function of temperature, pressure, site disorder, and alloying continues to be of interest in solid-state physics since it furnishes data that are the basis for testing band-structure calculations of increasing complexity (Ehrenreich and Hass, 1983).

Experimental and theoretical examination of the absorption edge of crystalline semiconductors led to the distinction between two kinds of optical transitions between the filled valence band and the unoccupied conduction band (Pankove, 1971; Wooten, 1972). The "direct transition" is one in which conservation of energy and momentum are met only by considering the incident photon energy E and the initial and final energy levels of the excited electron. Many crystalline semiconductors exhibit direct optical transitions, e.g., c-GaAs. There is, however, an equally numerous class of crystalline semiconductors that exhibit "indirect" optical transitions, e.g., c-Si. In these materials, excitation to the *lowest* energy electronic state is forbidden by the failure of conservation of crystal momentum. In a second-order process specific phonons participate in the optical transition and slightly modify the energy of the transition by the characteristic energy of the emitted or absorbed phonon. The added phonon momentum meets the overall requirements for crystal momentum conservation imposed by the spatial symmetry of the wave functions of the crystalline semiconductor (Wooten, 1972, p. 134). For both transitions the edge is sharp.

2. The Optical Absorption Edge of Amorphous Semiconductors

In contrast to crystalline semiconductors, the optical absorption edge of amorphous semiconductors is difficult to define experimentally. There is often a wide variation in the sharpness of the edge for the same material when subjected to annealing, as shown in Fig. 1 for a-Ge (Thèye, 1971). If we associate the effect of annealing with the removal of disorder, the absorption edge of amorphous semiconductors can sharpen considerably with "ordering" (Brodsky *et al.*, 1972; Brodsky, 1971). A further complication in defining the absorption edge is that materials synthesized under different conditions often exhibit large differences in their "as-grown" optical properties, as shown in Fig. 2 (Tsai and Fritzsche, 1979).

Figure 2 illustrates the dependence of the optical absorption on sample preparation for a specific amorphous semiconductor, a-Si : H_x. Despite this sensitivity, there remains a remarkable agreement among the shapes of the

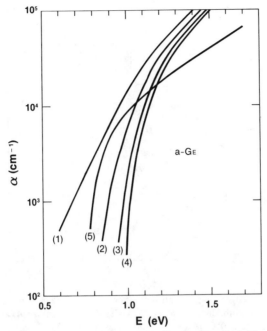

FIG. 1. Optical absorption of evaporated a-Ge film as deposited (1); and annealed at $T_A = 200°C$ (2); $300°C$ (3); $400°C$ (4); and for a recrystallized film (5). [After Thèye (1971). Adapted with permission from *Materials Research Bulletin,* Vol. 6, M. L. Thèye, Influence of annealing on the optical properties of amorphous germanium films, Copyright 1971, Pergamon Press, Ltd.]

optical absorption edges of different amorphous semiconductors. In Fig. 3, we show the exponential character of the absorption edges for several amorphous semiconductors (Mott and Davis, 1979, p. 279). In Fig. 4, we show the optical absorption edges for the well characterized materials a-Se and a-Si:H_x. Apart from the shift in the edge, there is a remarkable similarity in the shapes of the absorption edges of these materials. As generally described by Tauc (1970) and illustrated in Fig. 5, each material exhibits a shoulder in the optical absorption (region A) for $\alpha \lesssim 1$ cm^{-1}, an exponential rise (region B) for 1 cm$^{-1} \lesssim \alpha \lesssim 10^3$ cm^{-1}, and a slowly varying regime (region C) for $\alpha \gtrsim 10^4$ cm^{-1}.

We note from Figs. 3 and 4 that under specific preparation conditions many amorphous semiconductors exhibit absorption edges (regions B and C) that are remarkably similar in shape but considerably broader than the threshold absorption exhibited by either direct or indirect optical absorption edges in crystalline semiconductors. There is no pronounced feature in the absorption edge which can be directly related to an *optical gap.*

The data of Figs. 3 and 4 do suggest that the *differences* among the optical gaps of amorphous semiconductors may be easily obtained from a compari-

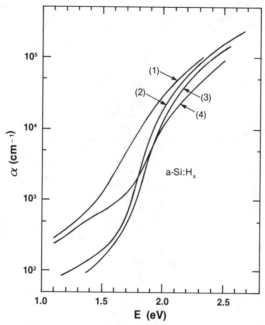

FIG. 2. Absorption edge of a-Si : H$_x$ films at different substrate temperatures and electrodes. Curves (1) and (3) are cathodic films at $T_S = 25°C$ and 270°C; curves (4) and (2) are anodic films at $T_S = 25°C$ and 270°C. [After Tsai and Fritzsche (1979).]

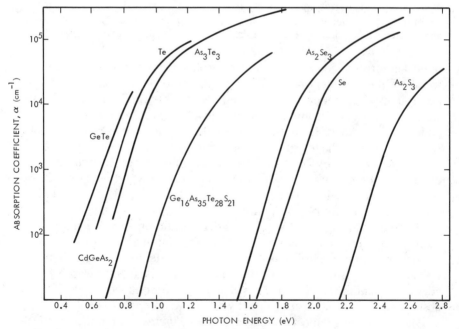

FIG. 3. Absorption edge of amorphous semiconductors. [After Mott and Davis (1979).]

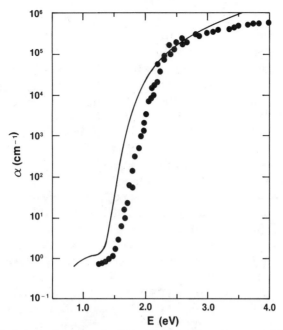

FIG. 4. Optical absorption edge of a-Se [solid circles, from Lanyon (1963)] and a-Si:H$_x$ (solid curve).

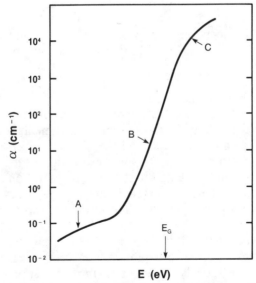

FIG. 5. Idealized absorption edge of amorphous semiconductors. [After Tauc (1970). Adapted with permission from *Materials Research Bulletin,* Vol. 5, J. Tauc, Absorption edge and internal electric fields in amorphous semiconductors, Copyright 1970, Pergamon Press, Ltd.]

son of the absorption edges for different materials , indeed more easily than the magnitude of the optical gap may be obtained from the absorption edge for one material. This observation is a dominant theme of experiments on the optical absorption edge for amorphous semiconductors. Changes in the inferred optical gap due to pressure, temperature, annealing, or preparation conditions may be more easily understood than would be the magnitude of the optical gap itself.

The absorption edge of amorphous silicon hydride (a-Si:H_x) is of great practical interest because of the importance of this amorphous semiconductor as the active material of thin film solar cells. For a given film thickness D and specular reflection, the quantity $A(E) = 1 - \exp[-2\alpha(E)D]$ is the maximum optical absorption at energy E. The position and shape of the absorption edge thus determines the maximum short-circuit current that can be obtained (Cody et al., 1980; Wronski et al., 1981) without optical enhancement (Yablonovitch and Cody, 1982).

The optical absorption edge of a-Si:H_x is also important for our understanding of optical absorption in amorphous semiconductors. One consequence of the worldwide interest in thin-film solar cells fabricated from a-Si:H_x is a plentiful supply of well characterized high purity thin films. Such films can be described as semiconductor grade, since device measurements show that they are homogeneous with electrically active defects at levels at or below 10^{16} cm^{-3} (Wronski et al., 1982). Films made by low-power glow-discharge decomposition of SiH_4 on substrates held at $250-300°$C exhibit no inhomogeneity and can have densities as high as 96% c-Si (Ruppert et al., 1981). Careful optical measurements made on such materials may be the most promising experimental starting point for a deep and quantitative understanding of the nature of the forbidden gap in amorphous semiconductors.

Remarkable advances have been achieved in our understanding of the transport properties of a-Si:H_x through such fundamental concepts as the mobility gap (Cohen et al., 1969) and the distinction between traps and free-carrier states (Rose, 1978). Despite these advances, the magnitude of the forbidden gap and its relation to the optical absorption edge remain a mystery. Until this mystery is resolved, advances in electronic and solar applications that require control of the forbidden gap will be seriously impeded.

In subsequent sections of this chapter, we shall briefly review the experimental techniques required to explore the absorption edge and broadly review the theoretical models that have been introduced to explain the shape and position of the absorption edge. With this theoretical framework as a guide, we shall review the experimental data on the absorption edge with respect to its dependence on preparation conditions, annealing, measurement temperature, and impurities. Our focus will be on regions B and C of Fig. 5. We shall be concerned about region A only as it contributes to the position and shape of the edge. Region A is the subject of Chapter 3 by Amer and Jackson.

II. Experimental Techniques

3. THIN FILM OPTICS

Thin films often exhibit inhomogeneities, for example "voids" in thick films, "islands" in the case of ultrathin films and in general, structure on a scale less than the wavelength of light. This section discusses some commonly used models which permit an experimentalist to estimate the effect of known or potential inhomogeneities on the measured complex dielectric constant, $\tilde{\varepsilon}(E)$. The experimentalist can then decide whether certain features of $\tilde{\varepsilon}(E)$ are either characteristic of a dominant phase, or strongly dependent on the presence of small amounts of impurities. In some cases, these models can be used to extract from the measured $\tilde{\varepsilon}(E)$ of an inhomogeneous material the dielectric function of a particular phase of interest.

The problem of constructing an average macroscopic dielectric function from the spatially varying dielectric function of the inhomogeneous material is an old one. It is currently being studied under the general title of effective medium theory (Landauer, 1978). Aspnes (1982) has recently reviewed the subject in the context of thin film optics.

Effective medium theory approaches the problem of averaging the dielectric properties of an inhomogeneous medium through a self-consistent average of the dipole fields. As summarized by Aspnes, we can define for a two-phase system of dielectric constant $\tilde{\varepsilon}_A(\tilde{\varepsilon}_B)$ and volume fraction $f_A(f_B)$ an average dielectric constant $\tilde{\varepsilon}$ in terms of a fictitious "host" dielectric constant $\tilde{\varepsilon}_H$ in which the dipoles contributing to $\tilde{\varepsilon}_A$ and $\tilde{\varepsilon}_B$ are embedded. Under these circumstances, within the dipole approximation,

$$\frac{\tilde{\varepsilon} - \tilde{\varepsilon}_H}{\tilde{\varepsilon} + 2\tilde{\varepsilon}_H} = f_A \frac{\tilde{\varepsilon}_A - \tilde{\varepsilon}_H}{\tilde{\varepsilon}_A + 2\tilde{\varepsilon}_H} + f_B \frac{\tilde{\varepsilon}_B - \tilde{\varepsilon}_H}{\tilde{\varepsilon}_B + 2\tilde{\varepsilon}_H}. \tag{1}$$

The Maxwell–Garnett approximation is to identify the B phase as an isolated phase entirely surrounded by A. Under these circumstances, $\tilde{\varepsilon}_H \equiv \tilde{\varepsilon}_A$ and the dielectric constant of the medium is given by the Maxwell–Garnett equation

$$\frac{\tilde{\varepsilon} - \tilde{\varepsilon}_A}{\tilde{\varepsilon} + 2\tilde{\varepsilon}_A} = f_B \frac{\tilde{\varepsilon}_B - \tilde{\varepsilon}_A}{\tilde{\varepsilon}_B + 2\tilde{\varepsilon}_A}. \tag{2}$$

An example in which Eq. (2) would be appropriate is the case for which phase B consists of a low percentage of isolated "voids" ($\tilde{\varepsilon}_B = 1$) in an otherwise homogeneous material of dielectric constant $\tilde{\varepsilon}_A$.

A solution that treats each phase in a similar manner and that is also easily generalized to multicomponent systems is the self-consistent effective-medium approximation (Bruggeman approximation), in which we define a self-consistent dielectric function $\tilde{\varepsilon}$ and set, in Eq. (1), $\tilde{\varepsilon}_H \equiv \tilde{\varepsilon}$. Under these circumstances, the dielectric constant is given by the Bruggeman formula

$$ f_A \frac{\tilde{\varepsilon}_A - \tilde{\varepsilon}}{\tilde{\varepsilon}_A + 2\tilde{\varepsilon}} + f_B \frac{\tilde{\varepsilon}_B - \tilde{\varepsilon}}{\tilde{\varepsilon}_B + 2\tilde{\varepsilon}} + \cdots = 0. \tag{3} $$

Equation (3) has been used with great success by Bagley et al. (1982) to obtain quantitative measures of the volume fraction of c-Si, a-Si, and voids in a-Si films prepared by chemical vapor deposition (CVD) and annealed above the solid-state recrystallization temperature. In this case, the presence of spectral signatures in the distinct phases gives confidence that the excellent agreement between Eq. (3) and experiment is more than fortuitous. Figure 6 exhibits the fit to the experimental spectra (imaginary part, ε_2, of $\tilde{\varepsilon}$) for a CVD film of polysilicon after laser annealing. Here the c-Si phase is readily identified by the double peak structure in ε_2, the a-Si by the broad peak in ε_2 below the double peak structure, and the presence of voids by the low value of ε_2 at the peak compared to either pure c-Si or pure a-Si. For comparison Fig. 7 exhibits the ε_2 and ε_1 spectra for the pure elements (Bagley

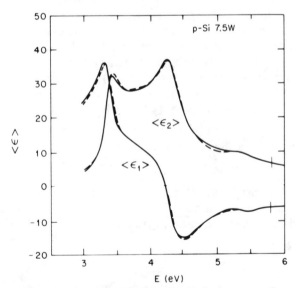

FIG. 6. Experimental ($\langle\tilde{\varepsilon}\rangle$) spectra for polycrystalline Si after laser annealing. —————, experimental; ————— 0.04 void + 0.88 c-Si + 0.08 a-Si. [From Bagley et al. (1982). Reprinted by permission of the publisher from "Laser and Electron Beam Interactions with Solids" by B. R. Appleton and G. K. Celler, eds., published by Elsevier Science Publ. Co., Copyright 1982.]

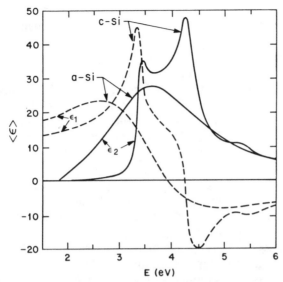

FIG. 7. Experimental ($\langle\tilde{\varepsilon}\rangle$) spectra for c-Si and a-Si (low-pressure CVD at a substrate temperature of 625 °C). [From Bagley *et al.* (1982). Reprinted by permission of the publisher from "Laser and Electron Beam Interactions with Solids" by B. R. Appleton and G. K. Celler, eds., published by Elsevier Science Publ. Co., Copyright 1982.]

et al., 1982). It is important to note that the solution of Eq. (3) for the effective-medium dielectric constant $\tilde{\varepsilon}$ requires knowledge of both the real *and* imaginary parts of $\tilde{\varepsilon}$ for the pure components.

The effective-medium theory in either its Bruggeman or Maxwell–Garnett form is an important tool for estimating the effects of uniformly distributed inhomogeneities on the measured optical properties. Given additional information in the form of spectral signatures, it can be used as a quantitative probe of the composition of a thin film specimen.

Interference fringes are not easily avoided in transmission measurements since attempts to reduce coherency by roughening the substrate, for example, may introduce scattering and other losses. For absorption measurements (photoconductivity or photoacoustic spectroscopy), roughening of the substrate can virtually eliminate fringes (Tiedje *et al.*, 1983a).

It is possible to combine reflectivity and transmission measurements on films of different thickness (Cody *et al.*, 1980) and obtain the real and imaginary parts of the dielectric constant for a-Si : H_x up to about 6 eV. This technique becomes insensitive in the region where $n \approx \kappa$. The method also suffers, as does ellipsometry, from the difficulty of preparing surfaces with high specular reflectivity and minimal surface oxide layers. It requires an additional assumption that films of different thickness are the same. An inhomogeneous region close to the substrate for example, would play an increasingly dominant role as film thickness is reduced. Ellipsometry is also sensitive to similar assumptions but is limited to the free surface of the film. For both techniques, judgment and sufficient measurements on a variety of films are required to obtain the magnitude and energy dependence of the dielectric constant.

In Figs. 8 and 9, we show data on n and κ for a-Si:H_x films made over a varying period of time under conditions of low and high contamination by water vapor and air. Figures 10 and 11 show similar data on ε_1 and ε_2. Finally, in Fig. 12, we compare ε_2 for a-Si:H_x ($x \approx 0.09$) with ε_2 for the high-structural-quality (high-density and homogeneity) low-pressure CVD (LPCVD) film of negligible hydrogen content shown in Fig. 7. We shall discuss the significance of the difference in Section 8c. In Fig. 13, we compare ε_1 data for the same two films.

In Fig. 14 we show on a semilog plot the optical edge as determined by optical absorption and photoconductivity for a variety of a-Si:H_x films made under nominally similar conditions. Above $\alpha \approx 3 \times 10^2$ cm^{-1} the measurements were made by transmission measurements which overlapped photoconductivity measurements made below $\alpha \approx 6 \times 10^2$ cm^{-1}. The photoconductivity measurements were done under conditions that ensured that the same recombination kinetics were maintained over the spectral

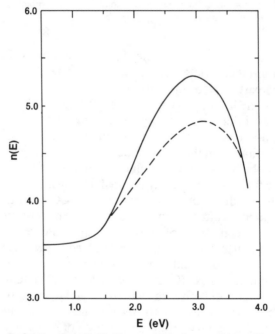

FIG. 8. Index of refraction for a-Si:H_x films grown under conditions of low (solid line) and high (dashed line) contamination by air and water vapor.

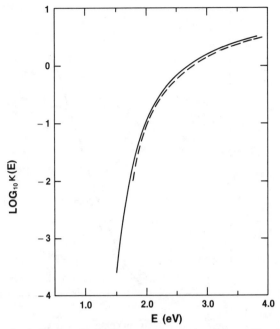

FIG. 9. Absorption index for a-Si : H_x films grown under conditions of low (solid line) and high (dashed line) contamination by air and water vapor.

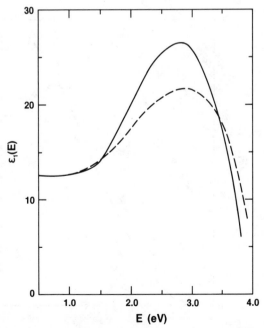

FIG. 10. Real part of the dielectric constant of a-Si : H_x grown under conditions of low (solid line) and high (dashed line) contamination by air and water vapor.

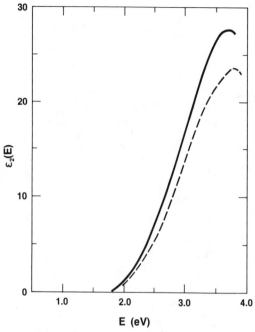

FIG. 11. Imaginary part of the dielectric constant of a-Si : H_x grown under conditions of low (solid line) and high (dashed line) contamination by air and water vapor.

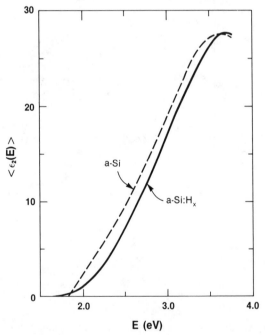

FIG. 12. Imaginary part of the dielectric constant of a-Si : H_x compared to that of a-Si [LPCVD at 625°C (Bagley *et al.*, 1982)].

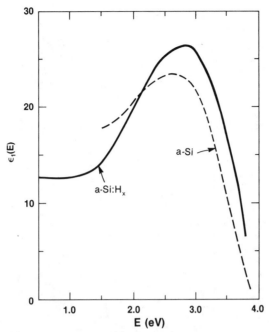

FIG. 13. Real part of the dielectric constant of a-Si : H$_x$ compared to that of a-Si [LPCVD at 625°C (Bagley *et al.*, 1982)].

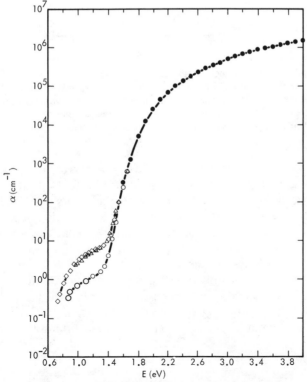

FIG. 14. Optical absorption coefficient determined by photoconductivity and transmission for a-Si : H$_x$ films made under same deposition conditions.

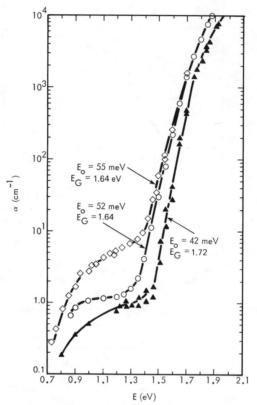

Fig. 15. Optical absorption coefficient measured by photoconductivity (Wronski, 1983) and transmission for several GD a-Si:H_x films. The notation E_G indicates the optical gap; E_0 indicates the slope of the exponential edge. The film with the highest optical gap was grown at a substrate temperature of 200°C compared with 250°C for the other films. It has a hydrogen concentration of about 14 at. % compared with about 9 at. % for the other films.

range. Above $\alpha \approx 10^3$ cm^{-1}, the points in Fig. 14 represent average data for 16 films. The absorption curve of high-density a-Si:H_x thus exhibits the characteristic behavior of amorphous semiconductors described by Tauc and others: a shoulder regime which is sample dependent for $\alpha \lesssim 1-10$ cm^{-1}, an exponential regime for $1-10$ cm$^{-1} \lesssim \alpha < 10^3$ cm^{-1}, and a slowly varying regime for $\alpha \gtrsim 10^4$ cm^{-1}.

In Fig. 15, in addition to the previous data, we show α for a sample of a-Si:H_x grown under slightly different conditions (substrate temperatures lower by 50°C and a hydrogen concentration about 50% higher). From Fig. 15, we note the sharpening of the edge that accompanies the shift in absorption to the blue as well as an apparent reduction in the "shoulder."

III. Theoretical Models for the Optical Absorption Edge of Amorphous Semiconductors

4. QUANTUM MECHANICS OF OPTICAL ABSORPTION

Throughout this chapter, we focus on the imaginary part of the dielectric constant $\varepsilon_2(E)$. The real part $\varepsilon_1(E)$ can be determined by Kramers–Kronig analysis from any theoretical model for ε_2. It is, however, of less experimental interest because its spectral variation, unlike that of ε_2, is not directly related to the density of states within the forbidden gap of the amorphous semiconductor. The theoretical models we shall consider for ε_2 permit us to make quantitative estimates of the density of states in the vicinity of the mobility edge.

Within the one-electron approximation, it is easily shown that

$$\varepsilon_2(E) = \left(\frac{2\pi e\hbar^2}{m}\right)^2 \left(\frac{2}{VE^2}\right) \sum |P_{o,u}|^2 \delta(E_o - E_u - E). \tag{4}$$

In Eq. (4), E is the energy of the incident light wave and V is the illuminated volume of the sample. The sum is over the occupied and unoccupied single-particle (one-spin) electronic states, which are normalized to unity over the volume V. The factor of 2 accounts for spin conservation in the transition and spin degeneracy in the occupation.

In the following, we consider $T = 0°K$, and define the occupied states by $\psi_v(\mathbf{r})$, a valence-band wave function, and the unoccupied states by $\psi_c(\mathbf{r})$, a conduction-band wave function. The matrix element $|P_{o,u}|^2$ can then be written as $|P_{v,c}|^2$, where

$$P_{v,c} = \int \psi_v^*(\mathbf{r}) \left(\frac{\tau \cdot \mathbf{P}}{\hbar}\right) \psi_c(\mathbf{r}) \, d\mathbf{r}, \tag{5}$$

where τ is the polarization vector of the light wave and $\mathbf{P}$ is the momentum operator. From the general relation between the momentum operator $\mathbf{P}$ and its conjugate coordinate $\mathbf{R}$, we have

$$\mathbf{P} = (im/\hbar)[H, \mathbf{R}]. \tag{6}$$

We can rewrite Eq. (4) in the simpler but equivalent form within the one-electron approximation as

$$\varepsilon_2(E) = (2\pi e)^2 \left(\frac{2}{V}\right) \sum_{v,c} |R_{v,c}|^2 \delta(E_c - E_v - E), \tag{7}$$

where

$$R_{v,c} = \int \psi_v^*(\mathbf{r})(\tau \cdot \mathbf{R})\psi_c(\mathbf{r}) \, d\mathbf{r}. \tag{8}$$

We shall use Eq. (7) to define ε_2 in this chapter. While there is no difference in principle between Eq. (7) and Eq. (4), there is experimental evidence for a-Si:H_x that the important and simplifying approximation of an energy-independent matrix element makes more physical sense when applied to Eq. (8) rather than to Eq. (5) (Cody et al., 1982).

If we multiply Eq. (7) by the energy E and integrate, we obtain

$$\int E\varepsilon_2(E)\, dE = (2\pi e)^2 \left(\frac{2}{V}\right) \sum_{v,c} |R_{v,c}|^2 (E_c - E_v). \qquad (9)$$

The summed quantity defines the oscillator strength $f_{v,c}$ as

$$|R_{v,c}|^2 (E_c - E_v) = f_{v,c}(\hbar^2/2m), \qquad (10)$$

and by the oscillator sum rule

$$\sum_c f_{v,c} = 1, \qquad (11)$$

we obtain the sum rule for ε_2 as

$$\int E\varepsilon_2(E)\, dE = \left(\frac{\pi}{2}\right)(\omega_p^2 \hbar^2). \qquad (12)$$

Equations (10) and (11) are useful in other ways as well. We note that the maximum matrix element is that which is associated with a *single* optical transition at $(E_c - E_v)$, and its magnitude is given by Eq. (10), with $f_{v,c} = 1$, as

$$|R_{v,c}|^2 = (\hbar^2/2m)[1/(E_c - E_v)] \qquad (13a)$$

or

$$|R_{v,c}|^2 = (3.8)/(E_c - E_v)\ (\text{Å}^2\ \text{eV}). \qquad (13b)$$

5. THE TAUC MODEL OF OPTICAL ABSORPTION IN AN AMORPHOUS SEMICONDUCTOR

We next explore the implications of Eq. (7) for a model of an amorphous semiconductor which was first introduced by Tauc et al. (1966) (Tauc, 1972, 1974). The optical absorption edge of amorphous semiconductors has received much attention since that time (Mott and Davis, 1979, p. 272). We focus on Tauc's model because it makes quantitative predictions which for a-Si:H_x turn out to be in quite reasonable agreement with experiment (Cody et al., 1982).

The Tauc model starts with a "virtual" crystal into which positional disorder is introduced (Ziman, 1979). The virtual crystal has the same nearest- and next-nearest-neighbor bonding in the amorphous phase as in the "real" crystal of the element or compound in question. The virtual crystal also has to simulate the topological disorder (Ziman, 1979) of the amorphous semiconductor, and its optical gap may not bear any obvious relation to that of the real crystal of the element or compound. The averaged features of the virtual crystal, however, may be closely related to the features of the real crystal. For example, the indirect absorption edge of c-Si is dominated by the long range symmetry of the diamond lattice, and detailed features of electronic states near this edge should not be expected to determine the optical absorption edge of a-Si:H_x. We might expect that certain optical properties of c-Si averaged over the valence and conduction bands would be closely related to similar averaged properties of a-Si:H_x. Examples of such properties include the "Penn gap" (Penn, 1962; Phillips, 1973) and the fundamental parameters of the single oscillator of Wemple and DiDominico (1971) and Wemple (1973, 1977). Such quantities depend more on density and chemistry than on artifacts of crystalline symmetry.

Following Tauc, we make these assumptions:

(1)   The amorphous semiconductor and its virtual crystal can be described by valence and conduction bands separated by a gap in which there is a very low density of states.

(2)   The valence- (conduction-) band wave functions for the amorphous semiconductor can be expanded in terms of the Bloch waves of the valence (conduction) band of the virtual crystal. There is no mixing of valence- and conduction-band wave functions due to positional disorder.

(3)   We model the *crystalline* semiconductor with an optical matrix element that is independent of energy as

$$(R_{v,c})_{crys} = R_{k_v, k_c'} \equiv Q\delta_{k_v, k_c'}. \tag{14}$$

In Eq. (14) the notation k_v and k_c' indicate Bloch wave functions appropriate to the valence and conduction bands with momentum k_v and k_c'. The Kronecker delta symbol describes electronic momentum conservation for the ordered virtual crystal.

We can derive from Eq. (7) under the above assumptions an expression for ε_2 in the virtual crystal which describes a *direct* optical transition since both energy and momentum are conserved. If we add to the above assumptions the additional one that there is *no* conservation of electron momentum in the disordered amorphous semiconductor, we derive an expression for ε_2 which describes an optical transition in which momentum is not conserved (Spicer and Donovan, 1970). This can be described as a *nondirect* transition to distinguish the absorption edge of the amorphous semiconductor with positional disorder from the *indirect* absorption edge of crystalline semiconductors.

Under the above assumptions, we can rewrite Eq. (7) as

$$\varepsilon_2(E) = (2\pi e)^2 (2/V) Q^2 \sum_{v,c} |\langle v|T|c \rangle|^2 \delta(E_c - E_v - E), \tag{15}$$

where

$$\langle v|T|c \rangle = \sum_{k_v, k_c} \langle v|k_v \rangle \delta_{k_v, k_c'} \langle k_c'|c \rangle. \tag{16}$$

If we note that the quantity $T^\dagger$ is a projection operator which takes a crystalline valence-band Bloch wave function and transforms it into a crystalline Bloch wave function in the conduction band with the same electronic momentum, that the quantity $TT^\dagger$ is a unit operator for the crystal valence band, and that $T^\dagger T$ is a unit operator for the crystal conduction band, we can define a sum rule due to Velicky (Tauc, 1972) as

$$\sum_{v,c} |\langle v|T|c \rangle|^2 = \sum_v \langle v|TT^\dagger|v \rangle = \sum_v 1 = N_0. \tag{17}$$

The quantity N_0 is the number of single spin states in the valence (conduction) band. The number of electrons that can occupy these states is $2N_0$.

We can now solve Eq. (15) under two conditions. The first is the *direct* optical transition where we choose for the valence- and conduction-band states the original Bloch wave functions of the virtual crystal. These wave functions have a specific crystal momentum. Under these circumstances

$$|\langle v|T|c \rangle|^2 = \delta_{k_v, k_c'} \tag{18}$$

and thus

$$\varepsilon_2(E) = (2\pi e)^2 Q^2 \left(\frac{2}{V} \right) \sum_k \delta[E_c(k) - E_v(k) - E]. \tag{19}$$

We solve Eq. (19) with the following density of states model: parabolic-energy-band edge with a density-of-states effective mass ratio M ($M = m_{\text{dos}}/m$) and an energy gap E_G. Then,

$$\varepsilon_2(E) = (1.4) Q^2 M^{3/2} (E - E_G)^{1/2}. \tag{20}$$

In Eq. (20), Q is in angstroms and E is in electron volts. Equation (20) has the expected form for the direct transition in a crystal (Pankove, 1971).

For the Tauc limit of Eq. (15), we replace the quantity $|\langle v|T|c \rangle|^2$ by a constant T_0^2. From the sum rule, Eq. (17), $T_0^2 = 1/N_0$, and then

$$\varepsilon_2(E) = (2\pi e)^2 Q^2 \left(\frac{2V}{N_0} \right) \int dZ \, N_c(Z) N_v(Z - E), \tag{21}$$

where

$$\sum_c \delta(E_c - Z) \equiv V N_c(Z) \tag{22}$$

defines the volume density of single spin states for the conduction band and equivalently for the valence band. Again, V is the illuminated volume of the sample to which all wave functions have been normalized. Comparing Eq. (21) with Eq. (19), we see that the assumption that $|\langle v|T|c\rangle|^2$ is a constant has replaced the electron momentum by the energy as the relevant quantum number. Equation (21) describes a *nondirect* optical transition.

We evaluate Eq. (21) by using a parabolic density of states and obtain

$$\varepsilon_2(E) = Q^2 M^3 (0.2/v)(E - E_G)^2, \tag{23}$$

where the units of Q^2 are angstroms squared and all energies are in electron volts. The quantity M is the ratio of the density-of-states effective mass to the free-electron mass for the amorphous semiconductor but differs slightly from the previous definition. For Eq. (23) M is the geometric mean of the conduction- and valence-band effective mass ratios. For Eq. (20), M is the inverse arithmetic average of the same quantities. The quantity v is the number of valence electrons per atom and is defined by

$$2N_0/V = v\rho_A, \tag{24}$$

where ρ_A is the atomic density.

In Fig. 16 we show a linear plot of $(\varepsilon_2)^{1/2}$ for a-Si:H$_x$ ($x \approx 0.09$) as a function of E, and we note the excellent agreement with Eq. (23) over 1.4 eV. The fit of Eq. (23) to the data has an interesting feature. The value of

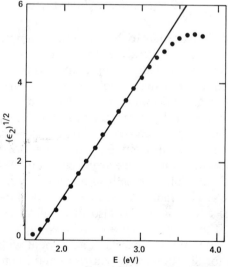

FIG. 16. The quantity $(\varepsilon_2)^{1/2}$ for a-Si:H$_x$ is plotted as a function of energy for 30 films. The straight line is a least squares fit to the data in the range 1.65–3.00 eV and defines an optical gap of 1.64 eV. [From Cody *et al.* (1982). Copyright North-Holland Publ. Co., Amsterdam, 1982.]

E_G, 1.64 eV, corresponds in Fig. 14 to a value of α of 6×10^2 cm^{-1}, which is just above the value of α at which the exponential absorption edge blends into the more slowly varying region (i.e., in Fig. 5, the transition between regions B and C). The quantity E_G determined by Eq. (23) thus exhibits an important correspondence with one of the few features of Fig. 14, the transition to an exponential absorption edge and presumably an exponential density of states. From the slope of the straight line in Fig. 16 we obtain from Eq. (23)

$$(Q^2 M^3/v) = 45 \text{ Å}^2. \tag{25}$$

In what follows we shall develop a technique for evaluating Q^2 and hence obtain from Eq. (25) estimates of M as a function of v.

A general sum rule for the quantity $\varepsilon_2(E)$ is (Wooten, 1972, p. 78)

$$\int_0^E \varepsilon_2(Z)Z \, dZ = \left(\frac{2\pi^2 e^2 \hbar^2}{m}\right) N(E), \tag{12'}$$

where $N(E)$ is the electron density involved in the transition up to energy E. This result is general and independent of any statement we might make about the matrix element and its energy dependence. If one assumes an energy independent matrix element, Eq. (14) leads to another sum rule. From Eq. (15), (19), or (21) we can easily show that

$$\int_0^{E_1} \varepsilon_2(Z) \, dZ = (2\pi e Q)^2 v \rho_A, \tag{26}$$

where E_1 is greater than or equal to the maximum energy difference between the valence and conduction bands. From Eq. (26) we note that when we make the constant matrix-element approximation, the area under the ε_2 curve is conserved between the virtual crystalline semiconductor and the disordered amorphous semiconductor derived from it.

In Fig. 7 we exhibited the ε_2 spectra for c-Si and a-Si. Up to 6 eV the areas under each curve are identical and equal to 62 eV. If we extend the curves to $\varepsilon_2 = 0$ (20 eV), we calculate from Eq. (26) $Q^2 = 0.9$ Å^2 for the atomic density of c-Si and four valence electrons per atom. When we insert this value of Q^2 in the oscillator-strength sum rule, we find a single-oscillator transition at 4.2 eV. This value is close to the energy of the maximum of the ε_2 spectra in c-Si and a-Si, and is also in excellent agreement with the single-oscillator fit to $n(E)$ for c-Si and a-Si (Wemple, 1977). Finally, the magnitude of Q^2 (0.9 Å^2) is in excellent agreement with values of optical matrix elements for Si estimated by Harrison and Pantelides (1976) from the analysis of the optical properties of groups IV, III–V, and II–VI tetrahedrally coordinated semiconductors.

We may use this self-consistent value for Q^2 in Eq. (25) to obtain the density-of-states mass ratio for a-Si:H_x when the conduction and valence bands are represented by parabolic energy bands. With $Q^2 = 0.9$ Å^2 and four electrons per atom ($v = 4$), we find $M = 6$. This is an implausibly high value of the density-of-states mass. From this value for M, we conclude that the free-electron equivalent to the a-Si:H_x density of states is a narrow band, with an effective density of states an order of magnitude larger than that usually assumed. If we had chosen $M = 1$ and $v = 1$ in Eq. (25), Q^2 would be about 45 Å^2, which is too large for a ground-state-to-excited-state transition on the basis of the oscillator-strength sum rule [Eq. (13a)] (Cody *et al.,* 1982).

If we naively apply Eq. (20) to the ε_2 spectra of c-Si, we can identify a direct optical transition at 3.2 eV and with $Q^2 = 0.9$ Å^2, a value for the density-of-states effective mass ratio of 12. This value of M is unphysically large. The large values of M for c-Si indicate the inadequacy of the constant matrix-element approximation for c-Si. The large value of M for a-Si:H_x cautions us to avoid attaching much significance to the precise numerical magnitude of Eq. (23) for a-Si:H_x. Indeed, this and other data to be presented suggest that, while the *form* of the fundamental equation of the Tauc model [Eq. (21)] is of general applicability, its *magnitude* can differ considerably from free-electron estimates.

We next give two examples which exhibit the usefulness of Eq. (21). The first is the analysis of the exponential regime (region B of Fig. 5). The second is the shoulder absorption (region A of Fig. 5). In each case, we describe the absorption edge through Eq. (21) and a simple density of states model.

We associate region C with a band-to-band optical absorption that we model by two parabolic densities of states of the form

$$N_{c,v}(Z) = N_{c,v}(Z - E_{c,v})^{1/2}, \tag{27}$$

where $E_G = E_c - E_v$, leads through Eq. (21) to

$$\varepsilon_2^{BB}(E) = (2\pi eQ)^2(4/v\rho_A)N_cN_v(\pi/8)(E - E_G)^2, \tag{28}$$

or generally,

$$\varepsilon_2^{BB}(E) = K_{BB}(E - E_G)^2, \tag{29}$$

where K_{BB} is a constant. The experimental value of K_{BB} is 9.36 eV^{-2} (refer to Fig. 23). From Eq. (28) with $Q^2 = 0.9$ Å^2, $v = 4$, and effective masses of unity, we calculate $K_{BB} = 4.7 \times 10^{-2}$ eV^{-2}. For reasonable choices of electronic parameters, Eq. (21) thus predicts about two orders of magnitude less optical absorption than is observed.

We next model region B by modifying the valence-band edge to include an exponential "tail" but continue to use Eq. (27) for the conduction band. We introduce for the *valence*-band density of states an exponential of the form

$$N_v(Z) = N_v(3E_0^v/2)^{1/2} \exp[(E_v^M - Z)/E_0^v].\qquad(30)$$

The parameters of Eq. (30) are defined by Fig. 17, where a parabolic density of states has been modified into an exponential distribution of characteristic energy E_0^v above E_M. From Fig. 17 and state conservation

$$E_v - E_v^M = \tfrac{3}{2}E_0^v.\qquad(31)$$

Since we assume no modification of the conduction-band density of states, the "mobility" gap E_G^M is simply given by

$$E_G^M = E_G + \tfrac{3}{2}E_0^v,\qquad(32)$$

where

$$E_G = E_c - E_v.$$

The contribution to ε_2 for the transition between the valence tail and the conduction-band parabolic states is thus

$$\varepsilon_2^{TB}(E) = (2\pi eQ)^2(4/v\rho_A)N_cN_v(E_0^v)^{3/2}(\sqrt{\pi}/2)\exp\{(E - E_G^M)/E_0^v\},\qquad(33)$$

or

$$\varepsilon_2^{TB}(E) = K_{TB}\exp\{E/E_0^v\},\qquad(34)$$

where

$$K_{TB} = K_{BB}(24/\pi)^{1/2}(E_0^v)^2\exp[-E_G^M/E_0^v].\qquad(35)$$

From Fig. 16, $K_{BB} = 9.36$ eV^{-2} and $E_G = 1.64$ eV. From Fig. 15, $K_{TB} = 1.87 \times 10^{-16}$ ($n = 4.0$, $\alpha = 3.0 \times 10^2$ cm^{-1} at $E = 1.60$ eV, and $E_0^v =$

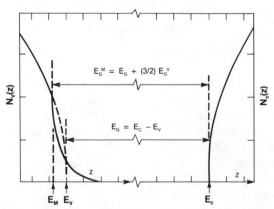

FIG. 17. Model density of states with parabolic conduction band and exponential band tail for valence band.

50×10^{-3} eV). The only undetermined quantity in Eq. (35) is then E_G^M, and solving for it, we obtain $E_G^M = 1.67$ eV. This result is in reasonable agreement with Eq. (32). It also suggests that the band-to-band and tail-to-band transitions can be scaled by the same factor.

A more stringent quantitative test of the Tauc model [Eq. (21)] is the shoulder optical absorption (region A in Fig. 9). We consider the valence-band density of states to include, in addition to the exponential tail, localized Si states deep in the gap. We thus replace Eq. (30) by

$$N_v(Z) = (\rho_D/2)\delta(Z - E_d), \tag{36}$$

where ρ_D is the volume density of "Si-like" deep states that we can think of as dangling bonds capable of being occupied by one or two electrons. For the conduction band, we again use Eq. (27). The contribution to ε_2 for the optical transition from these localized states to the conduction-band parabolic states is given by

$$\varepsilon_2^{SB}(E) = (2\pi e Q)^2 (4/v\rho_A)(\rho_D/2)N_c(E - E_D)^{1/2} \tag{37}$$

or

$$\varepsilon_2^{SB}(E) = K_{SB}(E - E_D)^{1/2}, \tag{38}$$

where

$$E_D = E_c - E_d \quad \text{and} \quad K_{SB} = K_{BB}(4/\pi)(\rho_D/N_v). \tag{39}$$

In Fig. 18, we show the optical absorption coefficient for a film of a-Si:H$_x$ grown from the rf decomposition of SiH$_4$ at a substrate temperature of 250°C. In this figure, we also show the effect on the optical absorption of a 30-min anneal at the indicated temperatures. The absorption coefficient plotted in Fig. 18 has been derived from optical transmission measurements combined with direct measurements of absorption through photodeflection spectroscopy (Roxlo *et al.*, 1984). As discussed in Section 7, heating a-Si:H$_x$ above its growth temperature drives out some of the bonded hydrogen. Dangling bonds are generated, but most are removed by reconstruction of the network. One consequence of this reconstruction is an increase in structural disorder. Another consequence is an increase in the density of neutral dangling bonds, which can be directly measured by electron spin resonance because they are singly occupied. Yoshida and Morigaki (1983) have reported spin densities for a film of a-Si:H$_x$ prepared in a manner similar to that for the film exhibited in Fig. 18, which had also been annealed at 550°C. Their value of the density of dangling bonds (5×10^{18} cm^{-3}) can be used to directly calculate K_{SB} from Eq. (37). The solid curves in Fig. 18 are a fit of the *sum* of Eq. (34) and Eq. (37) to the absorption coefficient below 10^3 cm^{-1}. The parameters determined by the fitting procedure for the film

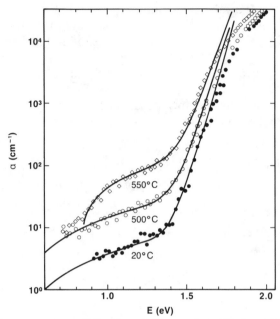

FIG. 18. Optical absorption coefficient for a-Si: H_x annealed a 500°C and 550°C for 30 min. As deposited, the film had $E_G = 1.64$ eV. The low-absorption data were obtained by photoacoustic absorption (Roxlo *et al.*, 1984) and the interference fringes have not been averaged out. Solid circles refer to as-grown film.

annealed at 550°C are $E_D = 0.8$ eV, $E_0^v = 60 \times 10^{-3}$ eV, and $K_{SB} = 7.7 \times 10^{-3}$ eV$^{-1/2}$. The *calculated* value for K_{SB} from Eq. (33) is 3.9×10^{-5} eV$^{-1/2}$ for $\rho_D = 5 \times 10^{18}$ cm^{-3}, $Q^2 = 0.9$ Å^2, $v = 4$, and an effective mass of one for the conduction-band density of states. The discrepancy between the simple Tauc theory and experiment is about two orders of magnitude, although the energy dependence of the optical absorption coefficient is fitted very well for reasonable estimates of the density of states. It is important to note that the discrepancy is of the same magnitude as that already noted for the band-to-band optical transition [Eq. (28)]. The similar scaling between the simple Tauc theory and experiment for the three regions of the optical absorption edge, shoulder, tail, and band, suggests a useful experimental paradigm. We may use Eq. (21) and experimental estimates of the density of states (e.g., Griep and Ley, 1983) to determine the *energy* dependence of the optical absorption coefficient. However, the *magnitude* of the absorption coefficient must be increased by almost a factor of 10^2 over that calculated using effective masses of one.

The power of this technique is illustrated in the interpretation of the absorption edge exhibited in Fig. 19. In this figure we show the optical

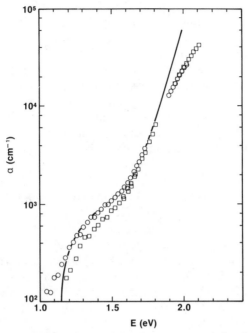

FIG. 19. Optical absorption coefficient determined by transmission for two phosphorus-doped a-Si:H_x films (2% PH_3 in SiH_4 at a substrate temperature of 250°C).

absorption coefficient of n-type a-Si:H_x that was prepared from a mixture of 2% PH_3 in SiH_4 at a substrate temperature of 250°C. The data were taken by optical transmission on two films. The data shown in Fig. 19 are essentially identical to that reported by Yamasaki *et al.* (1981b) for a film prepared from a mixture of 1% PH_3 in SiH_4. The large shoulder absorption shown in Fig. 19 can also be associated, from the analysis of the photoconduction, with an optical transition to the conduction band from an occupied (charged) dangling bond. Since the dangling bond in this case is *doubly* occupied, direct estimates of the dangling-bond density cannot be obtained from electron spin resonance. From photoconductivity measurements (Vanecek *et al.,* 1983; Wronski *et al.,* 1982) and from direct absorption measurements combined with the results of light-induced electron spin resonance or luminescence (Jackson, 1982; Jackson and Amer, 1982; Jackson *et al.,* 1983), we can estimate ρ_D to be between 10^{18} and 10^{19} cm^{-3} for a film doped with 2% PH_3. The solid curve in Fig. 19 is the best fit of the sum of Eq. (34) and Eq. (37). The parameters determined by the fitting procedure are $E_D = 1.1$ eV, $E_0^v = 80 \times 10^{-3}$ eV, and $K_{SB} = 7.7 \times 10^{-2}$ eV$^{-1/2}$. Based on the normalization derived from Fig. 18, we deduce a density of doubly occupied dangling bonds of 1×10^{19} cm^{-3}, about a factor of 3 higher than that extrapolated from photoconductivity data (Wronski *et al.,* 1982).

In summary, the Tauc model for the optical absorption is an accurate guide to the interpretation of the energy dependence of the absorption spectrum of a-Si: H$_x$ and, indeed, to those of other amorphous semiconductors. It permits the experimentalist to associate the absorption edge with fundamental features of the amorphous semiconductor. For a-Si: H$_x$, a parabolic density of states for the conduction and valence bands, the assumption of an energy independent dipole matrix element and an optical gap E_G reproduce the main features of the absorption spectra over a range of several volts. The additional assumption of an experimental broadening of the valence-band edge at a mobility gap E_G^M can account for the magnitude and position of the Urbach edge (Mott and Davis, 1979, p. 279; Tauc, 1970, 1972, 1974; Wood and Tauc, 1971; Kurik, 1971; Dunstan, 1982; Skettrup, 1978; Szczyrbowski, 1979) in a-Si: H$_x$. The Tauc model is thus in excellent agreement with transport measurements that lead to the same exponential density of trap states (Tiedje and Rose, 1981). Finally, the Tauc model gives an insight into the shape of the "shoulder absorption" in a-Si: H$_x$, associating it with a transition from a localized dangling bond at an energy E_D below the conduction band ($E_D \approx 0.8$ for Fig. 18, $E_D = 1.1$ for Fig. 19) to a parabolic distribution of conduction-band states.

There are difficulties in reconciling the numerical magnitude of Eq. (21) with experimental results. In general, the trend is that Eq. (21) may *underestimate* the magnitude of the absorption by as much as two orders of magnitude for physically reasonable values of the dipole matrix element and density of states. This is not a significant limitation on its use, since the normalization appears to be the same for a-Si: H$_x$ in the shoulder, tail, and band regions of the absorption edge. For the experimentalist, the importance of the model is in the parameters it introduces, particularly the optical gap E_G, which exhibits remarkable similarities to the traditional optical gap of crystalline semiconductors. In Section 7, we show the utility of the Tauc optical gap in understanding the electronic changes induced in the semiconductor through either structural or thermal disorder. Before we do, we shall examine a recent self-consistent model of optical absorption in a disordered semiconductor, which places the Tauc model, with its experimental utility, on a more consistent theoretical basis.

6. A Self-Consistent Theory of Optical Absorption in a Disordered Semiconductor

The agreement between the spectral dependence of Eq. (21) and experimental data on the absorption edge for reasonable model density of states confirms the physical insight contained in the Tauc model. However, the model remains inconsistent. The same disorder that removes momentum conservation as an ingredient of the band-to-band optical transition should also have a profound and unpredictable effect on the density of states at the band edge. In the preceding section, following Wood and Tauc (1971), we introduced an ad hoc modification of the density of states, e.g., the exponential tail, and then justified this modification by comparison with experiment. In this section, we consider a simple theory of optical absorption, due to Abe and Toyazawa (1981), that exhibits all the features of the absorption edge of a-Si:H_x and many other amorphous semiconductors in a self-consistent model.

Abe and Toyazawa, following Tauc, assume a virtual crystal with an energy gap E_x. They introduce Gaussian site disorder for the valence and conduction bands and permit the disorder to be correlated, uncorrelated, or anticorrelated between the bands. These authors again make the constant dipole matrix-element approximation (Q = const) but solve the disorder problem within the framework of the coherent potential approximation rather than the random phase approximation (*no* crystal momentum defined) of the Tauc model (Tauc, 1972; Hindley, 1970).

Following Abe and Toyazawa, we start with Eq. (15), rewritten as

$$\varepsilon_2(E) = (2\pi eQ)^2(2/V)I(E), \tag{40}$$

where

$$I(E) = \int\int dE_1 \, dE_2 \sum_{c,v} \delta(E_c - E_1)\delta(E_1 - E_2 - E)|\langle v|T|c\rangle|^2\delta(E_v - E_2). \tag{41}$$

We note that $VN_c(E_1) = \sum_c \delta(E_c - E_1)$ and that a similar expression holds for $VN_v(E_2)$. The Tauc model simplifies Eq. (41) through the random phase approximation. Abe and Toyazawa restate Eq. (41) in terms of an effective "matrix element" $M(E_1, E_2)$. Thus,

$$I(E) = \int dE_1 \, dE_2 \, S(E_1, E_2)\delta(E_1 - E_2 - E), \tag{42}$$

where

$$S(E_1, E_2) = \sum_{c,v} \delta(E_c - E_1)\delta(E_v - E_2)|\langle v|T|c\rangle|^2, \tag{43}$$

and they define

$$S(E_1, E_2) \equiv M(E_1, E_2)N_c(E_1)N_v(E_2), \tag{44}$$

where from Eqs. (43) and (17),

$$\int\int dE_1 \, dE_2 \, S(E_1, E_2) = N_0.$$

The generalization of Eq. (21) is thus

$$\varepsilon_2(E) =$$
$$(2\pi eQ)^2\left(\frac{2}{V}\right) \int \int dE_1 \, dE_2 \, N_c(E_1)M(E_1, E_2)N_v(E_2)\delta(E_1 - E_2 - E). \quad (45)$$

The Tauc model of Section 5 is thus a random phase approximation for which $M(E_1, E_2) = V^2/N_0$. Abe and Toyazawa, however, use the coherent potential approximation to calculate $M(E_1, E_2)$, $N_c(E_1)$, and $N_v(E_2)$ for a parabolic band model with varying degrees of Gaussian *site* disorder.

Before we present the chief results of this paper, we define the parameters of their model. For the virtual crystal without disorder, they utilize for simplicity a semielliptic density of states exhibiting a parabolic energy dependence at both edges of the bands. The band width of the conduction band $2B_c$ and the band width of the valence band $2B_v$ are important parameters that measure the strength, in this tight-bonding model, of interatomic transfer. The band gap of the virtual crystal E_x is a free parameter of this model, but its magnitude is determined by the competition between the bonding and antibonding levels of a chemical bond and the broadening of these levels due to intersite transfer. The measure of site disorder is taken to be the root-mean-square deviation of the conduction- or valence-band site energy from the mean site energy. If E_n^i is a random site energy in the conduction or valence band ($i = c$ or $i = v$), the measure of disorder is the energy W_i given by

$$W_i = (\langle\langle\langle(E_n^i - \bar{E}^i)^2\rangle\rangle\rangle)^{1/2}, \quad (46)$$

where we follow their notation and $\langle\langle\langle(\ )\rangle\rangle\rangle$ represents an ensemble average with respect to the Gaussian distribution. A correlation coefficient γ, defined by

$$\gamma = \frac{\langle\langle(E_n^c - \bar{E}^c)(E_n^v - \bar{E}^v)\rangle\rangle}{W_c W_v}, \quad (47)$$

defines the correlation of spatial fluctuation between the conduction and valence bands. The natural measure of disorder on the optical transition is a quantity proportional to the site disorder, e.g., W_i, and inversely proportional to some measure of intersite transfer, e.g., the band width $2B_i$. Dimensional conditions suggest the parameter

$$\Gamma_i = \text{const}(W_i^2/B_i), \quad (48)$$

and indeed Γ_i is the measure of disorder that emerges in the theory. Thus for the *same* site disorder, a narrower band with less intersite transfer has a higher effective disorder. For the *same* bonding chemistry a narrower band will have a higher energy gap independent of disorder.

We summarize the main results of this theory in Figs. 20–23 redrawn from the original article. In the first three figures, the quantity $V\varepsilon_2/2(2\pi eQ)^2 \equiv I(E)$ is plotted as a function of energy E for a direct transition between a valence band of unperturbed width 2 eV, a conduction band of unperturbed width 4 eV, and a band gap for the ordered virtual crystal of 3.0 eV. In Fig. 20, the quantity I is shown as a function of the parameter E (the photon energy is given by $E + 6.0$ eV) for different values of the disorder parameter $W(W_c = W_v \equiv W)$. The direct transition for $W = 0$ [Eq. (20)] is indicated, as is the nondirect transition of the Tauc model [Eq. (23)]. The curves for finite W show a broadening of $I(E)$ below the direct band gap E_x, as well as a reduction of $I(E)$ above E_x. This last result is, of course, a consequence of the constant dipole matrix-element sum rule [Eq. (26)]. It is apparent from Fig. 20 that the optical gap of the Tauc model, as defined in Section 5 [Eq. (29)], has to depend on disorder if it is to have any connection with the self-consistent coherent potential approximation result.

In order to relate the main experimental features of the optical absorption edge to the theoretical results summarized in Fig. 20, it is necessary to look at different energy regions in detail. In Fig. 21, again from the Abe and Toyazawa paper, we exhibit $\log_{10} I(E)$ as a function of E well below the optical gap at $E_x = 3.0$ eV ($E = -3.0$ eV). Over several orders of magnitude in ε_2, there is an approximately exponential variation with photon energy. Indeed, up to $W \approx 0.6$ the data is well represented by an expression of the "Urbach" form

$$\varepsilon_2 = \text{const} \exp[(E - E_x)/\Gamma(W)], \tag{49}$$

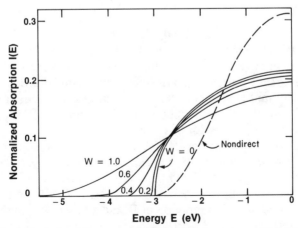

Energy E (eV)

FIG. 20. Normalized absorption as a function of photon energy ($E + 6.0$ eV) and site disorder W [Eq. (46)]. The correlation coefficient $\gamma = 0$. The nondirect transition [Eq. (28)] is shown by the dashed line. [From Abe and Toyazawa (1981).]

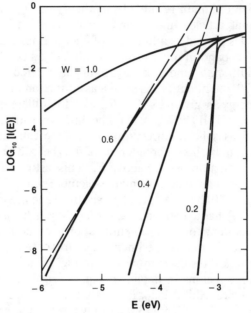

FIG. 21. The exponential tail of Fig. 20 exhibited as a function of photon energy $(E + 6.0 \text{ eV})$. The dashed curves are $\alpha = I_0 \exp[(E - E_x)/E_0]$, where $E_x = 3.0 \text{ eV}$ and $E_0 = (W^2/2.7) \text{ eV}$. [From Abe and Toyazawa (1981).]

where

$$\Gamma(W) = W^2/2.7, \tag{50}$$

as suggested by Eq. (48), and E_x, the Urbach focus, is the optical gap of the virtual crystal. The quantity $\Gamma(W)$ can be identified with E_0^x of Eq. (24). The coherent potential approximation thus exhibits an exponential absorption edge that, as is shown in Fig. 22, arises from an exponential density of states and a constant matrix element $M(E_1, E_2)$. The Urbach focus at E_x presumably arises from the constant dipole-moment sum rule [Eq. (26)], which implies a conservation law for ε_2.

Finally, close to the band edge, the Abe and Toyazawa model does exhibit square-law behavior characteristic of Eq. (23). In Fig. 23, $[I(E)]^{1/2}$ is plotted as a linear function of energy and there is a reasonable approximation to a straight line over a limited range. As expected, $E_G(W)$, defined as the intercept on the E axis, is a function of W and hence of Γ. Indeed from Fig. 23 and Eq. (50)

$$E_G(W) = -3.37W^2 + 3.0 \tag{51}$$

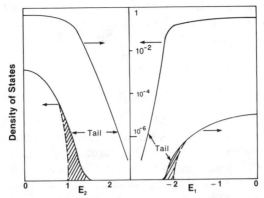

FIG. 22. The density of states as obtained by Abe and Toyazawa (1981) for $W = 0.4$ ($E_0 = 60 \times 10^{-3}$ eV) on a linear and exponential scale. The cross-hatched region shows the portion of the parabolic DOS that has been broadened into an exponential tail.

or

$$E_G(\Gamma) = -9.10\Gamma + E_x. \tag{52}$$

Thus from Eq. (52), $E_G(W = 0) = E_G(\Gamma = 0) = E_x$. From Fig. 23, even region C exhibits a focus at E_x.

From Eq. (51), we note that the Abe and Toyazawa model eliminates a major inconsistency of the Tauc model by permitting the derived optical

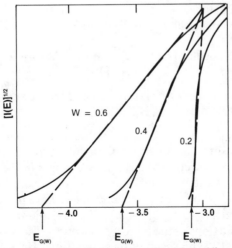

FIG. 23. The algebraic regime of Fig. 20. The extrapolated dashed lines define an optical gap [Eq. (28)] given by $E_G(W) = 3.0 - 3.37 \ W = 3.0 - 9.10 \ E_0$. [After Abe and Toyazawa (1981).]

energy gap to depend on the degree of disorder. As will be seen in Section 7, Eqs. (51) and (52) are in remarkable agreement with experiment.

We can summarize this and the preceding section in the equation

$$\varepsilon_2(E) = (2\pi e Q)^2 (2V/N_0) \int \int dE_1 \, dE_2$$
$$\times N_c(E_1)\overline{M}(E_1, E_2)N_v(E_2)\delta(E_1 - E_2 - E). \tag{53}$$

The quantity $\overline{M}(E_1, E_2)$ is defined as in Eq. (44) but is normalized to V^2/N_0. The Tauc model sets $\overline{M} = 1$ through the random phase approximation and utilizes a model density of states. The quantitative consequences of these assumptions will be discussed in Section 7. The coherent potential approximation of Abe and Toyazawa leads to values of $\overline{M}$ that are constant in the regions described in Figs. 21 and 23 but can be larger or smaller than one. As defined by Eq. (53), ε_2 satisfies the constant dipole-matrix-element sum rule [Eq. (26)] and exhibits a focus in both regions B and C of Fig. 5.

Finally, as shown in Fig. 22, the self-consistent density of states derived from this theory exhibits the required parabolic and exponential behavior required to explain the experimental utility of the Tauc expression for interpreting the absorption edge of a-Si:H_x and other amorphous semiconductors. In the Abe and Toyazawa theory, the exponential results from an interplay between transfer energy and Gaussian distributed site energies. This explanation of the Urbach edge should be distinguished from the exciton broadening mechanism of the Urbach edge phenomena in crystalline semiconductors and insulators for which a similar interplay may occur but specific phonons and electric fields are involved (Dow and Redfield, 1972). A major challenge for the experimentalist is to distinguish clearly between these two physically appealing models (Tauc, 1972).

IV. Experimental Data on the Optical Absorption Edge of a-Si:H_x

7. EFFECT OF THERMAL AND STRUCTURAL DISORDER ON THE ABSORPTION EDGE OF a-Si:H_x

It is sometimes the case in solid-state physics that changes in a quantity are easier to measure and interpret than the absolute magnitude of the quantity itself. In this section, we examine changes in the absorption edge due to thermal and structural site disorder. We define the position and shape of the absorption edge by the two parameters we have previously discussed. The first is the optical gap E_G defined by Eq. (29) and the second is E_0^v, the inverse logarithmic slope of the optical absorption edge, which is defined on the basis of Eq. (34). While each of these quantities lacks a firm theoretical foundation, changes in them are easily explained in terms of simple physical models.

Before we consider the experimental data, it is necessary to justify the choice of E_0^v as a fundamental parameter of the absorption edge. There is by now experimental evidence from drift mobility data (Tiedje et al., 1981, 1983b; Tiedje and Rose, 1981) that an exponential density of trapped states proceeding from the valence band that controls the hole drift mobility is also responsible for the exponential absorption edge. The inverse logarithmic slope of the trap distribution is about 50×10^{-3} eV for a-Si:H$_x$ at $T = 300°$K and is identical within experimental error to the quantity E_0^v (Tiedje et al., 1983a; Roxlo et al., 1983). Drift mobility measurements also demonstrate a similar exponential density of localized states proceeding from the *conduction* band but with a slope parameter E_0^c about equal to 25×10^{-3} eV at $T = 300°$K. From the form of Eq. (21), only the exponential associated with the valence-band-tail to conduction-band transition would be experimentally observed since $E_0^v \approx 2E_0^c$. In what follows, we shall write E_0 for E_0^v.

The simplest experimental change in a quantity is that due to temperature, and Fig. 24 exhibits the temperature shift in the optical gap of a-Si:H$_x$. The data was taken for $\alpha \gtrsim 10^4$ cm^{-1} either by isoabsorption or linear extrapolation of the quantity $(\alpha/E)^{1/2}$ as

$$(\alpha/E)^{1/2} = C'[E - E_G(T)], \tag{54}$$

where C' is a temperature-independent constant $[C' = 3.06$ eV$^{-3/2}$ μm$^{-1/2}]$.

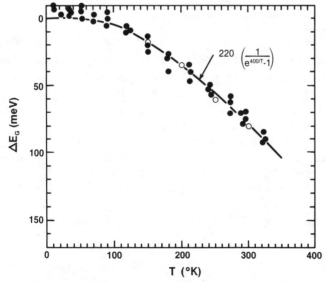

FIG. 24. Red shift of the Tauc optical gap of a-Si:H$_x$ as a function of measurement temperature. [From Cody et al. (1981).]

Equation (54) is derived by applying to Eq. (29) the well-known relation $\alpha(E) = 2\pi\varepsilon_2/n\lambda$.

An alternative expression for Eq. (54), and one commonly used, is a linear extrapolation of $(\alpha E)^{1/2}$ (Cody et al., 1982). This expression is derived from Eq. (29) when a constant *momentum* matrix element rather than a constant *dipole* momentum is used as the starting point for the theory of Sections 6 and 7. There is no *physical principle* that can determine the choice between these assumptions for amorphous semiconductors. Experimentally, the constant dipole matrix-element assumption is in considerably better agreement with experiment for a-Si:H_x, and more important, as shown by Cody et al. (1982), it defines an energy gap E_G that is independent of the energy range over which α is determined and hence independent of the thickness of the film. Moreover, the satisfaction of the sum rule [Eq. (26)] for reasonable values of Q^2, further suggests the fundamental basis for the assumption. In general, the distinction between $(\alpha/E)^{1/2}$ and $(\alpha E)^{1/2}$ as a definition of E_G is important only when films of widely differing thicknesses are compared. The two formulas give identical results if we consider only changes in E_G for the same film.

It is, of course, possible to plot any power of α against E and obtain, over some energy range, a linearly extrapolated E_G. Indeed, several recent papers suggest that the one-third power of α is linear over a *wider energy* range for a *specific* film of a-Si:H_x (Vorlicek et al., 1981; Klazes et al., 1982) than is the one-half power of α. There can be some ambiguity in defining the energy range of interest, since the high energy "algebraic" regime for α above 10^4 cm^{-1} becomes exponential for α below 10^3 cm^{-1}. Furthermore, there is no evidence that E_G defined by the extrapolation of $\alpha^{1/3}$ agrees with transport or luminescence measurements (Nitta et al., 1982) or that it is independent of film thickness (Cody et al., 1982). We thus choose for a-Si:H_x, Eq. (54) derived from Eq. (29), as the most physically meaningful fit to the absorption data above $\alpha = 10^4$ cm^{-1}. This choice is not general. For example, experimentally, $\varepsilon_2(E)$ appears to be a *linear* function of energy for a-Se (Mott and Davis, 1979, p. 522; Brooks and Stephens, 1982). For a-As_2Se_3 in the solid phase, ε_2 is a quadratic function of energy, but on melting it becomes a linear function (Andreev et al., 1976). The nature of the dimensionality of the free-carrier bands appears to play an important but still somewhat obscure role on the optical absorption edge (Weinstein et al., 1981; Zallen et al., 1981).

We return to Fig. 24. The plotted curve is of the form

$$\Delta E_G(T)|_P = E_G(0) - E_G(T)|_P = K/[\exp(\Theta_E/T) - 1], \qquad (55)$$

where $K = 220 \times 10^{-3}$ eV and $\Theta_E = 400°$K and we explicitly note that the measurements are made at constant pressure. In Fig. 25, we show a similar

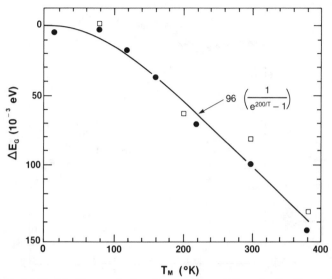

FIG. 25. Red shift of the Tauc optical gap for two sputtered films of a-Ge : H_x as a function of measurement temperature $[E_G(0) = 1.045$ eV]. [From Persans *et al.* (1983).]

plot for a-Ge : H_x, where $K = 96 \times 10^{-3}$ eV and $\Theta_E = 200°$K (Persans *et al.*, 1983). Both fits are excellent and should be compared to Fig. 26, in which the temperature dependence of the optical gap $E_G(T)$ for c-Si is shown fitted by an equation with the temperature dependence of Eq. (55).

It is surprising that the simple temperature dependence given by Eq. (55) has been overlooked in the experimental literature [see, e.g., Thurmond (1975)]. The temperature dependence of E_G is usually taken to be a semi-empirical formula, due originally to Varshni (1967) and later extended by Ravinda and Srivastava (1979). The physical basis for Eq. (55) is the Einstein representation of the total thermal energy of the solid, where $\Theta_E \approx \frac{3}{4}\Theta_D$ in order to normalize the zero point portion of the Einstein model to that of the Debye model and Θ_D is the "Debye–Waller" Debye temperature which is proportional to, but somewhat less than, the usual specific heat Debye temperature (Batterman and Chipman, 1962).

Understanding the physical basis for the form of Eq. (55) and estimating the magnitude of the constant K are simple matters. We first define a generalized coordinate Q that is coupled to the electronic states of the conduction and valence bands of the semiconductor. We define, as before, $E_G = E_c - E_v$ and define $\Delta E_G(T) = E_G(0) - E_G(T)$. For changes in temperature at constant volume

$$\Delta E_G(T)|_V = (D/2)(\langle q^2 \rangle_0 - \langle q^2 \rangle_T), \tag{56}$$

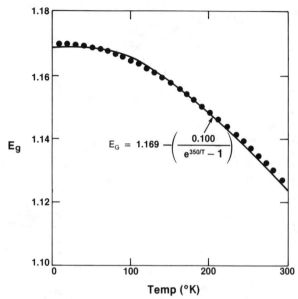

FIG. 26. Temperature dependence of the optical gap of c-Si. The curve is an Einstein oscillator fit [Eq. (64)]. [After Bludau *et al.* (1974).]

where

$$\langle q^2 \rangle_T = \langle (Q - \langle Q \rangle)^2 \rangle_T \tag{57}$$

is the mean square deviation of Q from its mean value $\langle Q \rangle$. For simplicity we use the notation $\langle \ \ \rangle$ to denote an ensemble average. The quantity D is an isotropic second-order deformation potential defined by

$$D = \langle \partial^2(E_c - E_v)/\partial Q^2 \rangle. \tag{58}$$

We anticipate D to be positive and with a magnitude of the same order as that of the spring constant of a vibrating atom, since $E_v - E_c$ is a measure of the binding energy. Thus, we estimate

$$D \approx M_a(k\Theta_E/\hbar)^2 \approx 8 \text{ eV/Å}^2,$$

where Θ_E is the Einstein temperature and M_a the atomic mass; for a-Si : H$_x$, $\Theta_E = 400°$K, and $M_a = 28$. Finally, by the virial theorem $\langle q^2 \rangle_T/\langle q^2 \rangle_0 = U(T)/U(0)$, where $U(T)$ is the total thermal energy, which can be expressed in the Einstein oscillator model by

$$U(T) = 3R\Theta_E\{[\exp(\Theta_E/T) - 1]^{-1} + \tfrac{1}{2}\}. \tag{59}$$

Equations (56)–(59) are easily shown to produce Eq. (55), but the argument is incomplete. We still have to include thermal expansion, since optical measurements are usually made at *constant pressure,* not *constant volume.* Fortunately, the effect of thermal expansion is small for tetrahedrally coordinated semiconductors and is easily incorporated in the model. We use the Grüneisen model for thermal expansion, which again relates the volume change due to thermal expansion to the total thermal energy $U(T)$ (Berard and Harris, 1983; Ashcroft and Mermin, 1976). We finally arrive again at Eq. (55) for constant pressure:

$$\Delta E_G(T)|_P = K/[\exp(\Theta_E/T) - 1]. \tag{55'}$$

The quantity K is given by $K_V + K_P$, where K_V is an intrinsic contribution (constant volume)

$$K_V = \langle q^2 \rangle_0 D \tag{60a}$$

and K_P arises from thermal expansion,

$$K_P = (\partial E_G/\partial P)_T(3R\Theta_E\Gamma/V_A). \tag{60b}$$

In the above equations, V_A is the gram atomic volume; $\hat{\Gamma}$ is the Grüneisen constant, defined as $\Gamma = -(\partial \ln \Theta/\partial \ln V)$; $(\partial E_G/\partial P)_T$ is the pressure coefficient of the optical gap at constant temperature; $\langle q^2 \rangle_0$ is the zero-point dispersion of the generalized coordinate Q, defined by Eq. (57); and D is the second-order deformation potential defined by Eq. (58). In Table I, we list for this simple model the experimental value of K and the quantity of K_V derived on the basis of indicated values of $(\partial E_G/\partial P)_T$, V_A, and Γ. The quantity Γ was determined from the indicated values of thermal expansion ε and compressibility χ.

The harmonic-oscillator model leads to $K_V = \langle q^2 \rangle_0 D \approx k\Theta_E$. We obtain, from Table I, $K_V = 7.7k\Theta_E - 0.08$, which is as much agreement with experiment as one should expect from such a naive model. The general problem of calculating the effects of temperature on the electronic band structure is complicated, since it involves, as discussed by Allen and Heine (1976), both intraband and interband self-energy and Debye–Waller contributions to K_V (Cohen and Chadi, 1980).

We have not attempted to improve on these exact theories but rather to show through Eq. (55) and Table I that temperature affects the Tauc optical gap of amorphous semiconductors defined by Eqs. (54) and (28) in the same manner as that in which it affects the optical gap of crystalline semiconductors. This conclusion, that thermal disorder affects crystalline and amorphous material in a similar way, should not be surprising at this point. The coherent potential model that was described in Section 6 is just as appropriate to Gaussian thermal site disorder as to Gaussian structural site disorder

TABLE I

THERMAL AND PRESSURE PARAMETERS FOR SELECTED CRYSTALLINE AND AMORPHOUS SEMICONDUCTORS

	K $(10^{-3}$ eV$)$	$K_V{}^a$ $(10^{-3}$ eV$)$	Θ_E $(°K)$	$\Theta_D{}^b$ $(°K)$	V_A (cm^3)	ε $(10^{-6} °K^{-1})$	χ $(10^{-12}$ cm^2 dyn$^{-1})$	$\partial E_G/\partial P$ $(10^{-12}$ eV dyn^{-1} cm$^2)$	Γ
c-C	740	820	1360[c]	2200	4.6	1.0	0.25	−0.50	1.6
c-Si	100[d]	105	350[d]	645	12.1	2.5	1.02	−1.50	0.4
c-Ge	95	80	230	374	13.5	5.7	1.33	+5.0	0.7
c-GaAs	90[e]	60	210[e]	344	13.6	6.0	1.33	+11.3	0.8
a-Si:H_x	220[d]	~220	400[d]	—	12.6[f]	1.3[g]	~1.0[h]	−1.0[i] − 0.8[j]	—
a-Ge:H_x	95[d]	~70	200[d]	—	14	17[k]	~1.3[h]	+3.5[l]	—

[a] From this table and Eq. (60).

[b] The value listed is the specific heat Debye temperature. A more relevant quantity might be the "Debye–Waller" Debye temperature (Batterman and Chipman, 1962).

[c] Clark et al. (1964).

[d] From Figs. 24–26.

[e] Panish and Casey (1969).

[f] Ruppert et al. (1981).

[g] Miyagi and Funakoshi (1981).

[h] Guttman (1977).

[i] Welber and Brodsky (1981).

[j] A. Hayes and H. Drickamer (1981). Private communication; 0.5 to 10 kbar.

[k] Persans and Ruppert (1984).

[l] Connell and Paul (1972).

(Abe and Toyazawa, 1981). Indeed, Eq. (46) can be rewritten as

$$W_i^2 = A_i \langle q^2 \rangle, \tag{61}$$

where A_i is a constant. Thus Eq. (51) of Abe and Toyazawa is entirely equivalent to Eq. (56). What may be surprising is the concept that structural and thermal disorder are additive through the Gaussian distribution law assumed for $\langle q^2 \rangle$ and that it is impossible to *separately* distinguish the effects of thermal and structural disorder on the absorption edge (Cody *et al.*, 1981).

This last point is illustrated by Figs. 27–29. In Fig. 27 we exhibit the optical absorption coefficient α for a-Si:H_x as a function of photon energy E. The data have been taken by optical transmission on one film and are accurate for $\alpha \gtrsim 2 \times 10^2$ cm^{-1}. The solid points indicate data obtained at measurement temperatures T_M below room temperature. The open points refer to measurements made on a similar film at $T = 293°$K, where the film

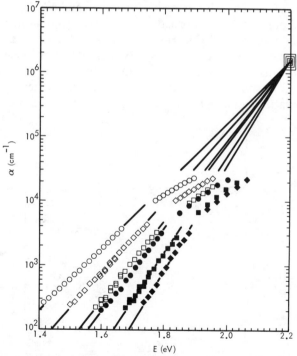

FIG. 27. The absorption edge of a-Si:H_x as a function of isochronal annealing temperature T_A and measurement temperature T_M. Open symbols ($T_M = 300°$K): $T_A = 575°$C (O), 525°C ($\Diamond$), 475°C ($\Box$). Solid symbols: $T_M = 293°$K (●), 151°K (■), 12.7°K (♦). [After Cody *et al.* (1981).]

had been isochronally annealed for 30-min periods at temperatures T_A up to 625°C. The film eventually crystallized at 700°C. The absorption edge in Fig. 27 exhibits the characteristic Urbach form

$$\alpha = \alpha_0 \exp(E - E_1)/E_0, \tag{62}$$

where E_1 and α_0 are independent of either thermal or structural disorder. In Fig. 28 we show the hydrogen content at each temperature of the film as inferred from measurements of evolved hydrogen pressure. We note that starting at about 300°C the hydrogen content of the film dropped from 13 at. % H to 2 at. % H. From Fig. 27 the large loss of hydrogen results in a red shift of the absorption curve ($\Delta E_G < 0$), and an increase in the inverse slope of the exponential edge ($\Delta E_0 > 0$).

Despite the remarkable difference between nondestructive changes in measurement temperature and the appreciable bond breaking represented by the loss of hydrogen in isochronal annealing at temperatures above 300°C, the different data plotted as shifts of ΔE_G [E_G is defined by Eq. (54) and positive ΔE_G is a red shift or *reduction* in E_G] and ΔE_0 [E_0 is defined by Eq. (62) and positive ΔE_0 is a broadening of the edge or an *increase* in E_0] are linear and indistinguishable as shown in Fig. 29 for several films.

The data on E_G, summarized in Fig. 29, were taken by linear extrapolation, as well as by interpreting isoabsorption measurements above $\alpha = 10^4$ cm^{-1} as shifts in E_G. The Urbach edge parameter E_0 was determined by measurements of the slope of log α and also by isoabsorption for

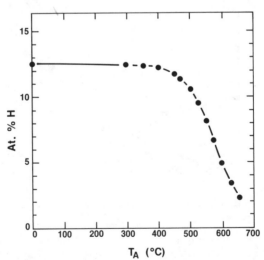

FIG. 28. Hydrogen content of a-Si:H$_x$ film after 30-min isochronal anneal at indicated temperature.

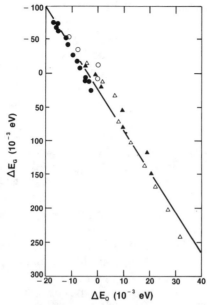

FIG. 29. Shift in Tauc optical gap E_G and Urbach edge parameter E_0 for three a-Si:H$_x$ films. The circles correspond to changes in measurement temperature. The triangles correspond to changes in E_G and E_0 observed at 300°K after isochronal annealing and subsequent loss of hydrogen. A positive ΔE_G is a decrease in E_G. A positive ΔE_0 is an *increase* in E_0. Sample 1: ●, temperature dependence; ▲, evolution. Sample 2: ○, temperature dependence. Sample 3: △, evolution. [From Cody *et al.* (1981).]

2×10^2 cm$^{-1} < \alpha < 5 \times 10^3$ cm^{-1}. Examination of Figs. 14 and 15 shows that this range is somewhat above the true exponential region and results in a value of E_0 perhaps 20×10^{-3} eV higher than would be determined over a wider range in α (Redfield, 1982; Roxlo *et al.*, 1983). The data have thus been plotted as ΔE_G with respect to ΔE_0. From the straight line we obtain

$$E_G = -6.2 E_0 + 2.0 \text{ eV}, \tag{63}$$

where we have used $E_G = 1.64$ eV from Fig. 16 and $E_0 = 50 \times 10^{-3}$ eV from Fig. 14. The Urbach focus, defined by Eq. (63) in the framework of the Abe and Toyazawa (1981) model of Section 6, is thus about 2.0 eV, about 0.2 eV below the approximate focus of 2.2 eV in Fig. 27, derived for $\alpha > 10^2$ cm^{-1}. The focus derived from the data of Fig. 15 for slight variations in growth conditions agrees well with that estimated from Eq. (63).

If we interpret the experimental Urbach focus in terms of the Abe and Toyazawa model of a virtual crystal with varying degrees of site disorder, we must ascribe to that virtual crystal a direct optical transition of about 2.0 eV.

This value is considerably less than the value of 3.6 eV for the direct transition of c-Si (Chelikowsky and Cohen, 1976).

Through Eq. (63), E_G and E_0 are linearly related, and it makes little difference whether the Tauc gap E_G or the Urbach slope E_0 is used as a measure of disorder. Experimentally, the effect on ΔE_G is larger than that on ΔE_0. But unlike E_0, shifts in E_G can also be caused by chemical or coordination effects on the band widths of the conduction and valence bands. For example, high hydrogen content a-Si: H_x ($x > 0.25$) prepared by homogeneous chemical vapor deposition (homoCVD) has a range of Tauc gaps depending on preparation, which are considerably higher than those of low hydrogen content a-Si: H_x prepared by glow-discharge decomposition of SiH_4 (see Chapter 2 by Scott in Volume 21A). One possibility for this difference is a reduction in atomic coordination at high hydrogen concentrations with subsequent band narrowing and hence a blue shift in E_G. If this hypothesis were correct, the Urbach focus for this interesting material would be perhaps 0.5 eV higher than for a-Si: H_x, but we should still expect Eq. (63) to be followed by *small* changes in thermal or structural site disorder.

An advantage in using E_0 rather than E_G as a measure of disorder is that this choice makes it possible to quantify the concept of structural disorder. For example, the definition of site disorder as the mean square dispersion of a generalized coordinate suggests the following general form for E_0:

$$E_0(T, X) = \left(\frac{\Theta_E}{\sigma}\right)\left\{\frac{1}{\exp(\Theta_E/T) - 1} + \frac{1 + X}{2}\right\}. \tag{64}$$

In Eq. (64) the only undefined parameters are σ, the Urbach edge parameter, and X, which is a measure of structural disorder.

We consider each in detail. As $T \to \infty$, $E_0(T, X) \to T/\sigma$, and hence σ represents a reduction of thermal disorder from kT. For crystalline materials in which an exciton mechanism defines the absorption edge, there is evidence that σ is significantly different from one. For example, c-GaAs exhibits $\sigma \simeq 2$ (Redfield and Afromowitz, 1967) and alkali halides exhibit $\sigma \simeq 0.5$ (Kurik, 1971). In these crystalline materials, the exciton broadening arises from the coupling to distinct phonons. In amorphous semiconductors, the edge is coupled to the total thermal disorder of the lattice. As we shall show, $\sigma \simeq 1$ appears to be a reasonable approximation for a-Si: H_x. The quantity X is defined by the relation

$$X = \langle q^2 \rangle_x / \langle q^2 \rangle_0, \tag{65}$$

where we have expressed the total disorder in terms of additive structural and thermal contributions as

$$\langle q^2 \rangle \equiv \langle q^2 \rangle_T + \langle q^2 \rangle_x. \tag{66}$$

Thus $\langle q^2 \rangle_x$ represents the frozen-in thermal disorder that has been so frequently used to describe modifications of the continuous random network (Tauc, 1970; Oheda, 1979; Street et al., 1974; Moss and Graczyk, 1970). Equation (65) scales this frozen disorder by the zero-point disorder of the generalized coordinate Q. From a harmonic oscillator model

$$\langle q^2 \rangle_0 = (6.40 \times 10^{-3} \text{ Å}^2)(400/\Theta_E)(28/M), \tag{67}$$

where M is the atomic mass and Θ_E is the Einstein equivalent to the Debye temperature.

From Eqs. (64) and (63) we identify $6.2\Theta_E/\sigma$ with the quantity K_V defined by Eq. (60a) and we obtain from Table I $\sigma = 0.95$. In what follows, we choose $\sigma = 1.0$. With this value of σ, we can by using $E_0(300, X) = 50 \times 10^{-3}$ eV (Fig. 15) obtain $X = 1.3$. From this value of X we deduce $[\langle q^2 \rangle_x]^{1/2} = 0.1 \text{Å}$. This value is considerably larger than would be expected for fluctuations in nearest-neighbor bonds, but it is of the right magnitude for the fluctuations of the next-nearest-neighbor distances due to root-mean-square fluctuations of bond angle by about 2° and is well within the variation one expects for a continuous random network (Guttman and Fong, 1982). This interpretation of X suggests that in a-Si : H_x, E_0, E_G, and presumably the top of the valence band are sensitive to bond angles.

We may use the quantity X to define a fictive temperature T_F from the width of the optical edge (Cody et al., 1983). By T_F we mean that temperature at which all the disorder associated with E_0 is entirely thermal. It is thus the temperature equivalent of the concept of frozen-in disorder and summarizes the data of Fig. 30 in terms that may convey structural insight. From Eq. (64) with $\sigma = 1$ we define T_F by

$$E_0(T_F, 0) \equiv E_0(T_M, X(T_A)), \tag{68}$$

where T_M indicates the measurement temperature and $X(T_A)$ denotes a structural site disorder induced by isochronal annealing at T_A. Of course, under certain circumstances $T_M = T_A$, but often T_M is fixed at 300°K and only T_A is varied. If we define the quantity B by $B = (E_0(T_M, X(T_A))/\Theta_E) - \frac{1}{2}$, the fictive temperature T_F is then given by

$$T_F = \Theta_E/\ln[(1 + B)/B]. \tag{69}$$

In Fig. 31 we show T_F for a-Si : H_x as a function of T_A for $T_M = 300°K$. Below $T_A \sim 700°K$, T_F is about 800°K, and just prior to crystallization T_F is about 1100°K. The maximum in T_F is thus about two thirds of the melting point of c-Si, in rough agreement with the simplest rule of thumb for interatomic diffusion controlled crystallization.

An Urbach edge is exhibited by many other amorphous semiconductors. Amorphous Se (Siemsen and Fenton, 1967) and As_2Se_3 (Andreev et al.,

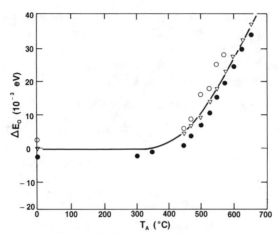

FIG. 30. Increase in E_0 at $T_M = 300°$K for three films of a-Si : H$_x$ after 30-minute isochronal anneals at T_A.

1976) are particularly interesting because it has been possible to measure the absorption edge both above and below the melting point. In Fig. 32 we show the absorption data for Se of Siemsen and Fenton (1967), which can be described by Eq. (62) with a focus point given by $E_1 = 2.4$ eV and $\alpha_0 \sim 10^7$ cm^{-1}. The edge parameter E_0 is a function of the measurement temperature T_M. Above the melting point, the quantity X increases rapidly with T_M. Presumably, the degree of structural site disorder X for a-Se films

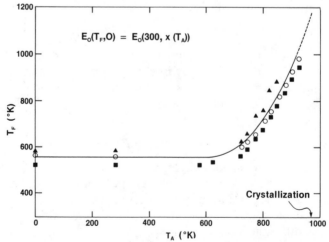

FIG. 31. The fictive temperature T_F of a-Si : H$_x$ defined by the Urbach edge parameter E_0 as a function of annealing temperature T_A.

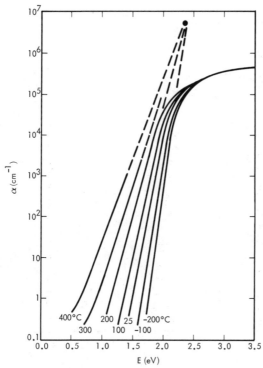

FIG. 32. Absorption edge of a-Se as a function of measurement temperature as measured by Siemsen and Fenton (1967). [After Mott and Davis (1979, p. 522).]

(a-Se normally self-anneals above 300°K) increases rapidly on melting. The representation of this disorder in terms of the fictive temperature T_F is shown in Fig. 33, and its temperature dependence is similar to that of a-Si:H_x. Bond breaking due to hydrogen desorption and bond breaking due to melting exhibit the same rapid rise in the parameter T_F.

There are interesting parallels that can be made between a-Se and a-Si:H_x. An important *difference* between a-Se and a-Si:H_x is the linear variation with energy of the band-to-band contribution to ε_2, which can be interpreted by a one-dimensional density of states for a-Se (Mott and Davis, 1979, p. 522; Davis, 1970). An interesting *similarity* to a-Si:H_x is the transport evidence that the broad exponential tail in the optical absorption of a-Se (Fig. 32) is associated with the band that contains the carrier with the *lowest* drift mobility. For a-Se, this low-mobility band is the conduction band (Enck and Pfister, 1976); for a-SiH_x, it is the valence band (see Chapter 4 by Tiedje in Volume 21C). For both materials, the exponential density of trap states deduced from transport measurements for the high-mobility

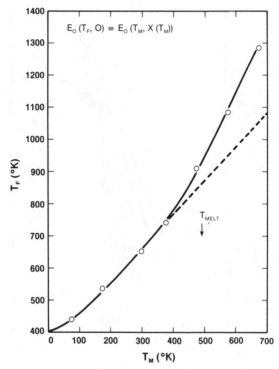

FIG. 33. Fictive temperature of a-Se derived from Fig. 32 as a function of measurement temperature T_M. The dashed line is the extrapolation of Eq. (69) for $X = \text{const} = X(373°K)$.

band has a width about one half that associated with the less mobile band. Another important difference is the dominant role of geminate recombination in a-Se and its small role in a-Si:H_x (Wronski *et al.,* 1981). These parallels between one of the oldest amorphous semiconductors and one of the newest are useful fixed points for any model that claims generality. Their explanation, in terms of fundamental concepts, should be rewarding, given the wealth of experimental data on each material.

Up to now we have discussed the effect of thermal and structural disorder on what might be considered an ideal system: a-Si:H_x grown from the glow-discharge decomposition of pure SiH_4 in deposition systems that operate under ultrahigh vacuum conditions at substrate temperatures T_S between 250 and 300°C. The data that has been discussed is, of course, only a part of a large body of literature. Since the pioneering studies of Tsai and Fritzsche (1979), the effects of annealing on the optical edge of a-Si:H_x, and the interplay among annealing temperature T_A, deposition temperature T_S, and measurement temperature T_M has been studied at many laboratories.

Similar studies have been made on the effects of annealing on sputtered films of a-Si: H_x (Paul and Anderson, 1981). Finally, there is extensive information on the effect of annealing and changes in measurement temperature on the optical edge of films made by the thermal decomposition of SiH_4 (CVD a-Si: H_x) (Janai and Karlsson, 1979; Divrechy et al., 1981; Yous et al., 1981). For these materials the hydrogen content is very low due to the high value of T_S (Hey and Seraphin, 1982).

In some cases, the films that have been studied are less ideal than what has been discussed until now. For example, there are often microscopic inhomogeneities in sputtered films or in glow-discharge films made from argon diluted SiH_4 as shown by low values of the index n when compared to Fig. 8 or ε_2 compared to Fig. 11 (Nitta et al., 1982). These films often exhibit hydrogen evolution curves that are considerably more complicated than that shown in Fig. 28 and that suggest a broad distribution in bonded hydrogen. Finally, under certain growth conditions, CVD, glow-discharge, or sputtered films may have large oxygen concentrations due to atmospheric leaks or water vapor, which may dominate certain features of the absorption edge (Lee and Lee, 1982).

Despite these differences in the films, there is broad general agreement in the literature on the effect of changes in T_M or T_A on the absorption edge. Lowering the temperature of measurement below room temperature leads to a blue shift in the optical gap of the same magnitude as that shown in Fig. 24. Raising the T_M has three effects on E_G [and through Eq. (97) on E_0]. For T_M above $300°K$ there is an intrinsic red shift due to increased thermal disorder, but since annealing may be occurring ($T_A \approx T_M$ for $T_M > 300°K$), a potential blue (red) shift can occur due to an ordering (disordering) reconstruction of the continuous random network. At higher measurement and/or annealing temperatures, a red shift in E_G is observed due to hydrogen evolution and bond breaking. Presumably this generation of point defects triggers reconstruction of the continuous random network and a consequent increase in site disorder. In most cases, measurements are done only at room temperature and only the effect of annealing is observed.

In all cases, the dominant feature of Fig. 29, the identical magnitude in ΔE_G for a temperature change of several hundred degrees celsius, or for the loss of 70 to 80% of the hydrogen, is observed. However, the relatively simple changes in the absorption coefficient α, exhibited in Fig. 34, are the exception rather than the rule. In this figure, we show the optical absorption edge as a function of 30-min isochronal anneals at the indicated temperatures for an a-Si: H_x film growth at $250°C$ under growth conditions that lead to high electronic quality (Roxlo et al., 1984; Cody et al., 1983). The measurements were made by a combination of optical transmission and photothermal absorption (see Chapter 3 by Amer and Jackson). The re-

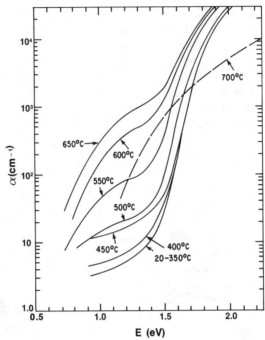

FIG. 34. Optical absorption edge of a film of a-Si:H$_x$ as determined by transmission and photodeflection spectroscopy. The dashed curve was obtained on crystallization and is about 20% higher than the optical absorption coefficient of c-Si as given by Aspnes and Studna (1982). [From Roxlo *et al.* (1984).]

markable stability of the absorption edge up to 400°C (Fig. 35), the monotonic shift of E_G to the red, the monotonic increase of E_0, and the rapid rise in the density of deep states (dangling bonds) prior to crystallization at 700°C are all characteristic of glow-discharge films made under these ideal conditions. To temperatures as high as 675°C, the films maintain very low scattering losses in absorption. An experimental problem with such measurements is separating the shoulder absorption from the tail absorption, so that $E_0[T_M, X(T_A)]$ can be estimated. One approach is to use isoabsorption, and Eq. (62), at high enough α, so that the shoulder can be neglected. Another approach is an optimized fit of Eq. (21) to the absorption curve with an assumed density of states for the optical absorption in the shoulder regime (Region A of Fig. 5) (Wronski *et al.*, 1982; Vanecek *et al.*, 1982; Jackson and Amer, 1982). There is no similar difficulty in extracting ΔE_G from high α data ($\geq 10^4$ cm^{-1}), and as shown in Fig. 35, the data is reproducible and monotonic with T_A.

As has been noted many times, glow-discharge films made at lower

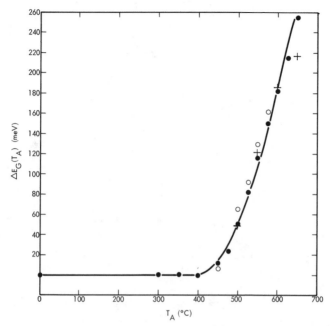

FIG. 35. Decrease in Tauc optical gap E_G of three films of a-Si:H_x as a function of annealing for 30 min at T_A. [From Cody et al. (1982). Copyright North-Holland Publ. Co., Amsterdam, 1982.]

substrate temperatures and higher hydrogen concentrations do not exhibit monotonic changes in the absorption edge with annealing temperature. In Fig. 36 we exhibit absorption curves as a function of isochronal annealing temperature T_A obtained by Yamasaki et al. (1981a) for material deposited at a substrate temperature of 100°C. As shown in Fig. 37, a net reduction of the spin density and of the shoulder absorption at 1.2 eV near $T_A = 300$°C is preceded by both a blue (ordering) and red (disordering) shift in E_G. For higher temperature anneals, an increase in spin density is accompanied by a continuous increase in disorder, as manifested by a red shifted E_G. This film had an initial hydrogen content of 22 at. % and an increase by almost a factor of two in ΔE_G at $T_A = 550$°C over that shown in Fig. 35 for a lower hydrogen content film. Reconstruction and consequent disordering of the network appears to be a function of net hydrogen desorption, as well as annealing temperature.

It is important to note that the spin density reported by Yamasaki et al. (1981b) for a film prepared at a substrate temperature of 100°C and annealed at 550°C is an order of magnitude *less* than that reported by Yoshida and Morigaki (1983) for a similarly annealed film initially prepared at a higher substrate temperature (300°C). This film has considerably less

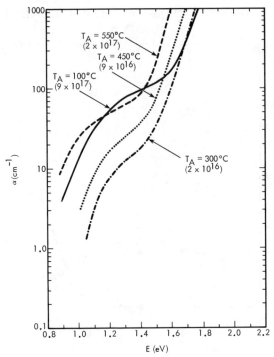

FIG. 36. Absorption edge for a GD film of a-Si:H_x prepared at a substrate temperature of $T_S = 100°C$ and annealed at the indicated temperatures T_A. [After Yamasaki *et al.* (1981a)]. The measured spin densities are indicated in parentheses.

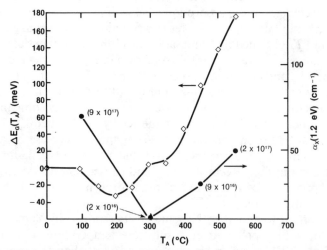

FIG. 37. Shift in the Tauc gap E_G and change in shoulder absorption at 1.2 eV (α_x) as a function of annealing temperature T_A for an a-Si:H_x film prepared at $T_S = 100°C$. Positive ΔE_G is a decrease in E_G. [From Yamasaki *et al.* (1981a).]

hydrogen than the film prepared at a substrate temperature of 100°C. One explanation of the discrepancy is that spin–spin interactions in the highly hydrogenated film have reduced the sensitivity of the spin resonance signal to the dangling-bond density.

This explanation is consistent with the identification of the shoulder absorption with the density of dangling bonds (see Chapter 3 by Amer and Jackson). As shown in Fig. 36, despite an identical shoulder absorption, there is an order of magnitude discrepancy between the spin density of the as-grown film and the film annealed at 550°C.

Even small changes in deposition conditions can lead to large changes in the optical absorption edge on annealing. In Fig. 38 we compare an a-Si : H_x film prepared at $T_S = 200$°C with 14 at. % H, $E_G = 1.72$ eV, and $E_0 = 40 \times 10^{-3}$ eV (film A) with a film of similar electrical quality, but prepared at $T_S = 250$°C with 9 at. % H, $E_G = 1.64$ eV, and $E_0 = 50 \times 10^{-3}$ eV (film B). For film B, there is little measurable change in E_G up to $T_A = 400$°C. For film A, E_G shifts to the red above $T_A = 250$°C, stays constant around 350°C, shifts to the blue up to 425°C, and then shifts to the red above this value. If we interpret blue and red shifts in ΔE_G as ordering and disordering of the continuous random network, we can tentatively identify (at least!) three processes: the first is perhaps the release of weakly bonded hydrogen, accompanied by a disordered reconstruction of the network (200–300°C); the second is an annealing of the network (300–425°C); and, finally, the generation of sufficient dangling bonds, through H desorption and perhaps

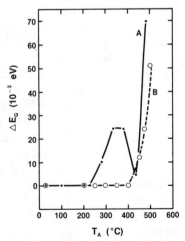

FIG. 38. Decrease of Tauc gap E_G as a function of annealing temperature T_A. Film A differs from film B in substrate temperature (200 rather than 250°C), hydrogen content (14 compared to 9 at. %), Tauc optical gap (1.72 compared to 1.64 eV), and Urbach edge parameter (42×10^{-3} compared to 50×10^{-3} eV).

other thermal processes, so as to lead to large increases in disorder through network reconstruction. Again, the net shift in ΔE_G in Fig. 38 is larger than that shown in Fig. 35, and again reflects the appreciably greater initial hydrogen concentration for film A.

Donnadieu et al. (1981), Berger et al. (1981), and Berger et al. (1983) examined a-Si:H_x films grown by glow-discharge decomposition of 50% SiH_4 in 50% Ar at various substrate temperatures and at measurement temperatures between 95 and 723°K. Berger et al. (1983) studied two of these films ($T_S = 350°C$, 8 at. % H, and $T_S = 280°C$, 18 at. % H) when they were isochronally annealed at temperatures up to 723°K. The optical data agree qualitatively with that shown in Figs. 30 and 35, but there are, however, significant quantitative differences. There is a considerably larger variation in ΔE_G and ΔE_0 with T_A than that shown in Figs. 30 and 35. In addition, this slope of the plot of ΔE_G with respect to ΔE_0 is closer to 3 than to 6, as shown in Fig. 29. Despite these quantitative differences, there remains fundamental agreement with the concept that isochronal annealing can be described within the context of a disorder-dependent Urbach edge, in which the shift and broadening of the edge is due to increased site disorder which is produced by the generation of dangling bonds due to hydrogen desorption and subsequent reconstruction of the network. The presence of trapped argon in these films may introduce additional processes in the annealing studies that prevent quantitative agreement with Fig. 29, or the discrepancy may be due to the previously noted problem in accurately measuring ΔE_0.

Matsumura and Furukawa (1980) and Hasegawa et al. (1978) have examined the effect of isochronal annealing on sputtered a-Si, a-Si:H_x, and a-Si:F_x films and again observed nonmonotonic behavior in E_G, corresponding to more than one annealing process. For a-Si sputtered films prepared at room temperature, Hasegawa et al. (1978) found a reduction in spin density by a factor of 10 and an increase in E_G for $T_A = 300°C$. This result is remarkably similar to that shown in Fig. 37 and again points to annealing out of dangling bonds and an ordering of the network if there is no hydrogen present between 200°C and 300°C. Annealing to 600°C leads to an increase in spin density by a factor of 10 and a red shift of ΔE_G by about 70×10^{-3} eV, about the same magnitude as that shown in Fig. 35 and presumably due to a similar thermal generation of dangling bonds as observed by Chik et al. (1980) for evaporated a-Si.

Finally, the comprehensive review of Paul and Anderson (1981) describes the effect of isochronal annealing on high electrical quality a-Si:H_x films made by reactive sputtering. There is general qualitative agreement with the glow-discharge-prepared material, but the red shifts in E_G with increase in T_A are considerably larger than that shown in Fig. 36 (i.e., at $T_A = 550°C$,

$\Delta E_G \sim 0.2$ eV for the sputtered films compared to 0.1 eV for the glow-discharge film shown in Fig. 35). Again, this may reflect the higher hydrogen content (~ 20 at. %) of these films.

We can summarize this section by noting that there is ample evidence that the Tauc gap E_G, the Urbach slope parameter E_0, and the shoulder in the optical absorption edge are sensitive to bond breaking induced by either hydrogen desorption or the thermal generation of dangling bonds. The data suggest that E_G and E_0 are sensitive functions of the disorder of the network and that hydrogen primarily affects the Tauc optical gap through its effect on disorder. Finally, analysis of the data suggests an Urbach focus of about 2.0 eV for a-Si:H_x ($x \lesssim 0.20$), which on the basis of the model described in Section 6 would be the direct optical gap of ordered a-Si:H_x. Differences in sample preparation, homogeneity, hydrogen concentration, and perhaps even optical roughness can have effects as large as a factor of 2 on ΔE_G and ΔE_0 but do not change the conclusion that the position and shape of the optical edge are sensitive and quantitative measures of the structural and thermal site disorder of a-Si:H_x and other amorphous semiconductors.

8. Effect of Film Preparation on the Absorption Edge of a-Si:H_x

To date, there are at least seven methods of film preparation (for further details see Volume 21A) for a-Si:H_x for which there are data on the optical absorption edge:

(1) glow discharge (GD) decomposition of pure SiH_4 or SiH_4 mixtures with inert or active gases (Cody et al., 1982; Berger et al., 1983),

(2) sputtering of Si by argon in the presence or absence of hydrogen (Paul and Anderson, 1981; Moustakas, 1982; Savvides et al., 1983),

(3) evaporation of Si (Pierce and Spicer, 1972; Brodsky, 1971; Chik et al., 1980),

(4) chemical vapor deposition of pure SiH_4 or SiH_4 mixtures with inert or active gas on a heated substrate (heterogenous CVD) (Hey and Seraphin, 1982; Bagley et al., 1982; Hirose, 1981),

(5) chemical vapor deposition of pure Si_2H_6 or Si_2H_6 mixed with inert or active gases (Akhtar et al., 1982; Delahoy, 1983),

(6) chemical vapor deposition of hot SiH_4 or SiH_4 mixtures with inert or active gases on a cooled substrate (homogeneous CVD) (see Chapter 2 in Volume 21A by Scott), and

(7) reactive ion beam sputtering of a-Si:H_x (RIBS) (Kasdan et al., 1984).

Up to this point we have concentrated on a-Si:H_x films made by the GD decomposition of pure SiH_4 at substrate temperatures between 200 and

300°C, in bakable UHV deposition systems, and under growth conditions that produce films of high electrical quality as measured by thin film solar cell performance and high structural perfection as measured by high density and high dielectric constant. We have also emphasized undoped films since we are primarily interested in the intrinsic optical properties of a-Si : H_x. In this section we briefly summarize the general features of the absorption edge for representative films made by some of the above techniques and focus on important differences and similarities. A general conclusion of the comparison is that the similarities far outweigh the differences. In the band-to-band region (C of Fig. 5), the coefficient K_{BB} is identical, within experimental error, for the evaporated a-Si films of Pierce and Spicer (1972) ($E_G = 1.4$ eV) and a-Si : H_x ($E_G = 1.6$ eV). In the tail region (B of Fig. 5), the parameter E_0 can vary by as much as a factor of two. For high electronic quality a-Si : H_x, E_0 can be as low as 43×10^{-3} eV at $T = 300°K$ (Tiedje *et al.*, 1983a; Roxlo *et al.*, 1983). For high temperature CVD films (Janai and Karlsson, 1979) or evaporated films, E_0 varies from 100 to 200×10^{-3} eV, and there is rough agreement with Eq. (63) for the corresponding change in E_G. The largest differences in the absorption edge for films made by different techniques occur in the shoulder region (A of Fig. 5) where the residual deep-gap absorption varies by as much as a factor of 10^4 (cf. Fig. 1).

All of these differences are compatible with the central hypothesis of this chapter: that the position (E_G) and shape (E_0) of the absorption edge are quantitative measures of site disorder. Figure 39 exhibits the trends that

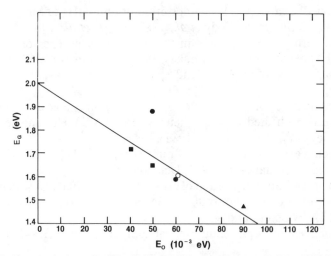

FIG. 39. Tauc optical gap E_G [Eq. (54)] and Urbach parameter E_0 [Eq. (62)] for a-Si : H_x films prepared by reactive sputtering (O), reactive ion beam sputtering (●) (Kasdan *et al.*, 1984), high-pressure CVD (▲) (Hey *et al.*, 1983), and GD (■). The straight line is Eq. (63).

occur when a-Si:H_x films that have been prepared by GD, reactive sputtering, ion-beam sputtering, and high-temperature CVD are compared. The data are referenced to glow-discharge films previously discussed, with $E_G = 1.64$ eV and $E_0 = 50 \times 10^{-3}$ eV. The scatter around the trend line [Eq. (63)] for films made by different techniques may reflect differences in growth morphology, extrinsic impurities, or perhaps the connectivity of the network. Despite the scatter of Fig. 39, the correlation between the broadening of the edge and the disorder-induced red shift of E_G is considerably stronger than any correlation of E_G with hydrogen content (0–20 at. %) for as-grown films (Fig. 40). In this range of concentration, hydrogen appears to influence the shape and position of the absorption edge of a-Si:H_x chiefly through its effect on the disorder of the network.

In this concluding section of this chapter on the optical absorption edge, we briefly indicate the present dominant features of a-Si:H_x films prepared by the techniques given in this chapter. In principle, all techniques should converge on a homogeneous, high-density, continuous, random, tetrahedrally coordinated Si network with sufficient hydrogen to produce very low levels of residual dangling bonds ($\lesssim 10^{15}$ cm^{-3}) and that may contain additional bonded hydrogen due to the kinetics of the specific growth mechanism. In this sense the technique of film growth that has received the most attention in this chapter, films produced by GD of pure SiH$_x$, is singled out only because of the large, world-wide effort on this technique.

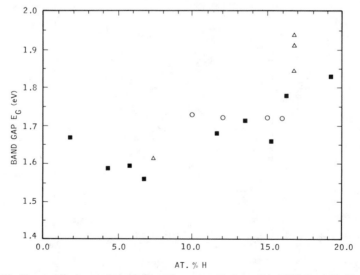

FIG. 40. Tauc optical gap derived from the extrapolation of $(\alpha E)^{1/2}$ for a-Si:H_x films prepared by a variety of techniques as a function of hydrogen concentration. ■, RIBS; ○, rf glow discharge; △, reactive sputtering. [From Kasdan et al. (1984).]

a. The Optical Properties of Films of a-Si : H_x Produced by Glow Discharge

If we define an *optically good* film as one with high values of ε_1 and ε_2, indicating homogeneous material of maximum density, GD material made from pure SiH_4 exhibits excellent optical properties. Dilution of the SiH_4 by inert gases such as argon or helium can reduce the index of refraction at 0.5 eV by as much as 10%, which suggests argon-containing cavities. The absorption edge of electrically optimized intrinsic material is similar to that shown in Fig. 15, where the film with the steepest edge and lowest shoulder corresponds to the material with the highest efficiency (Deckman *et al.*, 1983). As shown in Fig. 41, additions of small amounts of boron increase the shoulder absorption but do not appreciably shift the position of the edge. Larger amounts of boron, as first noted by Tsai (1979), give large changes in the position and shape of the edge and indicate a Si–B alloy (Fig. 42). As shown in Figs. 19 and 43, the effect of phosphorus appears to increase the shoulder absorption but has no appreciable effect on the optical gap.

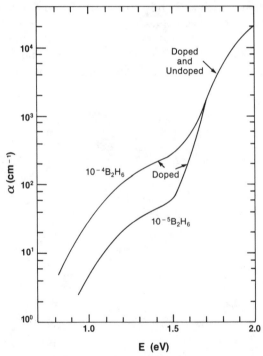

FIG. 41. Optical absorption edge of lightly boron-doped a-Si : H_x. The indicated concentrations refer to the gas phase. Below $\alpha = 4 \times 10^2$ cm^{-1}, the absorption edge was determined by photoconductivity (Wronski, 1983).

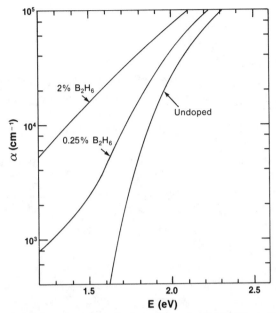

FIG. 42. The absorption edge of heavily boron-doped a-Si:H$_x$. The indicated concentrations refer to the gas phase. The undoped curve is taken from Fig. 14.

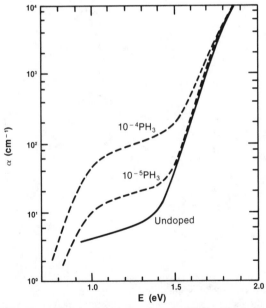

FIG. 43. Absorption edge of lightly phosphorus-doped a-Si:H$_x$. The indicated concentrations refer to gas-phase concentrations. Below $\alpha = 10^2$ cm^{-1}, the absorption edge was determined by photoconductivity (Wronski, 1983); at $\alpha \approx 10^2$ cm^{-1} and above, this value was determined by optical transmission. (For strong phosphorus doping, see Fig. 19.)

b. *The Optical Properties of Films of* a-Si:H$_x$ *Produced by Reactive Sputtering*

These data have been extensively reviewed by Paul and Anderson (1981). In this subsection we consider only the comparison between the absorption edge of a-Si:H$_x$ films of high electronic quality prepared by GD and the absorption edge of similarly high electronic quality reactively sputtered films.

In Fig. 44 we compare a reactively sputtered film containing 20 at. % H, with an E_G of 1.89 eV and an E_0 of 50×10^{-3} eV [subtraction of the shoulder absorption through Eq. (38) would reduce this value by almost 20%], with film B of Fig. 38, containing 9 at. % H with $E_G = 1.64$ eV and

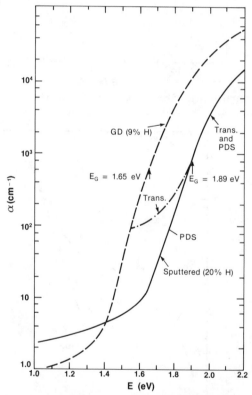

FIG. 44. Absorption edge of high-hydrogen-content sputtered a-Si:H$_x$ compared with GD a-Si:H$_x$. The arrow in each case indicates the Tauc optical gap defined by Eq. (54). For the sputtered film the data for $\alpha \lesssim 10^3$ cm^{-1} was taken by photodeflection spectroscopy. The high-absorption data were obtained by transmission. The discrepancy between transmission and PDS around $\alpha \approx 10^2$ cm^{-1} is ascribed to scattering.

again an $E_0 = 50 \times 10^{-3}$ eV. The shapes of the absorption edges of these films (E_0) are identical but there is almost a 0.3 eV blue shift (E_G) in the edge for the sputtered film. This film departs considerably from Eq. (63) as shown in Fig. 39 and suggests that for hydrogen concentrations close to 20 at. % disorder is not the sole factor in determining E_G. Despite this difference, it is interesting to note that the similarity in the shape of the curve is such that the value of $\alpha(E_G)$ is the same for the two films $[\alpha(E_G) \approx 7 \times 10^2$ cm$^{-1}]$. A major difference between the films is the considerably lower value of the index n. The index n for the GD film is about 3.55 at $E \approx 0.5$ eV (Fig. 8), while for the sputtered film at the same energy, $n \approx 3.00$. In Fig. 44, the saturation in α at low energies for optical transmission is due to scattering, as can be seen from the comparison with α determined by photothermal deflection spectroscopy. Sputtered films often exhibit a surface region of scattering centers which can lead to a spurious shoulder in the derived α that is not seen when absorption is directly measured by either photoconductivity or by photothermal deflection spectroscopy. For the reactively sputtered film, the slope of the linear plot of $(\alpha/E)^{1/2}$ is 2.84 eV$^{-3/2}$ μm$^{-1/2}$, about 5% less than the value of 3.00 eV$^{-3/2}$ μm$^{-1/2}$ obtained for the GD films. This discrepancy is in the opposite direction from the change due to n and may be ascribed to a void density of about 10% (cf. Section 3).

In Fig. 45 we show the absorption edge of a reactively sputtered film containing 3 at. % H with $E_G = 1.58$ and $E_0 = 65 \times 10^{-3}$ eV. The GD film exhibited for comparison is the same as that shown in Fig. 44. The refractive index of the sputtered film is 3.5 at $E = 0.5$ eV, and the shape of the linear plot of $(\alpha/E)^{1/2}$ is 3.10, in reasonable agreement with the GD data. This sputtered film agrees well with the GD standard in both position (E_G) and shape (E_0), and the small shift in E_G is in agreement with Eq. (63).

In Fig. 46, we present some very recent data on ε_2 for a-Si sputtered on a substrate held at 200°C (Savvides et al., 1983). In this figure, we also show for comparison ε_2 for a CVD film (Bagley et al., 1982) and an a-Si:H$_x$ film (Fig. 12). Apart from the displacement of the CVD and sputtered films to low energies, there is remarkable agreement between the three curves.

In Fig. 47, we show a linear plot of $(\varepsilon_2)^{1/2}$ for each of the films shown in Fig. 46. When the data is expressed in the form of Eq. (29), the quantities K_{BB} and E_G, as determined by least squares fits over a range of about 1 eV, are 5.64 eV^{-2}, 1.030 eV for the sputtered film; 7.68 eV^{-2}, 1.391 eV for the CVD film; and 10.02 eV^{-2}, 1.661 eV for the GD film. The common focus for the high α data shown in Fig. 47 is about 3.6 eV. Figure 47 should be compared to Fig. 23 [drawn from Abe and Toyazawa (1981)]. At face value, the data exhibited in Fig. 47 suggest that the focus of region C (~ 3.6 eV) is considerably different from the focus of region B (~ 2.0 eV).

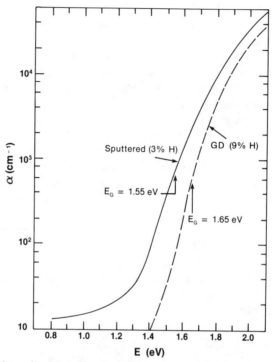

FIG. 45. The absorption edge of low-hydrogen-concentration sputtered a-Si : H_x compared to that of glow-discharge a-Si : H_x. For the sputtered film the low-absorption data were obtained by photothermal deflection spectroscopy. High-absorption data were taken by transmission.

c. *The Optical Properties of Films of* a-Si : H_x *Produced by Chemical Vapor Deposition of Silane (Heterogeneous CVD of* SiH_4)

Atmospheric pressure CVD of SiH_4 is a well-established technology for the growth of epitaxial crystalline thin films. At temperatures below 650°C it leads to amorphous silicon films. The optical properties of such amorphous films have been reported by Janai and Karlsson (1979), Divrechy *et al.* (1981), Yous *et al.* (1981), and Hirose (1981). Perhaps the most complete optical data are those of Bagley *et al.* (1982) and Aspnes and Studna (1981) on films of a-Si grown by *low-pressure* CVD. Both high-pressure and low-pressure CVD films have a hydrogen content of less than 1%. The difference between the techniques lies in the growth kinetics and chemistry.

The optical data on the low-pressure films were obtained by ellipsometry and has already been compared to GD a-Si : H_x, in Figs. 12 and 13. A least

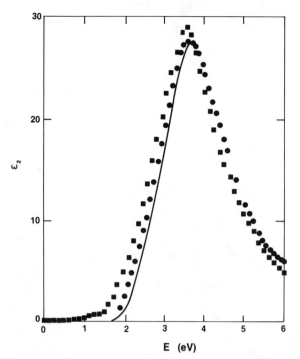

FIG. 46. Imaginary part of the dielectric constant as a function of photon energy for sputtered a-Si [solid squares, Savvides *et al.* (1983)], CVD a-Si [solid circles, Bagley *et al.* (1982)], and GD a-Si : H$_x$ [line, Cody *et al.* (1982)].

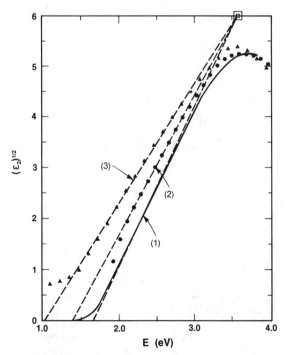

FIG. 47. The quantity $(\varepsilon_2)^{1/2}$ obtained from Fig. 46 plotted as a linear function of energy for the data of Savvides *et al.* (1983) (curve 3), Bagley *et al.* (1982) (curve 2), and Cody *et al.* (1982) (curve 1).

squares fit to $(\varepsilon_2)^{1/2}$ for this film in the range 1.9–3.0 eV, leads to

$$(\varepsilon_2)^{1/2} = 2.80(E - 1.40). \tag{70}$$

Equation (70) implies a significant red shift in the optical gap of low-pressure CVD films prepared at 600°C, compared to GD a-Si:H_x, of about 0.2 eV. The systematic dependence of K_{BB} and E_G exhibited in Fig. 47 is suggestive in light of Fig. 23. However, considering the widely differing measuring techniques (ellipsometry compared to transmission and reflection), it is difficult to assess the significance of the trend.

In Fig. 48, we show data for four high-pressure CVD films prepared by B. O. Seraphin's group at the University of Arizona (Hey *et al.*, 1984). The deposition (substrate) temperatures (550, 600, and 650°C) are indicated in the figure. All the films shown in Fig. 48 have a Tauc optical gap E_G of

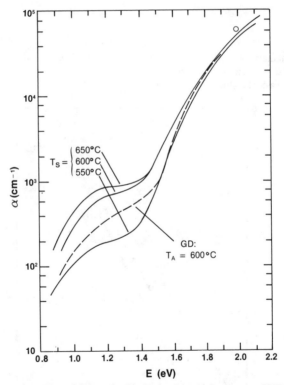

FIG. 48. Absorption edge of films of a-Si obtained by high-pressure CVD of SiH$_4$ at the indicated substrate temperatures (Hey *et al.*, 1983). The absorption edge was obtained by transmission and photothermal deflection spectroscopy (Roxlo *et al.*, 1984). The CV absorption edge is compared to that of GD a-Si:H_x annealed at a temperature of 600°C. The open circle is derived from the low-pressure CVD film of Bagley *et al.* (1982).

1.50 eV, although there are, as shown, major differences in the shoulder absorption. In Fig. 48 one point was derived from Figs. 12 and 13 for a low-pressure CVD film. The agreement between the two CVD films is quite good, as suggested by the similar magnitude of their optical gaps. In Fig. 48 we also show data from Fig. 34 for a GD film isochronally annealed at the same temperatures as the deposition temperature for the CVD films. There is reasonable agreement in the shoulder region of the edge, but, as shown in Fig. 36, there is, for the annealed GD films, a red shift that is a function of T_A and that is absent in the CVD films.

For the annealed GD films, the red shift in E_G is about 0.12 eV between $T_A = 550$ and $650°C$. For the CVD films, the red shift in E_G is less than 0.03 eV. This difference in the behavior of E_G, despite the similar magnitude of the shoulder absorption, is similar to the decoupling of the optical gap E_G from the shoulder absorption in the series of lightly doped GD films shown in Fig. 41 and Fig. 43.

A possible explanation for this difference may lie in the distinction between defects induced *during growth* and defects induced *after growth.* The first kind of defect, e.g., dangling bonds associated with phosphorus or boron or even thermally ruptured bonds, can be healed as growth continues. Indeed, the presence of such defects may relax the random network. On the other hand, defects induced after growth, e.g., through dangling bonds generated by hydrogen desorption, can only strain the network. In the first case, there is no change in site disorder and hence no change in E_G. In the second case, there is an increase in site disorder and a red shift in E_G.

d. The Optical Properties of Films of a-Si:H_x Produced by Chemical Vapor Deposition of Si_2H_6

An interesting new chemical approach to either CVD preparation of a-Si:H_x (Akhtar *et al.,* 1982) or GD preparation (Scott *et al.,* 1980) is the substitution of Si_2H_6 for SiH_4 (Scott *et al.,* 1980). There do not appear to be very large differences in the optical properties of GD films produced from Si_2H_6. However, CVD films prepared from this material appear to exhibit improved electrical properties compared to their SiH_4 counterparts, and we might anticipate significant optical differences.

The limited optical data we have to date suggest that despite the more rapid deposition rates, these films do not differ in an essential way from the films that have been discussed up to now under Section 8c.

In Fig. 49, we show the high-energy, band-to-band absorption edge for a film prepared by disilane CVD at a substrate temperature of $400°C$ with a cold wall (Delahoy, 1983). We again reference these data to the "standard" a-Si:H_x film prepared by GD ($E_G = 1.64$ eV, $E_0 = 50 \times 10^{-3}$ eV). The optical gap, obtained from Eq. (54), for the CVD film is $E_G = 1.40$ eV and

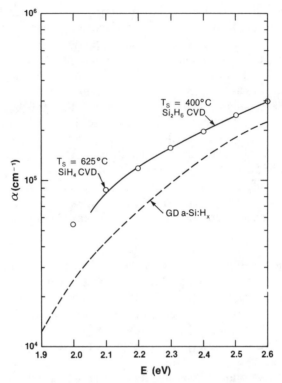

FIG. 49. High-energy absorption edge of a-Si: H_x prepared by CVD of Si_2H_6 at $T_s = 400°C$. The open circles are derived from ε_1 and ε_2 data on a low-pressure CVD film prepared at a substrate temperature of 625°C (Bagley *et al.*, 1982). The optical absorption data for GD a-Si: H_x (Fig. 14) are shown for comparison. [From Delahoy (1983).]

the constant C' is 2.8 $eV^{-3/2}$ $\mu m^{-1/2}$. For the GD films $E_G = 1.64$ eV and $C' = 3.1$ $eV^{-3/2}$ $\mu m^{-1/2}$. There is thus no essential optical difference between a film prepared by disilane CVD and the silane CVD films of the previous section (Section 8c).

The relatively low optical gap of the CVD films suggests high structural disorder. However, within the theory discussed in Section 7, thermal and structural disorder are additive, and hence the temperature dependence of E_G and E_0 for these films should be the same as for the GD material. The high electrical quality of the Si_2H_6 CVD films makes it possible to accurately probe the temperature variation of both E_G and E_0. Figure 50 (Delahoy, 1983) gives the results for a CVD film prepared from Si_2H_6 at 400°C. In agreement with the experiments and model of Section 7, $\partial E_G/\partial E_0 = -7.2$, and from Eq. (63), the derived Urbach focus for this film is ~2.1 eV in fair agreement with Fig. 27.

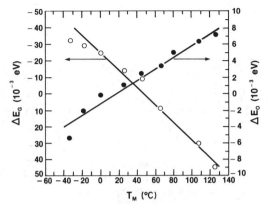

FIG. 50. Shift in E_G and E_0 with measurement temperature as determined by photoconductivity measurements for a-Si:H_x prepared by CVD from disilane at substrate temperature of 450°C. The temperature coefficients are $\partial E_G/\partial T = -5.2 \times 10^{-4}$ eV/°C and $\partial E_0/\partial T = 7.2 \times 10^{-5}$ eV/°C implying $\Delta E_G/\Delta E_0 = -7.2$ [compare Eq. (63)]. Positive ΔE_G is a *decrease* in E_G. Positive ΔE_G is an *increase* in E_0. [From Delahoy (1983).]

e. The Optical Properties of Films of a-Si:H_x Produced by Chemical Vapor Deposition of Hot SiH_4 on a Cold Substrate (Homogeneous CVD)

Homogeneous CVD is an ingenious method of film growth of a-Si:H_x that permits a wide range of substrate temperatures T_S (see Chapter 2 by Scott in Volume 21A). Films obtained by this technique exhibit significant differences from GD (plasma-enhanced CVD) films prepared at similar substrate temperatures. In general, at low T_S, these films have very high hydrogen concentrations, low spin densities, and a very large optical gap. The low density of nonradiative recombination centers in these films leads to very efficient "white" photoluminescence at room temperature. As shown in Fig. 51, for low T_S these films exhibit large blue shifts in the absorption curve and presumably in E_G. The high hydrogen content of the films prepared at low T_S suggests that the reduced connectivity or dimensionality of the network may dominate site disorder in the electronic structure.

f. The Optical Properties of Films of a-Si:H_x Produced by Reactive Ion Beam Sputtering (RIBS)

Ion beam sputtering replaces the target–plasma–substrate sputtering system by a high-flux ion beam bombarding a target in a high vacuum system. It differs from traditional sputtering in several ways. First, the substrate can be placed outside the plasma environment; second, the low

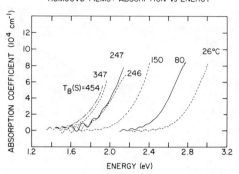

FIG. 51. Absorption coefficient versus energy for homoCVD films prepared at different substrate temperatures. Film compositions range from < 1 at. % H at a growth temperature of 454°C to almost 40 at. % H at room temperature (see Chapter 7 by Scott in Volume 21A).

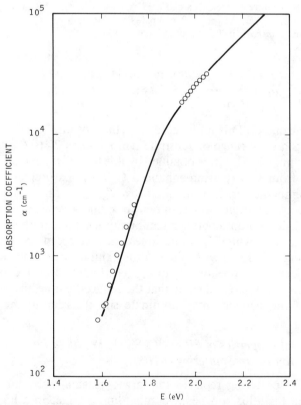

FIG. 52. The absorption edge of a film of a-Si : H_x containing 15 at. % H prepared by reactive ion beam sputtering. The solid line is averaged data for a-Si : H_x films prepared by GD (Kasdan et al., 1984). The open circles indicate data taken from films prepared by RIBS.

pressure of the system eliminates using gas-phase reactions; and, finally, the reactive gas pressure can be varied over a very wide range compared to that of the argon sputtering beam. This technique thus offers considerably greater flexibility in growth parameters compared to that of GD of SiH_4 or reactive sputtering techniques.

Ion beam sputtering is a new technique for film growth which followed the recent development of intense ion beam sources. Despite its novelty, it has been applied to a-SiH_x growth by Weissmantel et al. (1978) in Germany, Kobayashi et al. (1981) in Japan, and in the U.S., by Ceasar et al. (1981) and Kasdan et al. (1984). The remarkable changes in the surface morphology of the film under different growth conditions has been described by Kasdan and Goshorn (1982).

In Fig. 52, we show the absorption edge of a film of a-Si:H_x prepared by RIBS containing about 15 at. % H. Again, we compare the absorption data with a-Si:H_x prepared by glow discharge ($E_G = 1.64$ eV, ~9 at. % H) and the agreement is excellent over the indicated range in α.

Reactive ion beam sputtering can produce films with large variations in hydrogen concentrations. As shown in Fig. 40, the Tauc gap E_G does not correlate well with the hydrogen concentration.

V. Summary and Conclusion

In this chapter, we have explored the effect of temperature, isochronal annealing, and preparation conditions on the optical absorption edge of a-Si:H_x films. We attempted to understand the data on the basis of a simple model, due originally to Tauc (1974) and recently generalized by Abe and Toyazawa (1981). The Tauc model defines the absorption edge of an amorphous semiconductor through the concept of a nondirect optical transition in which electron momentum is not conserved (total disorder). By utilizing the random phase approximation and parabolic bands, Tauc defined an optical gap E_G that is widely used by experimentalists to define the position and the shape of the high-energy side of the absorption edge for a-Si:H_x and other amorphous semiconductors. The quantity E_G is not a significant feature of the absorption edge but can be successfully correlated with the magnitude of the mobility gap deduced from transport measurements. Abe and Toyazawa (1981) generalized the Tauc model to include a variable degree of Gaussian site disorder on the basis of the coherent potential approximation. From this generalization, the Tauc optical gap E_G becomes an additive function of structural and thermal disorder, in agreement with experiment. Furthermore, the exponential character of the edge below E_G is also related to structural disorder through the inverse logarithmic slope of the exponential or Urbach edge E_0, which describes an exponential density of gap states. In the Abe and Toyazawa model (1981), a third parameter E_x occurs, which is the zero-disorder limit of E_G. The three parameters E_G, E_0, and E_x are approximately linearly related by an equation of the form

$$E_G(T, X) = E_x - GE_0(T, X), \qquad (71)$$

where T, the absolute temperature, is a measure of thermal disorder and X, defined by Eq. (65), is a measure of structural disorder. The quantity G is a number of order 6–10 and reflects conservation of states under the effect of disorder. We find that all of the published experimental data for the absorption edge of a-Si : H_x with a hydrogen concentration below 20 at. % can be understood on the basis of Eq. (71). For higher hydrogen concentrations, there can be significant departures from Eq. (71), which suggest changes in E_x, possibly due to the hydrogen reducing the connectivity of the network. From Eq. (71), changes in the state of order of the a-Si : H_x can be measured by shifts in E_G and complementary shifts in E_0, revealing several annealing stages. An important conclusion that emerges from the experimental data is that defects induced *during* growth have a considerably smaller effect on E_G and E_0 than do defects induced *after* film growth.

In this chapter we have emphasized the role of site disorder in determining the position and shape of the absorption edge of a-Si : H_x. The role of topological order and connectivity of the network in determining the magnitude of the optical gap in the limit of small site disorder (i.e., the Urbach focus energy E_x) has not been discussed. The magnitude of the optical gap remains an unsolved problem despite several recent imaginative attempts at physical models (Brodsky, 1980; Brodsky and DiVincenzo, 1982).

The theoretical models discussed in this chapter supply a conceptual framework that correlates many of the features of the optical absorption edge with the structure of a-Si : H_x and are very useful to the experimentalist despite (or because of!) their simplicity. The underlying physics behind the regularities exhibited in Figs. 27 and 47 have yet to receive a fundamental explanation. Recent reviews of the theoretical situation (Yonezawa and Cohen, 1981; Cohen, 1982) suggest the complexity of the optical and transport behavior of amorphous semiconductors.

Acknowledgments

I am indebted to my colleagues at Exxon for excellent materials, stimulating experimental results, good ideas, and strong criticisms. F. R. Gamble and A. Rose asked many fundamental questions on the optical absorption in amorphous semiconductors, and this chapter is an attempt to answer some of them. I am deeply indebted to B. G. Brooks for a rewarding technical collaboration and to B. A. Palmer for preparation of the manuscript. Professor Y.

Goldstein of the Hebrew University played an important role in initiating the experiment on the optical effects of isochronal annealing. Professor H. Ehrenreich of Harvard University has been helpful in establishing the common ground between the physics of the optical properties of amorphous semiconductors and that of crystalline semiconductors.

REFERENCES

Abe, S., and Toyazawa, Y. (1981). *J. Phys. Soc. Jpn.* **50,** 2185.

Akhtar, M., Dalal, V. L., Ramaprasad, K. R., Gau, S., and Cambridge, J. A. (1982). *Appl. Phys. Lett.* **41,** 1146.

Allen, P. B., and Heine, V. (1976). *J. Phys. C* **9,** 2305.

Andreev, A. A., Kolomiets, B. T., Mazets, T. F., Manukyan, A. L., and Pavlov, S. K. (1976). *Sov. Phys. Solid State* **18,** 29.

Ashcroft, N. W., and Mermin, N. D. (1976). *"Solid State Physics,"* p. 462. Saunders, Philadelphia, Pennsylvania.

Aspnes, D. E. (1982). *Thin Solid Films* **89,** 249.

Aspnes, D. E., and Studna, A. A. (1981). *In* "Optical Characterization Techniques for Semiconductor Technology." *Proc. Soc. Photo-Opt. Instrum. Eng.* **276,** 227.

Aspnes, D. E., and Studna, A. A. (1982). *Phys. Rev. B* **27,** 985.

Bagley, B. G., Aspnes, D. E., Celler, G. K., and Adams, A. C. (1982). *In* "Laser and Electron Beam Interactions with Solids" (B. R. Appleton and G. K. Celler, eds.), p. 483. Elsevier, Amsterdam.

Batterman, B., and Chipman, D. R. (1962). *Phys. Rev.* **127,** 690.

Berard, M., and Harris, R. (1983). *Phys. Lett. A* **94,** 89.

Berger, J. M., Ferraton, J. P., Yous, B., and Donnadieu, A. (1981). *Thin Solid Films* **86,** 337.

Berger, J. M., DeChelle, F., Ferraton, J. P., and Donnadieu, A. (1983). *Thin Solid Films* **105,** 107.

Bludau, W., Onton, A., and Heinke, H. (1974). *J. Appl. Phys.* **45,** 1846.

Brodsky, M. (1971). *J. Vac. Sci. Technol.* **8,** 125.

Brodsky, M. H. (1980). *Solid State Commun.* **36,** 55.

Brodsky, M. H., and DiVincenzo, D. P. (1982). *Proc. Int. Conf. Phys. Semicond., 16th, Montpellier, France.* To be published.

Brodsky, M., Kaplan, D. M., and Ziegler, J. F. (1972). *Proc. Inst. Conf. Phys. Semicond., 11th, Warsaw.*

Brooks, B. G., and Stephens, R. B. (1982). Private communication.

Ceasar, G. P., Grimshaw, S. F., and Okomura, K. (1981). *Solid State Commun.* **38,** 89.

Chelikowsky, J. R., and Cohen, M. L. (1976). *Phys. Rev. B* **14,** 556.

Chik, K. P., Feng, S. Y., and Poon, S. K. (1980). *Solid State Commun.* **33,** 1019.

Clark, C. D., Dean, P. J., and Harris, P. V. (1964). *Proc. R. Soc. London, Ser. A* **277,** 312.

Cody, G. D., Wronski, C. R., Abeles, B., Stephens, R. B., and Brooks, B. (1980). *Solar Cells* **2.** 227.

Cody, G. D., Tiedje, T., Abeles, B., Brooks, B., and Goldstein, Y. (1981). *Phys. Rev. Lett.* **47,** 1480.

Cody, G. D., Brooks, B. G., and Abeles, B. (1982). *Solar Energy Mat.* **4,** 231.

Cody, G. D., Abeles, B., Brooks, B., Persans, P., Roxlo, C., Ruppert, A., and Wronski, C. (1983). *J. Non-Cryst. Solids* **59 & 60,** 325.

Cohen, M. H. (1982). *In* "Melting Localization and Chaos" (R. K. Kalia and P. Vashishta, eds.), p. 125. Elsevier, Amsterdam.

Cohen, M. L., and Chadi, D. J. (1980). *In* "Handbook of Semiconductors" (M. Balkanski, ed.), Vol. 2, p. 155. North-Holland, Amsterdam.

Cohen, M. L., Fritzsche, H., and Ovshinsky, S. R. (1969). *Phys. Rev. Lett.* **22,** 1065.
Connell, G. A. N., and Paul, W. (1972). *J. Non-Cryst. Solids* **8–10,** 215.
Davis, E. A. (1970). *J. Non-Cryst. Solids* **4,** 107.
Deckman, H. W., Wronski, C. R., and Witzke, H. (1983). *J. Vac. Sci. Technol. A* **1,** 578.
Delahoy, A. E. (1983). *Proc. Soc. Photo. Opt. Instram. Eng.* **407,** 47.
Divrechy, A., Yous, B., Berger, J. M., Ferraton, R. J., and Donnadien, A. (1981). *Thin Solid Films* **78,** 235.
Donnadieu, A., Yous, B., Berger, J. M., Ferraton, J. P., and Robin, J. (1981). *J. Phys. Colloq. Orsay, Fr.* **42,** C4-655.
Dow, J. P., and Redfield, D. (1972). *Phys. Rev. B* **5,** 594.
Dunstan, D. J. (1982). *Solid State Phys.* **30,** L419.
Ehrenreich, H., and Hass, K. C. (1983). *J. Vac. Soc.* **105,** 107.
Enck, R. G., and Pfister, G. (1976). *In* "Photoconductivity and Related Phenomena" (J. Mort and D. M. Pai, eds.), p. 278. Elsevier, Amsterdam.
Griep, S., and Ley, L. (1983). *J. Non-Cryst. Solids* **59 & 60,** 253.
Guttman, L. (1977). *Solid State Commun.* **24,** 211.
Guttman, L., and Fong, C. Y. (1982). *Phys. Rev. B* **26,** 6756.
Harrison, W. A., and Pantelides, S. T. (1976). *Phys. Rev. B* **14,** 691.
Hasegawa, S., Yazaki, S., and Shimizu, T. (1978). *Solid State Commun.* **26,** 407.
Hey, P., and Seraphin, B. O. (1982). *Solar Energy Mat.* **8,** 215.
Hey, P., Seraphin, B. O., Brooks, B. G., Roxlo, C., and Cody, G. D. (1984). To be published.
Hindley, N. (1970). *J. Non-Cryst. Solids* **5,** 17.
Hirose, M. (1981). *J. Phys. Colloq. Orsay, Fr.* **42,** C4-705.
Jackson, W. B. (1982). *Solid State Commun.* **44,** 477.
Jackson, W. B., and Amer, N. M. (1982). *Phys. Rev. B* **25,** 5559.
Jackson, W. B., Biegelson, D. K., Nemanich, R. J., and Knights, J. C. (1983). *Appl. Phys. Lett.* **42,** 105.
Janai, M., and Karlsson, B. (1979). *Solar Energy Mat.* **1,** 387.
Kasdan, A., and Goshorn, D. P. (1982). *Appl. Phys. Lett.* **42,** 36.
Kasdan, A., Goshorn, D. P., and Lanford, W. A. (1984). To be published.
Kasdan, A., Goshorn, D. P., and Lanford, W. A. (1983). *J. Non-Cryst. Solids,* to be published.
Klazes, R. H., Van den Broek, M. H. L. M., Rezemer, J., and Radelaar, S. (1982). *Philos. Mag. B* **45,** 377.
Kobayashi, M., Saraie, J., and Matsunami, H. (1981). *Appl. Phys. Lett.* **38,** 696.
Kurik, M. V. (1971). *Phys. Status Solidi A* **8,** 9.
Landauer, R. (1978). *AIP Conf. Proc.* **40,** 2.
Lanyon, H. P. D. (1963). *Phys. Rev.* **130,** 134.
Lee, H. K., and Lee, C. (1982). *Solid State Commun.* **42,** 71.
Matsumura, H., and Furukawa, S. (1980). *Jpn. J. Appl. Phys.* **20,** Suppl. 1, 275.
Miyagi, M., and Funakoshi, N. (1981). *Jpn. J. Appl. Phys.* **20,** 289.
Moss, S. C., and Graczyk, J. F. (1970). *In Proc. Int. Conf. Phys. Semicond., 10th, Cambridge, Massachusetts* (S. Keller, J. C. Hensel, and F. Stern, eds.), p. 658.
Mott, N. F., and Davis, E. A. (1979). "Electronic Processes in Non-Crystalline Materials," 2nd ed. Oxford Univ. Press (Clarendon), London and New York.
Moustakas, T. (1982). *Solar Energy Mat.* **8,** 187.
Nitta, S., Itoh, S., Tanaka, M., Endo, T., and Hatano, A. (1982). *Solar Energy Mat.* **8,** 249.
Oheda, H. (1979). *Jpn. J. Appl. Phys.* **18,** 1973.
Panish, M. B., and Casey, H. C., Jr. (1969). *J. Appl. Phys.* **40,** 163.
Pankove, J. (1971). "Optical Processes in Semiconductors," p. 34. Prentice-Hall, Englewood Cliffs, New Jersey.

Paul, W., and Anderson, D. A. (1981). *Solar Energy Mat.* **5,** 229.
Penn, D. R. (1962). *Phys. Rev.* **128,** 2093.
Persans, P., and Ruppert, A. F. (1984). To be published.
Persans, P., Cody, G. D., Ruppert, A. F., and Brooks, B. G. (1983). *Bull. Am. Phys. Soc.* **28,** 532.
Phillips, J. C. (1973). "Bonds and Bands in Semiconductors." Academic Press, New York.
Pierce, D. T., and Spicer, W. E. (1972). *Phys. Rev.* **135,** 3017.
Ravinda, N. M., and Srivastava, V. K. (1979). *J. Phys. Chem. Solids* **40,** 791.
Redfield, D. (1982). *Solid State Commun.* **44,** 1347.
Redfield, D., and Afromowitz, M. A. (1967). *Appl. Phys. Lett.* **11,** 138.
Rose, A. (1978). "Concepts in Photoconductivity and Allied Problems," 2nd ed. Krieger Publ., Huntington, New York.
Roxlo, C. B., Abeles, B., Wronski, C. R., Cody, G. D., and Tiedje, T. (1983). *Solid State Commun.* **47,** 985.
Roxlo, C., Cody, G. D., and Brooks, B. G. (1984). To be published.
Ruppert, A., Abeles, B., deNeufville, J. P., and Schriesheim, R. (1981). *Bull. Am. Phys. Soc.* **26,** 387.
Savvides, N., McKenzie, D. R., and McPhedran, R. C. (1983). *Solid State Commun.,* **48,** 189.
Scott, B. A., Brodsky, M. H., Green, D. C., Kirby, P. B., Plecenik, R. M., and Simonye, E. E. (1980). *Appl. Phys. Lett.* **37,** 725.
Siemsen, K. J., and Fenton, E. W. (1967). *Phys. Rev.* **161,** 632.
Skettrup, T. (1978). *Phys. Rev. B* **18,** 2622.
Spicer, W. E., and Donovan, T. M. (1970). *Phys. Rev. Lett.* **24,** 595.
Street, R. A., Searle, T. M., Austin, I. G., and Sussmann, R. S. (1974). *J. Phys. C* **7,** 1582.
Szczyrbowski, J. (1979). *Phys. Status Solidi B* **96,** 769.
Tauc, J. (1970). *Mater. Res. Bull.* **5,** 721.
Tauc, J. (1972). *In* "Optical Properties of Solids" (F. Abeles, ed.), p. 279. North-Holland Publ., Amsterdam.
Tauc, J. (1974). *In* "Amorphous and Liquid Semiconductors" (J. Tauc, ed.), Chap. 4. Plenum, New York.
Tauc, J., Grigorovici, R., and Vancu, A. (1966). *Phys. Status Solidi* **15,** 627.
Thèye, M. L. (1971). *Mater. Res. Bull.* **6,** 103.
Thurmond, C. D. (1975). *J. Elec. Chem. Soc.: Solid State Sci. Technol.* **122,** 1134.
Tiedje, T., and Rose, A. (1981). *Solid State Commun.* **37,** 49.
Tiedje, T., Cebulka, J. M., Morel, D. L., and Abeles, B. (1981). *Phys. Rev. Lett.* **46,** 1425.
Tiedje, T., Abeles, B., Cebulka, J. M., and Pelz, J. (1983a). *Appl. Phys. Lett.* **42,** 712.
Tiedje, T., Abeles, B., and Cebulka, J. M. (1983b). *Solid State Commun.* **47,** 493.
Tsai, C. C. (1979). *Phys. Rev. B* **19,** 2041.
Tsai, C. C., and Fritzsche, H. (1979). *Solar Energy Mat.* **1,** 29.
Vanecek, M., Kocka, J., Stuchlik, J., Kozisek, T., Stika, O., and Triska, A. (1983). *Solar Energy Mat.* **8,** 411.
Varshni, Y. P. (1967). *Physica* **39,** 149.
Vorlicek, V., Zavetova, M., Pavlov, S. K., and Patasova, L. (1981). *J. Non-Cryst. Solids* **45,** 289.
Weinstein, B. A., Zallen, R., Slade, M. L., and De Lozanne, A. (1981). *Phys. Rev. B* **24,** 4652.
Weissmantel, C., Bewilogua, K., Dietrich, D., Erler, H. J., Hinneburg, H. J., Klose, S., Nowick, W., and Reisse, G. (1978). *Thin Solid Films* **72,** 19.
Welber, B., and Brodsky, M. (1981). *Phys. Rev. B* **23,** 787.
Wemple, S. H. (1973). *Phys. Rev. B* **7,** 1973.
Wemple, S. H. (1977). *J. Chem. Phys.* **67,** 2151.

Wemple, S. H., and DiDomenico, M., Jr. (1971). *Phys. Rev. B* **3**, 1338.

Wood, D. L., and Tauc, J. (1971). *Phys. Rev. B* **5**, 3144.

Wooten, F. (1972). "Optical Properties of Solids." Academic Press, New York.

Wronski, C. R. (1983). Private communication.

Wronski, C. R., Cody, G. D., Abeles, B., Stephens, R. B., Brooks, B., and Sherrier, R. (1981). *Proc. IEEE Photovoltaic Specialists Conf., 15th,* p. 1973.

Wronski, C. R., Abeles, B., Tiedje, T., and Cody, G. D. (1982). *Solid State Commun.* **44,** 1423.

Yablonovitch, E., and Cody, G. (1982). *IEEE Trans. Electron Devices* **ED-29,** 300.

Yamasaki, S., Hata, N., Yoshida, T., Oheda, H., Matsuda, A., Okushi, H., and Tanaka, K. (1981a). *J. Phys. Colloq. Orsay, Fr.* **42,** C4-297.

Yamasaki, S., Nakagawa, K., Yamamoto, H., Matsuda, A., Okushi, H., and Tanaka, K. (1981b). *AIP Conf. Proc.* **73,** 258.

Yonezawa, F., and Cohen, M. H. (1981). *In* "Fundamental Physics of Amorphous Semiconductors" (F. Yonezawa, ed.), p. 119. Springer-Verlag, Berlin and New York.

Yoshida, M., and Morigaki, K. (1983). *J. Non-Cryst. Solids* **59** & **60,** 357.

Yous, B., Berger, J. M., Ferraton, J. P., and Donnadieu, A. (1981). *Thin Solid Films* **82,** 279.

Zallen, R., Weinstein, B. A., and Slade, M. L. (1981). *J. Phys. Colloq. Orsay, Fr.* **42,** C4-241.

Ziman, J. M. (1979). "Models of Disorder." Cambridge Univ. Press, London and New York.

SEMICONDUCTORS AND SEMIMETALS, VOL. 21, PART B

CHAPTER 3

Optical Properties of Defect States in a-Si:H

Nabil M. Amer

LAWRENCE BERKELEY LABORATORY
UNIVERSITY OF CALIFORNIA
BERKELEY, CALIFORNIA

and

Warren B. Jackson

XEROX CORPORATION
PALO ALTO RESEARCH CENTER
PALO ALTO, CALIFORNIA

I. Introduction

One of the effects of disorder and doping on the electronic structure of amorphous semiconductors is the introduction of states in the pseudogap. These states, commonly classified as defects, significantly affect the properties of the material. Thus, information about their nature, density, and energy levels is of great interest.

Defects have been extensively studied by luminescence, conductivity, photoconductivity, electron spin resonance, and capacitance techniques. However, these techniques are sensitive to the Fermi-level position, detect only the radiative component of deexcitation, or require special doping; hence they do not measure all defects. Subgap absorption, on the other hand, is a fundamental process that measures the energy and number of all defects, thus providing a unique and versatile tool for their study.

Unlike the case of chalcogenide glasses, for which optical spectra were obtained with conventional techniques and the absorption was attributed to transitions involving defect centers (Mott, 1979), no direct or reliable spectra were measured in the case of hydrogenated amorphous silicon (a-Si:H). For transmission and reflection techniques to measure the weak absorption associated with gap states, an accuracy of better than one part in 10^5 in the difference between the incident and transmitted intensities is required. This is clearly beyond the capability of conventional transmission and reflection methods. Consequently, the minimum absorption coefficient α that could be measured reliably is $50-100$ cm^{-1} for a 1-μm-thick film (Cody *et al.*, 1980; Abeles *et al.*, 1980; Tsai and Fritzsche, 1979; Freeman and Paul, 1979; Zanzucchi *et al.*, 1977; Connell and Pawlik, 1976). The absorption associated with gap states can be smaller by a factor of 100.

In an attempt to circumvent these difficulties, photoconductivity was employed to deduce the nature of optical absorption in the fundamental gap (Wronski *et al.*, 1982; Crandall, 1980a,b; Moddel *et al.*, 1980; Loveland *et al.*, 1973). Photoconductivity σ is given by

$$\sigma \propto \eta\mu\tau(1 - e^{-\alpha l}), \tag{1}$$

where η is the efficiency of generating mobile carriers, μ their mobility, τ their lifetime, and l the material thickness. However, τ is a function of both the generation rate and α, which results in difficulties in deducing α from Eq. (1). Furthermore, to extract α from such measurements requires making the heretofore unverified assumption that the product $\eta\mu\tau$ is a constant independent of the energy of the exciting photons. As will be seen, we have shown in recent experiments that, for example, in the case of undoped material, photoconductivity measurements do not yield accurate absorption spectra below 1.5 eV. Localized-to-localized transitions do not contribute to the photocurrent and cause $\eta\mu\tau$ to depend on photon energy.

The limitations of these other methods motivated us to devise a new approach for measuring small optical absorption coefficients. This approach, photothermal deflection spectroscopy (PDS), is based on measuring the thermal energy deposited in the material of interest as electromagnetic radiation is absorbed. This direct method for measuring the optical absorption proved to be a highly sensitive ($\alpha l \sim 10^{-7}$ for solids) and a relatively simple tool for material characterization, particularly for thin films.

In Part II of this chapter we describe the technique of photothermal deflection spectroscopy. A condensed review of the physics of photothermal generation is given first, with emphasis on those features that are most applicable to the case of a-Si:H films. A detailed description of the experimental arrangement follows. Finally, we compare PDS with techniques such as conventional absorption and photoacoustic spectroscopies.

In Part III, the optical properties of states in the gap of a-Si:H are reviewed. The effects of deposition parameters, doping, and photoinduced fatigue (Staebler–Wronski effect) on these states and on the Urbach edge are discussed. Optically derived total correlation-energy measurements will be described and compared with those obtained from other techniques, and the contribution of surface defects is reviewed.

In Part IV, the relationship between photoconductivity and optical absorption is presented, and a discussion of the implications of the energy dependence of the electron and hole $\eta\mu\tau$ products is given. Then, by combining the results presented in Parts III and IV, we sketch in Part V a density-of-states diagram for a-Si:H. Finally, in Part VI we summarize the main points of the chapter and speculate on future directions.

II. Photothermal Deflection Spectroscopy

In this section the technique of photothermal deflection spectroscopy (PDS) (Jackson *et al.*, 1981; Boccara *et al.*, 1980a,b) is briefly described. The steps of a comprehensive calculation of the relationship between the optical and thermal properties of the material and the measured signal are outlined. Formulas for the special case of thin films are presented and their limitations are discussed. Details of the experimental setup, signal optimization, and data analysis are given. Finally, we compare PDS with other techniques.

1. PHYSICAL AND THEORETICAL BASIS OF PDS

The physical principle underlying PDS is straightforward. When an intensity-modulated beam of light (pump beam) is absorbed by a given medium, periodic heating will occur (Fig. 1). The heat from the absorbing material flows into the surrounding medium causing a corresponding modulation in the index of refraction near the material surface. A weak laser (probe) beam essentially grazing the material will experience a periodic deflection synchronous with the intensity modulation. The amplitude and phase of this periodic deflection can be measured with a position sensor and a differential ac synchronous detection scheme. Thus, if the wavelength of

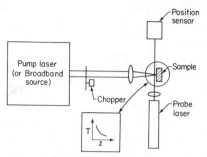

FIG. 1. The experimental arrangement.

the pump beam is varied, the deflection of the probe beam becomes a measure of the optical absorption spectrum of the material of interest.

In order to obtain the optical absorption coefficient, the theory of PDS has been developed (Jackson *et al.,* 1981). We concentrate here on those formulas for temperature, deflection, and detector response that are applicable to the case of thin films.

The derivation of the relation between the optical absorption of the sample and the measured voltage output consists of four steps:

(1) The spatial distribution of the optical energy $I(\mathbf{r}, t)$ is determined within the sample and in the surrounding media. The effect of interference and scattering on the resulting intensity distribution is included at this stage (Cody *et al.,* 1980; Yasa *et al.,* 1982).

(2) The optical energy is absorbed, and the generated heat per unit volume per second is given by $\alpha_f I(\mathbf{r}, t)$, where α_f is the extinction coefficient of the film. The heat energy is the source term for the temperature equations that are solved for the material and the surrounding media.

(3) The temperature rise induces an index-of-refraction gradient that deflects the probe beam. The beam deflection is proportional to the heat flow perpendicular to the sample surface across the probe beam.

(4) The detector converts the deflection into a voltage signal, which depends on the type of the detector and the experimental geometry.

Following these steps, it can be shown that the deflection signal S is given by (Jackson *et al.,* 1981; Yasa *et al.,* 1982)

$$S = T_r(1/n_0)(dn/dT)L[dT(z_0)/dz]e^{i\omega t} + \text{c.c.,} \qquad (2)$$

where T_r is the detector transduction factor (in volts per radian of deflection), which depends on the experimental configuration and the detector type. A maximum value is $1 \times 10^3 \text{ V/rad}^{-1}$. The product $(1/n_0) \, dn/dT$ is the relative index-of-refraction change with temperature for the deflecting medium, L the length of interaction between the optically heated region and the probe beam, T the amplitude of the ac temperature rise above the average temperature, z_0 the distance of the probe beam from the sample surface, and c.c. denotes the complex conjugate. Since the heat flow is proportional to dT/dz, Eq. (1) demonstrates that the deflection measures the heat flow across the probe beam.

The periodic solution for the temperature in the deflecting medium, the a-Si:H film, and the substrate has the form $T_j \exp(\pm ik_j z - i\omega t) + \text{c.c.}$, where $k_j^2 = (i\omega/\lambda_j)$, in which ω is the modulation frequency, λ_j is the thermal diffusivity for the jth medium, $i = \sqrt{-1}$, and T_j is the complex amplitude of the thermal wave determined from the boundary conditions. The subscript j stands for d, s, or f, which refers to the deflecting medium, the substrate, and the film, respectively. The quantity λ_j can be written as $\kappa_j/\rho_j C_j$, where κ_j is the thermal conductivity, ρ_j the density, and C_j the specific heat.

The physical interpretation of these solutions is that the temperature can be represented by highly damped waves that decay over the thermal length $1/|k_j|$. These waves experience both reflection and transmission when they propagate through a boundary. The quantity $\kappa_j k_j$ is the "index of refraction" of the thermal wave in the jth medium. Since the deflection signal is proportional to the heat flow, the deflection measures the flux of thermal waves across the probe beam path. Specifically, if the film is thermally thin, the temperature gradient is given by

$$\frac{dT(z_0)}{dz} = \frac{k_d \exp(-k_d z_0)I}{\kappa_d k_d + \kappa_s k_s} \begin{cases} A & \text{(film absorption)} & \text{(3a)} \\ -\alpha_d/k_d & \begin{array}{l}\text{(deflecting medium} \\ \text{absorption)}\end{array} & \text{(3b)} \\ \alpha_s/k_s & \text{(substrate absorption),} & \text{(3c)} \end{cases}$$

where I is the pump beam intensity (W cm^{-2} sec^{-1}), A is the absorptance of the film, and α_j is the absorption coefficient (if any) of the jth medium. Equations (2) and (3a) show that the photothermal deflection is directly related to the optical properties of the film. The terms preceding the brace in Eqs. (3a)–(3c) are independent of photon energy and depend only on the thermal properties of the deflecting medium and the substrate. At high photon energy, the a-Si:H films are completely absorbing. Consequently, $\alpha_f l \gg 1$, $A = 1 - R_F$, and $S_{sat} = C(1 - R_F)$, where l is the film thickness, $R_F(R_B)$ the reflectance of the front (back) interface, S_{sat} the saturated PDS signal, and C a constant composed of factors in Eqs. (2) and (3). For lower photon energies

$$\langle S \rangle = C[1 - \exp(-\alpha_f l)]\{[(1 - R_F)(1 + R_B)]/(1 - R_F R_B)\}, \quad (4)$$

where $\langle \ \rangle$ denotes averaging of the interference fringes on a logarithmic scale (equivalent to geometric averaging). Hence

$$1 - \exp(-\alpha_f l) = \langle S/S_{sat} \rangle. \qquad (5)$$

This formula gives the absorptance for an incoherent pump source. In Eq. (5), we assume that the reflectivity factor in the braces of Eq. (4) does not vary significantly with photon energy. The error associated with this assumption is less than 20%. Equation (5) enables one to determine α_f from PDS alone once the film thickness is known. Additional optical measurements are not needed.

It should be noted that Eq. (5) is valid under the following conditions:

(1)   The film is thermally thin ($|k_f|l \ll 1$). This approximation is valid in the case of a-Si : H since the thermal diffusion length is greater than 50 μm at the intensity modulation used in performing the measurement (7 Hz).

(2)   The pump beam should have a radius larger than the thermal diffusion length $1/|k_j|$.

2. THE EXPERIMENTAL ARRANGEMENT

The experimental setup is shown in Fig. 1. Typically, the pump beam is a 1-kW Xe arc lamp focused on the slit of a $\frac{1}{4}$-m monochromator. Various order-sorting filters are used to remove unwanted diffraction orders, and the intensity modulation is accomplished with a mechanical chopper operating between 7 and 18 Hz. These modulation frequencies represent a trade-off between a high frequency, which is desirable to eliminate the effects of mechanical vibrations and turbulence within the deflecting medium, and a low frequency, which increases both the thermal diffusion length and the signal amplitude. The pump beam should be focused as tightly as possible on the film since from Eq. (3a) the signal is proportional to the intensity (power per unit area) of the pump beam. The probe beam is typically a weak ($\sim$ 1-mW) He – Ne laser, which is focused roughly to 75 μm and which grazes the film surface. The focal length of the probe beam lens represents a trade-off between a large confocal distance and a small spot size. The deflection of the probe beam is measured by a position sensor (from Silicon Detector Corporation, Newbury Park, California). The outputs of the position sensor are amplified by current amplifiers and fed into the (A – B) input of a computer-interfaced lock-in amplifier. The intensity of the pump beam is monitored by splitting off part of the pump beam before it reaches the film and detecting it with a pyroelectric detector. The amplitude of the PDS signal is then averaged and normalized to the incident light intensity. The effects of dispersion and residual nonlinearities of the system are removed by utilizing a PDS spectrum of graphite to normalize the a-Si:H spectra. The correction is typically only a factor of two in the 2.1 – 2.6 eV range of the spectrum where the dispersion is largest.

There are a number of other points to be considered in maximizing the sensitivity of PDS:

(1)   Equations (2) and (3a) demonstrate that a figure of merit for various deflecting media is given by

$$F = 1/(n_0 \kappa_d) \, dn/dT. \tag{6}$$

For a well-aligned system, $\exp(-k_d z_0) \simeq 1$ and $\kappa_d k_d > \kappa_s k_s$ if the substrate is SiO_2. In addition, the deflecting medium should have no significant absorption in the photon energy range of interest. Carbon tetrachloride (CCl_4), which has a large F, has an extremely low absorption in the 0.4–2.3 μm range, and ESR measurements show that it does not alter the properties of the a-Si:H films. These considerations make CCl_4 the preferred deflecting medium for a-Si:H studies.

(2)   The signal is also significantly larger if both the substrate and the deflecting medium as low a thermal conductivity as possible. Consequently, SiO_2 is superior to Al_2O_3 or crystalline Si.

(3)   The probe beam should be as close to the film surface as possible so that $\exp(-k_d z_0) \simeq 1$. This requires that the sample be placed on a micrometer-driven translation stage to adjust the distance between the probe beam and the sample. In addition, it is desirable that the sample holder have a rotational degree of freedom to allow the probe beam to be close to the sample, without being eclipsed.

(4)   Particular care should be taken to keep the deflecting fluid free from particulates since they can be a significant source of noise.

(5)   The pointing stability of the probe laser can be the factor limiting the sensitivity of the technique.

(6)   To eliminate air currents, the optical paths should be suitably enclosed.

3. PDS SENSITIVITY AND COMPARISON WITH OTHER TECHNIQUES

Photothermal deflection spectroscopy is a highly sensitive technique with numerous advantages over traditional reflection and transmission and photoacoustic methods. A detailed comparison is discussed by Jackson et al. (1981). In this section we discuss the sensitivity of PDS and compare it to other techniques that have been employed in studies of amorphous materials.

The PDS sensitivity that can be achieved routinely with monochromatic broadband sources is 1×10^{-8} W of absorbed power. Since typical available powers are on the order of ~ 1 mW at 0.8 eV, an absorptance of 10^{-5} can be readily measured. For a 1-μm film, values of α as low as 0.1 cm^{-1} at 0.8 eV can be determined. The sensitivity is limited by the probe beam stability, turbulence, and scattering from particles in the deflecting medium. Furthermore, at this level of sensitivity, absorption by the substrate material can be a sensitivity-limiting factor. Typically SiO_2 substrates have a surface absorbing region with an absorptance of $\sim 10^{-5}$ (Parker, 1973), in addition to the O–H overtone absorption band at 1.38 μm. This background may be minimized by using OH-free glass and monitoring the signal phase. Nevertheless, it does place a limit on the ultimate sensitivity achieved. Note that if the pump beam is a tunable laser, the signal-to-noise ratio increases because of the higher power and smaller beam spot size.

It is instructive to write the PDS signal in terms of film surface temperature rise $\tau(0)$ integrated over the optically heated area so that the sensitivity may be compared with that of photoacoustic spectroscopy. Equation (2) can be evaluated to give $S = 10$ V cm^{-2} °C^{-1} $\tau(0)$. In the case of photoacoustic detection, the corresponding factor is 0.1 V cm^{-2} °C^{-1} $\tau(0)$. Hence, PDS is approximately 100 times more sensitive than photoacoustic spectroscopy.

One of the strengths of PDS over other techniques is that the background signal (e.g., signal due to substrate absorption) is small and can easily be separated from that due to the film absorption. This is not the case for conventional absorption techniques. Equations (3b) and (3c) indicate that one can distinguish between the substrate and deflecting medium absorption by utilizing the signal phase information. The substrate produces a signal with a lagging phase of 45°, while the deflecting fluid signal will be leading by a phase of 135° with respect to that of the mechanical chopper. This ability to determine the source of absorption is extremely important in low-absorption measurements since substrates such as quartz and pyrex glass are known to have weak surface absorption (Parker, 1973).

The effects of elastically scattered light are much less significant for PDS than for both photoacoustic spectroscopy and transmission techniques. In transmission techniques, the scattered and absorbed light are not separable, so the measured extinction coefficient is often considerably larger than the actual coefficient. In PDS, light scattered elastically from the bulk will not contribute to the signal unless it is scattered along the probe beam and is subsequently absorbed. Since this event is unlikely, PDS is highly insensitive to scattering (Yasa et al., 1982). Also, PDS is less sensitive than photoacoustics to scattering from surface inhomogeneities, since the scattered light must be absorbed near the probe beam. Photoacoustic spectroscopy also has a background signal due to scattered light striking the photoacoustic cell and/or the microphone detector. In PDS, the probe beam may be blocked to determine whether stray light is falling on the position sensor. Such light may be filtered out by the use of a He–Ne laser line filter. Throughout our investigations on various a-Si : H films, we found that PDS is largely insensitive to scattering except in the case of mechanically roughened samples.

III. Optical Spectroscopy of Defects in a-Si:H

As in crystalline semiconductors, factors determining the electronic structure and the density of states in a-Si:H are of fundamental and practical interest. While some states, localized and extended, occur in the material regardless of the method and parameters employed in making the films, others are introduced during the deposition process. The nature and number of these extrinsic states, usually called *defects*, is a strong function of the deposition parameters and the type and level of doping. Furthermore, even after the deposition of the material, exposure to light introduces metastable defects that are annealed away by heating. Although the intrinsic states are of interest, we concentrate our coverage in this chapter on those extrinsic states that we defined earlier as defects.

4. THE EFFECTS OF DEPOSITION AND DOPING PARAMETERS

Traditionally, the convention is to divide the optical absorption edge of amorphous semiconductors into three regions: (a) a power law region; (b) an exponential absorption edge, often called the Urbach edge; and (c) a tail, which extends beyond the Urbach edge into the pseudogap. The latter two regions are known to be sensitive to defects in crystalline semiconductors.

Figure 2 shows the effect of substrate temperature on the optical absorption spectrum in regions (b) and (c) as obtained by photothermal deflection spectroscopy. The subgap absorption is lowest for material deposited at 230°C and increases for both lower and higher deposition temperatures. The effect of increasing the deposition power density on the subgap absorption for a constant substrate temperature is given in Fig. 3. Again, deposition conditions known to produce large defect densities and columnar growth yield a discernible shoulder at ~1.25 eV. A second pronounced effect is the

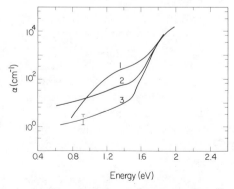

FIG. 2. The substrate temperature dependence of gap-state absorption (undoped material). (1) 100°C, (2) 330°C, (3) 230°C. [From Jackson and Amer (1981a,b).]

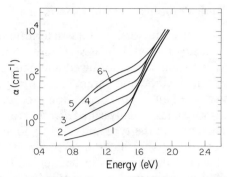

FIG. 3. The dependence of gap-state absorption on deposition power (undoped material). (1) 1 W, (2) 2 W, (3) 5 W, (4) 15 W, (5) 30 W, (6) 40 W. [After Jackson and Amer (1982).]

progressive decrease in the slope of the exponential edge as the deposition conditions depart from optimum (Jackson and Amer, 1981a,b). Finally, if undoped films, deposited at low substrate temperatures, are annealed, subgap absorption first decreases as the annealing temperature is increased to ~300°C, then increases again for temperatures higher than 300°C (Yamasaki *et al.*, 1981).

The effects of single doping and compensation (Jackson and Amer, 1982) are shown in Fig. 4. The spectra for various phosphorus doping levels are summarized in Fig. 4a. The gap-state absorption increases and the Urbach edge broadens as the phosphorus concentration is increased. Boron doping has a similar qualitative effect on the absorption spectrum (Fig. 4b). However, the change in the slope which differentiates between the Urbach edge and the subgap absorption becomes less pronounced than is the case for phosphorus-doped material.

The changes due to the introduction of dopants can be caused either by a shift of the Fermi level or by the bonding configurations of the dopant atoms. These effects can be separated by investigating compensated materials, i.e., materials into which n- and p-type dopants have been simultaneously incorporated. The results from a series of studies on compensated material are shown in Fig. 4c. The films were prepared by fixing the PH_3 concentration and gradually increasing the B_2H_6 vapor pressure. As the boron concentration increases, the magnitude of the gap-state absorption *decreases*. Concurrently, the Urbach edge broadens and shifts to lower energies.

The results in Figs. 2–4 demonstrate that the subgap region is highly sensitive to doping and to deposition conditions. From the subgap absorption, information about the nature, energy, and number of defects in a-Si : H can be obtained. The number of defects can be estimated by separating the

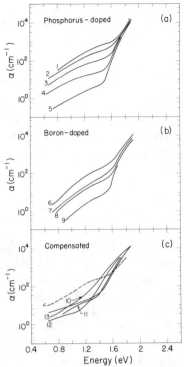

FIG. 4. The dependence of gap-state absorption on the type and level of doping. (a) PH_3 doping concentration: (1) 10^{-2}, (2) 10^{-3}, (3) 3×10^{-4}, (4) 10^{-5}, and (5) 10^{-6}; (b) B_2H_6 doping concentration: (6) 10^{-3}, (7) 3×10^{-4}, and (8) 10^{-4}, (9) 10^{-5}; (c) compensated material: all samples have 10^{-3} PH_3, and the B_2H_6 concentrations are (10) 2×10^{-4}, (11) 4×10^{-4}, (12) 2×10^{-3}, and (13) 4×10^{-3}. All doping concentrations refer to gas-phase concentration. Substrate temperature is 230°C and the rf deposition power is 2 W. [After Jackson and Amer (1982).]

subgap defect absorption from the exponential band-tail absorption. The excess optical absorption α_{ex} due to gap-state defect absorption is given by

$$\alpha_{ex} = \alpha - \alpha_0 \exp[\hbar\omega/E_0], \qquad (7)$$

where α_0 and E_0 are obtained from a fit to the exponential absorption.

From an optical sum rule, the number of defects N_s is related to the absorption by

$$N_s = \frac{cnm}{2\pi^2\hbar^2 e_s^{*2}} \int \alpha_{ex}(E) \, dE, \qquad (8)$$

where c is the speed of light, n (≈ 3.8) the index of refraction of Si, m the electron mass, and e_s^* the effective charge of the defect (Brodsky *et al.*, 1977). The integration limits extend from zero to the energy at which the exponential absorption terminates ($\sim 1.3-1.6$ eV). Using the local field corrections employed in interpreting the infrared absorption spectra of a-Si:H (Brodsky *et al.*, 1977; Dexter, 1956), e_s^{*2} is given by

$$e_s^{*2} = \frac{9n^2}{(1+2n^2)^2} e^2 f_{0j}, \qquad (9)$$

where f_{0j} is the oscillator strength of the absorption transition and e is the electron charge. By assuming that $f_{0j} = 1$ and noting that the local field corrections have been determined to overestimate the correct local field by a factor of 2 in a-Si:H (Brodsky *et al.*, 1977; Freeman and Paul, 1978), we obtain

$$N_s = 7.9 \times 10^{15} \int \alpha_{ex} \, dE. \qquad (10)$$

Equation (10) provides a useful measure of the relative number of defects in the a-Si:H. The assumptions that $f_{0j} = 1$ and that the integration terminates at ~ 1.5 eV was verified by comparing the number of defects obtained from Eq. (10) with those measured by other techniques such as electron spin resonance (ESR), light-induced ESR (LESR), and luminescence. The dominant defect in undoped a-Si:H is an ESR active defect with a *g*-value of 2.0055, which is characteristic of a silicon dangling bond (Street, 1981). By comparing the number of dangling bonds determined by ESR with N_s as measured by PDS, it is found that the numbers agree extremely well over three orders of magnitude (see Fig. 5). This excellent agreement shows that the subgap absorption in undoped material is due to singly occupied silicon dangling bonds. It should be noted that, in addition to yielding the absolute number of dangling bonds, PDS is sensitive to smaller dangling-bond densities than is ESR. The linear relation between the absorption and the dangling-bond density is unlike that for the case of unhydrogenated amorphous silicon in which the number of the absorption-derived defects varies as the square of the number of ESR-measured spins (Brodsky *et al.*, 1972).

In the case of phosphorus doping, the silicon dangling bonds are doubly occupied, whereas for boron-doped material the dangling bond is unoccupied. Consequently, the defects are LESR-active with a dangling-bond *g*-value of 2.0055. The number of these defects may be determined from LESR or deduced from the quenching of the luminescence (Street, 1981). There is a good correlation between N_s and the number of doubly occupied and unoccupied dangling bonds as shown in Fig. 6. This is clear evidence that the silicon dangling-bond defect is the source of gap-state absorption in doped a-Si:H as well. Equation (10) appears to underestimate the number of defects in boron-doped material. This is most likely due to the fact that

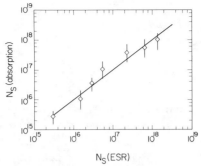

FIG. 5. The correlation between the number of defects deduced from PDS and ESR, undoped material. [After Jackson and Amer (1982).]

subgap absorption, in the case of boron doping, overlaps with the band-to-band transitions responsible for the exponential tail absorption. To compensate for this, the limits of integration should be extended by ~0.1 eV.

Figure 6 demonstrates that the defect density varies as the dopant concentration is increased. From PDS, LESR, and photoconductivity measurements, it is found that the defect density varies as the square root of the dopant concentration (Street, 1982, and references therein). Consequently, by combining the sensitivity of PDS with the square root dependence, very low doping concentrations can be measured readily. Because of uncertainties in photoconductivity, luminescence, and LESR measurements, PDS gives the most reliable estimates of defect densities in doped material. Recently, PDS results from doped material were invoked in a model to explain doping in a-Si:H (Street, 1982, and references therein).

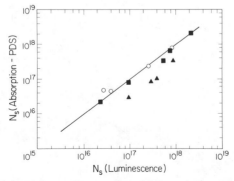

FIG. 6. The correlation between the number of defects deduced from PDS and from luminescence quenching. ▲, β-doped; ○, compensated; ■, P-doped. [After Jackson and Amer (1982).]

In addition to measuring the number of defects, PDS may also be used to determine the energy of the silicon dangling-bond defects. The absorption coefficient is given by

$$\alpha\hbar\omega \propto M^2 \int_{E_F-\hbar\omega}^{E_F} N_V(E)N_C(E + \hbar\omega)\, dE, \tag{11}$$

where M is the transition matrix element, $N_V(N_C)$ the density of states below (above) the Fermi level E_F, and M is assumed to be independent of energy. Using the density of states discussed in Part V, the energy dependence of the absorption coefficient may be fitted to the data with an adjustable peak position. Then best fit is obtained for a peak position of approximately 1.25 eV and 0.9 eV below the conduction band for undoped and phosphorus-doped materials, respectively.

5. The Correlation Energy of the Silicon Dangling-Bond Defect

By using PDS, the energies of the singly and doubly occupied dangling-bond defects can be compared to give a measure of the electronic correlation energy of the defect.

The total energy required to place a second electron on the dangling bond is the sum of the Coulomb repulsion and the dielectric and atomic relaxation energies. If the sum is positive, the doubly occupied dangling bond will be closer to the conduction band. Consequently, gap-state absorption will exhibit a threshold at a lower energy; otherwise it remains unchanged. By comparing the singly and doubly occupied dangling bonds in undoped and phosphorus-doped material, respectively, it is found that the absorption of the doped films is shifted to lower energies by ⁓0.35 eV (Fig. 7). Hence the

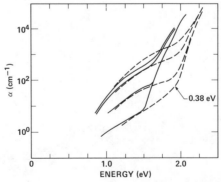

FIG. 7. Photon energy dependence of the absorption for phosphorus-doped (dashed curves) and undoped (solid curves) material with a wide range of defects. The doped curves have been shifted to the right by 0.38 eV and multiplied by a factor of 6. [From Jackson (1982).]

correlation energy U of the silicon dangling-bond defect is positive and has a value of -0.35 eV (Jackson, 1982), which agrees with the value deduced from the ESR study of the dependence of the number of spins on the Fermi-level position (Dersch *et al.*, 1981b). The fact that U is positive implies that the Coulomb repulsion dominates. The implications of these results for the density of states are discussed in Part V.

6. SURFACE DEFECTS

There is significant evidence for the existence of surface and interface defects in a-Si:H films. Such evidence comes from ESR (Knights *et al.*, 1977), photoconductivity (Ast and Brodsky, 1980), and conductivity (Fritzsche, 1980; Solomon *et al.*, 1978) measurements. However, ESR requires the use of relatively thick films, and both ESR and conductivity measurements are inherently unable to differentiate among actual changes in the density of states and band bending and/or Fermi-level shifts. Photothermal deflection spectroscopy, with its high sensitivity, enables one to measure the total density of defects over a wide range of film thicknesses. This would provide a measure of the density of surface defects independent of the Fermi-level position.

By combining Eq. (10) with the PDS results of a thickness dependence study (Jackson *et al.*, 1983a), it was found that for low-defect material the film surface (or its interface with the substrate) has a defect density of $\sim 10^{12}$ silicon dangling bonds (per cm^2) (Fig. 8). The results also indicate that the

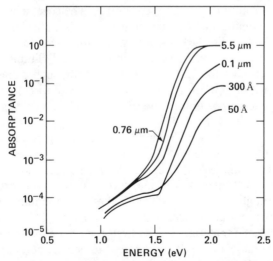

FIG. 8. Thickness dependence of the absorption in undoped a-Si:H at 230°C. [From Jackson *et al.* (1983a).]

defects are confined to a very thin (5-nm) surface or interface layer. The absorption spectra of the defective layer are similar to those obtained from the highly defective bulk a-Si : H (high gap-state absorption and broadened Urbach edge). This shows that the bulk density of states is different from that at the surface (or interface). It also explains why field-effect measurements, which are surface sensitive, consistently yield larger densities of gap-states than those obtained by transient capacitance measurements.

7. OPTICAL PROPERTIES OF PHOTOINDUCED DEFECTS

The reversible photoinduced changes (Staebler–Wronski effect) in the properties of a-Si : H have attracted much attention (Staebler and Wronski, 1977, 1980). Illumination creates metastable defects that are annealed away by heating the films above 150°C. However, the exact mechanism responsible for this effect remains to be elucidated.

An interesting question is how the so-called Staebler–Wronski effect affects the absorption spectrum of states residing in the gap of a-Si: H. Furthermore, although ESR measurements suggest an increase in the number of silicon dangling bonds after illumination, these studies detect unpaired spins only. Therefore, it is possible that the density of dangling bonds remains constant, and the observed increase would then be due to a shift in the Fermi level deeper in the gap, into a region of increased dangling-bond states. In this case, the number of unpaired spins would increase even though the total number of dangling-bond defects remains constant.

Figure 9 shows the effect of illumination with broadband light on the gap-state absorption of an undoped sample. The spectra show a clear and reproducible enhancement in gap-state absorption after illumination. Annealing at 150°C in the dark restores the magnitude of the absorption to its original value. The spectra of the annealed and illuminated films are qualitatively similar, the only difference being the higher magnitude of the subgap absorption. By using the procedures described in Section 4, the density and energy position of the optically induced defects can be determined (Amer et al., 1983).

In the case of undoped material, the increase in gap-state defect density ΔN_s is found to be constant ($\sim 10^{16}$ defects cm^{-3}), in agreement with ESR results (Dersch et al., 1981a), irrespective of the preparation conditions. The dependence of ΔN_s on doping level is shown in Fig. 10. The ratio $\Delta N_s / N_s$ is found to be roughly constant, indicating that the photoinduced defects scale with the doping-induced defect concentration (Fig. 11). In the case of the fully compensated material, ΔN_s showed the least enhancement. By using Eq. (11), the energy of the defect is found to be 1.25 and 0.9 eV below the conduction band for undoped and phosphorus-doped films, respectively.

The PDS spectra show no change in the Urbach edge, which in the case of

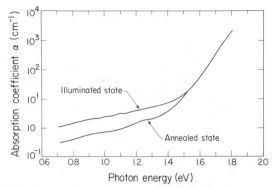

FIG. 9. The effect of illumination on the absorption spectrum of undoped a-Si : H. [From Amer *et al.* (1983).]

a-Si : H is dominated by the valence-band edge (Tiedje *et al.,* 1981). Another experimental tool for probing the edge is photoinduced absorption, in which the shape of the valence-band tails determines the rate of the photoexcited carrier decay (Wake and Amer, 1983). No change was found in the decay rate between the illuminated and the annealed states. One can then conclude that the photoinduced defects do not alter the shape of the valence-

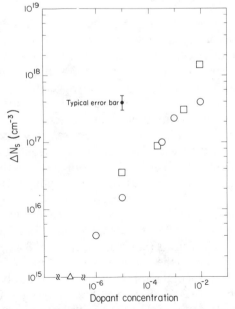

FIG. 10. The dependence of ΔN_s on doping concentration. □, boron doped; ○, phosphorus doped; △, compensated. [From Amer *et al.* (1983).]

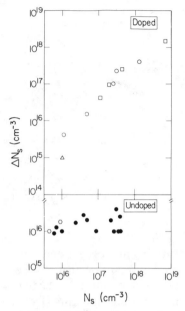

FIG. 11. The dependence of ΔN_s on N_s for doped and undoped materials. □, boron doped; ○, phosphorus doped; △, compensated; ○, oxygenated. [From Skumanich *et al.* (1983a).]

band tail (Skumanich *et al.*, 1983b). Neither the PDS nor the photoinduced absorption results are consistent with an increase in the number of states near the valence-band edge as suggested by Lang *et al.* (1982).

Several significant conclusions can be drawn from these results. The observed enhancement in dangling-bond defects on illumination is due to the creation of new defects and is not caused by a Fermi-level shift toward midgap. A shift in the Fermi level to lower energy alone would result in a decrease in absorption rather than an increase, as is observed experimentally. Furthermore, the successive increase in dopant concentration shifts the Fermi level into the band tails. It will then be in a region of the density-of-states diagram that has smaller dangling-bond density; consequently, the detected effect should decrease with increased doping. However, as seen in Fig. 11, the photoinduced change in defect density increases with doping. One is then led to conclude that illumination creates new silicon dangling-bond defects (Skumanich *et al.*, 1983a).

Since ΔN_s is constant in the undoped material, independent of deposition parameters, this implies that photoinduced defects may be related to impurities rather than to weak Si–Si bonds. Preliminary results (Skumanich *et al.*, 1983a) with films of a-Si:H grown with large oxygen content tend to confirm the conclusion that impurities may be involved. The constant ratios

$\Delta N_s/N_s$ for singly doped films, independent of the type and concentration of the dopant and of the film thickness, indicate that the photoinduced states are themselves related to defects associated with doping. Finally, compensation with equal amounts of phosphorus and boron has the striking effect of drastically minimizing ΔN_s. This suggests that the position of the Fermi level may be important in determining the number of photoinduced defects.

The results can also be used to shed some light on the mechanism responsible for creating the photoinduced defect. The data imply that such a mechanism does not involve breaking Si–Si bonds. The scaling of ΔN_s with doping can be attributed to lowering the defect creation energy as caused by the change in the position of the Fermi level. Furthermore, high defect-density undoped material has a large number of weak Si–Si bonds due to strain disorder. Yet ΔN_s is constant, independent of the degree of disorder. Thus the photoinduced defects are not related to strain disorder, and this in turn suggests that weak Si–Si bonds are not primarily involved in defect creation.

Finally, it has been reported that when air is leaked into the deposition chamber, solar cells exhibit an even greater drop in efficiency after illumination, which can be annealed away by heating (Carlson, 1982). However, by using PDS we have found that, for an a-Si:H film containing up to 2000 ppm of oxygen, the density of light-induced defects is independent of the oxygen concentrations.

8. URBACH EDGE STUDIES BY PDS

A full treatment of the exponential edge absorption involving band-to-band transitions is given in Chapter 2 by Cody. In this section we review results obtained with PDS (Jackson and Amer, 1981) in conjunction with the work discussed in Section 4. From Figs. 2–4 it is evident that the slope of the exponential edge depends on the conditions under which a-Si:H is prepared as well as on the type and degree of doping. As the number of defects increases, the Urbach edge broadens.

The absorption edge in that region can be fitted to the form

$$\alpha = \alpha_0 \exp(\hbar\omega/E_0), \tag{12}$$

where E_0 is the width of the exponential edge. To test the degree of correlation between the width of the edge and the number of defects, E_0 is plotted in Fig. 12 as a function of the equilibrium spin density. A strong correlation is found, with the functional form $N_s \propto \exp(E_0/10 \text{ meV})$, that could be interpreted as evidence that the disorder-induced fields and strains responsible for controlling the slope of the Urbach edge are caused by dangling-bond defects. Alternatively, the defects and the structural disorder may be produced concurrently while the material is being deposited.

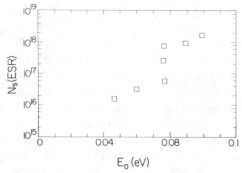

FIG. 12. The correlation between the Urbach edge energy and the ESR-deduced spin density. [From Jackson and Amer (1981b).]

The optical absorption edge of a-Si:H is dominated by the band with the broader exponential tail, i.e., the valence-band tail (Tiedje *et al.*, 1981). Photoinduced absorption studies (Wake and Amer, 1983) have been assumed to measure this part of the density-of-states diagram. This induced absorption $\Delta T_r/T_r$ is given by

$$\Delta T_r/T_r \propto t^{-\beta}, \tag{13}$$

where $\beta = T/T_0$, $T_0 = E_0/k_B$, and k_B is the Boltzmann constant. As shown in Fig. 13, an excellent correlation is found between E_0 from PDS and from photoinduced absorption. Therefore, this strong correlation verifies the assumption that the shape of the valence-band tail determines the rate of photoinduced absorption decay.

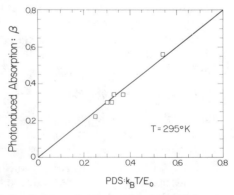

FIG. 13. The correlation between photoinduced and PDS-deduced optical absorption for undoped material with varying defect density. [After Wake and Amer (1983).]

IV. Carrier Transport

In the previous sections, PDS was used to determine the optical absorption of a-Si:H. This technique may be employed to measure transport properties as well. Because photoconductivity (PC) depends on both the absorption and carrier transport, the combination of PDS and PC provides information about the dependence of the carrier transport on excitation energy (Jackson *et al.*, 1983b). In addition PDS also provides a way of normalizing the photoconductivity to account for effects of intensity on the carrier lifetime (Jackson *et al.*, 1983b). A complete review of the various aspects of photoconductivity is given in Chapter 8 by Crandall.

Disorder significantly alters the transport of carriers. Within the bands, the carriers occupy extended states, whereas within the gap the states are localized. Consequently, the distance a carrier moves in a field (the range) should exhibit a decrease as the energy of the exciting light is decreased. The range per unit field is given by $\mu\tau$, where μ and τ are the carrier mobility and lifetime, respectively. Comparison of PC and PDS has shown that there is a fairly rapid transition from extended to localized transport.

9. TYPES OF PHOTOCONDUCTIVITY

Three different configurations are used for measuring PC, and they depend on different aspects of carrier transport. If PC is measured with a gap electrode configuration, the photocurrent depends on the transport of both electrons and holes (Fig. 14). The electrodes can inject both electrons and holes to maintain the charge neutrality of the film. The resulting photocurrent ΔI_{ss} is given by

$$\Delta I_{ss} \propto [(\eta\mu\tau)_e + (\eta\mu\tau)_h]f, \qquad (14)$$

where η is the efficiency of carrier generation and the subscripts e and h refer to the electron and hole, respectively. The average number of photons f absorbed per cubic centimeter per second is given by

$$f = \Phi S/lS_{sat}, \qquad (15)$$

where Φ is the flux of photons per square centimeter per second, S is the photothermal deflection signal, and l is the film thickness.

If the photocurrent is measured on a forward-biased Schottky barrier, the electrons can be injected by the back contact, whereas the holes cannot be injected by the Schottky barrier. Consequently, the photocurrent ΔI_{sp} depends only on the electron transport as

$$\Delta I_{sp} \propto f(\eta\mu\tau)_e. \qquad (16)$$

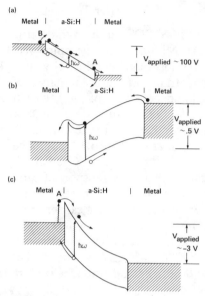

FIG. 14. Types of photoconductivity: (a) secondary gap photoconductivity (SSPC), (b) secondary forward-bias photoconductivity (SPPC), and (c) primary reverse-bias photoconductivity (PPPC). [From Jackson *et al.* (1983b).]

Finally, if the photocurrent is measured on a reverse-biased Schottky barrier, neither electrode can inject carriers. When the film is thick and the carriers are excited by subgap light, the photocurrent is given by

$$\Delta I_{pp} \propto (\eta\mu\tau)_h f. \tag{17}$$

Equations (16) and (17) indicate that in principle $(\eta\mu\tau)_e$ and $(\eta\mu\tau)_h$ can be determined by using photoconductivity. Unfortunately, $\eta\mu$ may depend on the photon energy and τ depends on the generation rate f, which also depends on photon energy. The PDS measurements allow one to adjust the incident light intensity to hold f constant and hence to eliminate the f dependence of τ. Dividing the constant generation rate normalized PC spectra by the PDS spectra gives the energy dependence of $(\eta\mu\tau)_e$ and $(\eta\mu\tau)_h$ at equilibrium due to changes in the photon energy.

10. PHOTOCONDUCTIVITY RESULTS

The results of the three different kinds of photoconductivity measurements and PDS for an undoped film are shown in Fig. 15. At the high energies above 1.6 eV, the spectral dependence of the four measurements is the same. Below ∼ 1.5 eV, the reverse-biased Schottky photocurrent begins

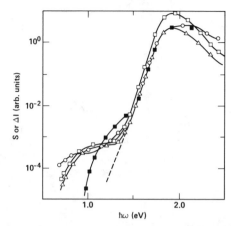

FIG. 15. The dependence of constant generation rate photoconductivity and absorptance on photon energy. ○, PDS; △, SSPC; □, SPPC, ■, PPPC. The dashed line shows the primary reverse-bias photoconductivity without the internal photoemission contribution. [From Jackson *et al.* (1983b).]

to deviate from other measurements. Finally, at ⁓ 0.9 eV, the absorption differs from the gap electrode and forward-biased Shottky photocurrents.

These results are explained by considering the density of states discussed in Part V. For photon energies greater than ⁓ 1.6 eV, both photogenerated carriers either are mobile or can thermalize at room temperature during their lifetime. The spectral dependence of the three photocurrents follows the absorption, indicating that $(\eta\mu\tau)_e$ and $(\eta\mu\tau)_h$ are independent of energy (Fig. 16). Below ⁓ 1.5 eV, a hole cannot thermalize to the valence-band edge. The photocurrent of a reverse-biased Schottky barrier decreases relative to the absorption. The dotted line in Fig. 16 shows the dependence of the reverse-biased Schottky barrier after corrections for internal photoemission (Jackson *et al.*, 1983b). Because the electron current dominates the hole current in undoped and phosphorus-doped films, the gap cell and forward-biased Schottky photocurrents will follow the absorption since the electron is still mobile. The quantity $(\eta\mu\tau)_h$ shows a rapid decrease, whereas $(\eta\mu\tau)_e$ remains constant (see Figs. 16 and 17). For energies less than 1.0 eV, neither the electron nor the hole is mobile, leading to a deviation between the gap electrode and forward-biased Schottky photocurrents and the absorption. Here, $(\eta\mu\tau)_e$ exhibits a decrease, especially in the phosphorus-doped case.

The fact that $(\eta\mu\tau)_e$ is roughly constant to 0.9 eV, demonstrates that the defect absorption transition results in the same final state as band-to-band transitions. Consequently, the defect absorption is due to transitions from the defect to the conduction band rather than from the valence band to the

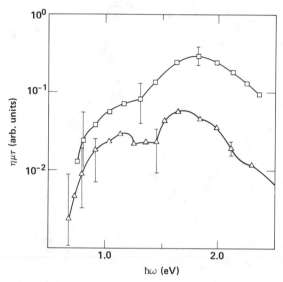

FIG. 16. $\eta\mu\tau$ derived from photoconductivity and absorptance. $\triangle$, $(\eta\mu\tau)_e + (\eta\mu\tau)_h$ from SSPC; $\square$, $(\eta\mu\tau)_e$ from SPPC. The curves are offset vertically for clarity. [From Jackson *et al.* (1983b).]

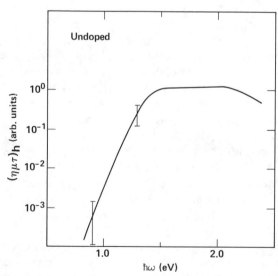

FIG. 17. $(\eta\mu\tau)_h$, relative to the value at 2.0 eV, as a function of photon energy. [From Jackson *et al.* (1983b).]

defect. The energy threshold of the defect absorption measures the energy of the defect relative to the conduction band.

The results also indicate that geminate recombination is not the dominate recombination path; otherwise $(\eta\mu\tau)_e$ and $(\eta\mu\tau)_h$ would show a large decrease for photon energies less than 1.6 eV. Due to the large error bars, the possibility that a small fraction of the photocarriers recombine geminately cannot be excluded.

The comparison of PDS with PC indicates that although in principle PC can be used to determine the absorption, in practice the variation of the carrier lifetime with intensity causes distortions in the energy dependence of the photoconductivity. Empirically, it has been found that after careful normalization, the defect absorption may deviate by as much as a factor of 2 to 3 relative to the above band value. Hence, the photoconductivity-derived absorption slopes are uncertain by about 15 meV.

Photothermal deflection spectroscopy can be used to normalize photoconductivity since the lifetime is a function of the generation rate. If the photocurrent is measured as a function of light intensity, one obtains the functional dependence of the lifetime on generation rate. By using the absorption from PDS and the measured light flux, one can determine the generation rate to derive the photocurrent at a constant generation rate (and hence lifetime). This technique enables one to use photoconductivity in situations in which PDS is unsuitable due to background absorption.

The results presented in this section demonstrate the versatility and utility of PDS. Not only can the optical absorption be measured but also the range of photoexcited carriers can be determined as well.

V. PDS-Deduced Features in the Density of States of a-Si:H

In this section, the contributions of PDS to the understanding of the density of states in a-Si:H are described, and the results discussed in the previous sections will be placed in a more comprehensive picture.

The density-of-states diagram, as deduced from a variety of techniques, including photoemission, dispersive transport, photoinduced absorption, deep level transient spectroscopy (DLTS), luminescence, and ESR, as well as PDS, is shown in Fig. 18 for undoped and phosphorus-doped a-Si:H. First consider the conduction band. Little is known about the bottom of the conduction band near the band edge other than that it is probably close to a parabolic density of states and possesses no sharp structure (Pierce and Spicer, 1972). Comparison of derived absorption spectra with PDS spectra suggests that the conduction band may be somewhat broadened compared with $(E - E_c)^{1/2}$; however, more experiments need to be performed.

Near the conduction-band edge, the states become localized below the

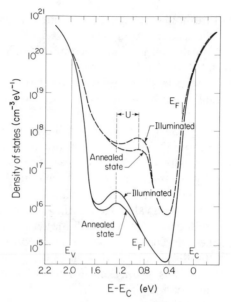

FIG. 18. Density of states for undoped (——) and phosphorus-doped (— — —) a-Si:H. E_F denotes the Fermi level for single occupancy.

conduction-band mobility edge E_c. The comparison of the absorption with the photoconductivity supports the view that the states become localized near the conduction-band edge. Below the mobility edge, dispersive transport has shown that the states tail exponentially into the gap. The edge is extremely sharp, decreasing by five orders of magnitude in 0.25 eV. An important consequence of the rapid decrease of the conduction-band tail is that the spectral dependence of the absorption is dominated by the energy distribution of valence-band and defect states, rather than being a mixture of features of the conduction- and valence-band densities of states.

In the energy range 0.7–1.5 eV below the conduction band, the density of states is dominated by silicon dangling-bond defects. The correlation of the spin density with the gap-state absorption provided the first evidence that a peak in the density of states, located slightly below midgap at ~1.25 eV below the conduction band, was specifically associated with the silicon dangling bonds (Jackson and Amer, 1981a,b). This peak grows as the rf power increases, if the substrate temperature deviates from 180–250°C, or if the film is annealed to remove hydrogen. The integrated magnitude of the dangling-bond peak ranges roughly from 10^{15} to 10^{19} states cm^{-3}. The integrated magnitude of the dangling-bond peak increases to ~10^{17} states cm^{-3} near the interface and/or surface. The spatial inhomogeneity explains the discrepancy between the interface-sensitive field-effect measurements of the density of states and the bulk-sensitive capacitance DLTS measurements. The singly occupied dangling-bond defect peak increases by 10^{16} states cm^{-3} if the films are exposed to illumination above the band gap for a protracted time.

Phosphorus doping causes the dangling-bond defect peak to increase in magnitude as the square root of the doping concentration and to move to a higher energy ($+0.38$ eV) due to the Coulomb repulsion between the two localized electrons on the dangling bond. Boron doping also increases the defect concentration as the square root of the doping concentration. Note that since the spectral dependence of the photoconductivity and PDS has not been determined for boron-doped material, it is not known whether the peak lies close to the conduction band or near the valence band. Finally, compensation decreases the dangling-bond defect peak.

The effect of illumination is to increase the magnitude of the dangling-bond peak by $\sim 30\%$ in both phosphorus- and boron-doped samples. By a mechanism not yet understood, the number of photoinduced defects is related to the number of defects that existed prior to illumination. The process apparently involves the position of the Fermi level since the compensated material exhibits a very small increase.

The valence-band edge decreases exponentially into the gap from the valence-band mobility edge. The width of the valence band is very sensitive to deposition conditions and doping. Many parameters that increase the dangling-bond defect density also increase the width of the valence-band tail, roughly according to the relation $N_s \propto \exp(E_0/10 \text{ meV})$. However, if the dangling bonds are photoinduced, an increase in the valence-band tail does not occur. This fact suggests the possibility that disorder, as measured by the valence-band width, causes dangling bonds during deposition. Apparently, the photoinduced dangling bonds do not by themselves cause disorder of the tail above the valence mobility edge.

Comparison of the spectral dependence of the hole current with the absorption indicates that at least over the range of tenths of an electron volt the hole states make a transition from extended to localized behavior. The width of the transition cannot be determined accurately because the measurements were performed at high temperatures and the internal photoemission causes a large background. The photoconductivity measurements on Schottky barriers demonstrate that the optical transitions occur between filled defect states and the conduction band for undoped and phosphorus-doped material.

The density of states discussed in this section has been used to calculate the absorption in the vicinity of the band edge (Jackson, 1982). The calculated and experimental spectra are compared in Fig. 19. The good agreement between the two is indicative of the consistency between the

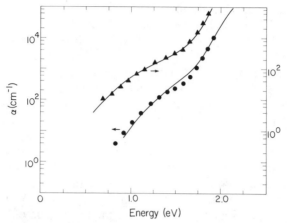

Fig. 19. Calculated and PDS-deduced absorption spectra of undoped and phosphorus-doped materials. ●, 30 W undoped; ▲, 10^{-3} PH$_3$. [From Jackson (1982).]

experimental absorption data and the density of states model shown in Fig. 18. Thus PDS can be employed as a check on the density of states models derived from other techniques.

VI. Concluding Remarks

The preceding discussion demonstrates the utility of PDS in obtaining information about the density of states in a-Si:H. The power of measuring optical absorption arises from the fact that it is a fundamental and well-understood process occurring in all solids. Typically, absorption measurements do not require contacts, doping conditions, Fermi-level positions, special temperatures, device structures, etc. The effect of almost any material parameter on the density of states can be assessed by measuring the effect of the parameter on the optical absorption.

As shown in this chapter, photothermal deflection spectroscopy has proven to be a powerful tool for the investigation of very small optical absorptions. By overcoming the inherent limitations of conventional absorption techniques, PDS enables the direct and reliable determination of the energy and number of defects in a-Si:H. Furthermore, it is an excellent tool for the characterization of material parameters such as doping concentration, defect level, and film homogeneity.

As to future directions, an obvious extension of the work reviewed in this chapter is the study of intrinsic defects and other impurities. Another possibility is the investigation of the effects of alloying amorphous silicon with other materials such as germanium, carbon, or tin. Photothermal deflection spectroscopy is a unique tool for probing nonradiative transitions. When combined with luminescence studies, a complete picture of the branching ratios of radiative and nonradiative deexcitation should emerge. The study of the nature of defects in microcrystalline and CVD-deposited silicon is also a possible direction for future research.

Given the relative ease with which PDS can be performed, this technique has the potential of becoming an important addition to the tools employed in materials characterization, process control, and device diagnostics.

ACKNOWLEDGMENTS

It is a pleasure to acknowledge the contributions of A. C. Boccara and D. Fournier to the development of photothermal deflection spectroscopy. This work was supported by the Assistant Secretary for Conservation and Renewable Energy, Photovoltaic Systems Division of the Department of Energy under contract No. DE-AC03-76SF00098, and by the Solar Energy Research Institute under contract No. XJ-0-9079-1.

REFERENCES

Abeles, B., Wronski, C. R., Tiedje, T., and Cody, G. D. (1980). *Solid State Commun.* **36,** 537.
Amer, N. M., Skumanich, A., and Jackson, W. B. (1983). *Physica Utrecht* **117B & 118B,** 897.
Ast, D. G., and Brodsky, M. H. (1980). *Philos. Mag. B* **24,** 273.
Boccara, A. C., Fournier, D., and Badoz, J. (1980a). *Appl. Phys. Lett.* **36,** 130.
Boccara, A. C., Fournier, D., Jackson, W. B., and Amer, N. M. (1980b). *Optics Lett.* **5,** 377.
Brodsky, M. H., Kaplan, D. M., and Ziegler, J. F. (1972). *In Proc. Int. Conf. Phys. Semicond., 11th, Warsaw, 1972* (Polish Academy of Sciences, eds.), p. 529. Elsevier, Amsterdam.
Brodsky, M. H., Cardona, M., and Cuomo, J. J. (1977). *Phys. Rev. B* **16,** 3556.
Carlson, D. E. (1982). *Solar Energy Mat.* **8,** 129.
Cody, G., Wronski, C. R., Abeles, B., Stephens, R., and Brooks, B. (1980). *Solar Cells* **2,** 227.
Connell, G. A. N., and Pawlik, J. R. (1976). *Phys. Rev. B* **13,** 787.
Crandall, R. S. (1980a). *Phys. Rev. Lett.* **44,** 749.
Crandall, R. S. (1980b). *Solar Cells* **2,** 319.
Dersch, H., Stuke, J., and Beichler, J. (1981a). *Appl. Phys. Lett.* **38,** 456.
Dersch, H., Stuke, J., and Beichler, J. (1981b). *Phys. Status Solidi B* **105,** 265.
Dexter, D. L. (1956). *Phys. Rev.* **101,** 48.
Freeman, E. C., and Paul, W. (1978). *Phys. Rev. B.* **18,** 4288.
Freeman, E. C., and Paul, W. (1979). *Phys. Rev. B* **20,** 716.
Fritzsche, H. (1980). *Solar Cells* **2,** 289.
Jackson, W. B. (1982). *Solid State Commun.* **44,** 477.
Jackson, W. B., and Amer, N. M. (1981a). *AIP Conf. Proc.* **73,** 263.
Jackson, W. B., and Amer, N. M. (1981b). *J. Phys. Colloq. Orsay Fr.* **42,** C4-293.
Jackson, W. B., and Amer, N. M. (1982). *Phys. Rev. B* **25,** 5559.
Jackson, W. B., Amer, N. M., Boccara, A. C., and Fournier, D. (1981). *Appl. Opt.* **20,** 1333.
Jackson, W. B., Biegelsen, D. K., Nemanich, R. J., and Knights, J. C. (1983a). *Appl. Phys. Lett.* **42,** 105.
Jackson, W. B., Nemanich, R. J., and Amer, N. M. (1983b). *Phys. Rev. B* **27,** 4861.
Knights, J. C., Biegelsen, D. K., and Solomon, I. (1977). *Solid State Commun.* **22,** 133.

Lang, D. V., Cohen, J. D., Harbison, J. P., and Sergent, A. M. (1982). *Appl. Phys. Lett.* **40**, 474.
Loveland, R. J., Spear, W. E., and Al-Sharbaty, A. (1973). *J. Non-Cryst. Solids* **11**, 55.
Moddel, G., Anderson, D. A., and Paul, W. (1980). *Phys. Rev. B* **22**, 1918.
Mott, N. F. (1979). "Electronic Processes in Non-Crystalline Materials," 2nd ed. Oxford Univ. Press, London and New York.
Parker, G. (1973). *Appl. Opt.* **12**, 2974.
Pierce, D. J., and Spicer, W. E. (1972). *Phys. Rev. B* **5**, 3017.
Skumanich, A., Amer, N. M., and Jackson, W. B. (1983a). (Submitted.)
Skumanich, A., Wake, D. R., and Amer, N. M. (1983b). *Bull. Am. Phys. Soc.* **28**, 257.
Solomon, I., Dietl, T., and Kaplan, D. (1978). *J. Phys. Paris* **39**, 1241.
Staebler, D. L., and Wronski, C. R. (1977). *Appl. Phys. Lett.* **31**, 292.
Staebler, D. L., and Wronski, C. R. (1980). *J. Appl. Phys.* **51**, 3262.
Street, R. A. (1981). *Phys. Rev. B* **24**, 969.
Street, R. A. (1982). *Phys. Rev. Lett.* **49**, 1187.
Tiedje, T., Cebulka, J. M., Morel, D. L., and Abeles, B. (1981). *Phys. Rev. Lett.* **46**, 1425.
Tsai, C. C., and Fritzsche, F. (1979). *Solar Energy Mat.* **1**, 29.
Wake, D. R., and Amer, N. M. (1983). *Phys. Rev. B* **27**, 2598.
Wronski, C. R., Abeles, B., Tiedje, T., and Cody, G. D. (1982). *Solid State Commun.* **44**, 1423.
Yamasaki, S., Hata, N., Yoshida, T., Oheda, H., Matsuda, A., Okushi, H., and Tanaka, K. (1981). *J. Phys. Colloq. Orsay Fr.* **42**, C4-297.
Yasa, Z. A., Jackson, W. B., and Amer, N. M. (1982). *Appl. Opt.* **21**, 21.
Zanzucchi, P. J., Wronski, C. R., and Carlson, D. E. (1977). *J. Appl Phys.* **48**, 5228.

CHAPTER 4

The Vibrational Spectra of a-Si:H

P. J. Zanzucchi

RCA/DAVID SARNOFF RESEARCH CENTER
PRINCETON, NEW JERSEY

I. Introduction

In this chapter, the vibrational properties of hydrogenated amorphous silicon (a-Si:H) are reviewed. Present understanding of these properties has evolved with the development of amorphous silicon materials and devices (Carlson and Wronski, 1976; Carlson, 1977; Wronski, 1979; Hamakawa, 1981) and with the formulation of new models for local and structural vibrational states in amorphous and crystalline solids (Newman, 1969; Alben *et al.*, 1975; Barker and Sievers, 1975; Pollard and Joannopoulos, 1981).

When first prepared, and for some time thereafter, the properties of amorphous silicon produced by glow discharge were a puzzle (Brodsky *et al.*, 1970). Unlike a-Si films prepared by evaporation or sputtering, the discharge-produced films had good photoconductivity and carrier lifetimes

113

(Chittick *et al.*, 1969). Within this same time period, infrared data reported by Taft (1971) showed that there was extensive incorporation of bonded hydrogen species in silicon nitride films produced by decomposition of silane in glow discharge. In retrospect, these data suggested that bonded hydrogen was a significant factor in the properties of amorphous silicon when prepared by glow discharge. However, at that time the role of bonded hydrogen was not well understood. As is now known, bonded hydrogen reduces the number of incomplete silicon bonds that can act as carrier traps (Spear and LeComber, 1975; Lewis, 1976; Connell and Pawlik, 1976; Paul *et al.*, 1976).

Understanding of the properties of bonded hydrogen in the variously prepared a-Si:H films came, in part, from detailed studies of the vibrational spectra of these films (e.g. Brodsky *et al.*, 1977). Typically, these spectra cover the wavelength range 2.5–40 μm, which is the mid-infrared region, and show the magnitude of dipole absorption. Note from Table I that the frequency range in which infrared absorption occurs is broad. In this review, we shall be concerned with fundamental infrared absorption in the mid-infrared.

As will be discussed (Part IV), the frequency of a fundamental infrared absorption mode depends on both the mass of the oscillating dipole and the strength of the bond between the atoms comprising the dipole. For a model

TABLE I

COMPARISON OF FREQUENCY AND THE TYPE OF MEASUREMENT IN THE INFRARED[a]

FREQUENCY Hz	WAVELENGTH μm	MEASUREMENT
		⌈ INFRARED
10^{14}	1	overtone vibrations
10^{13}	10^1	{ fundamental vibrations
10^{12}	10^2	bending mode
10^{11}	10^3	⌈ ESR
		{ rotational states
10^7	10 m	
10^6	10^2	⌈ NMR
		{ ION CYCLOTRON
10^5	10^3	{ MASS SPEC
		INTERFERO- METRY
10^4	10^4	{ ELECTRO- CHEMISTRY

[a] Other common analytical methods, and their corresponding frequency ranges, are listed for comparison.

of the vibration of a dipole, let two spheres of unequal mass be connected by a spring. The vibration of the model will follow Hooke's law [see Eq. (9) of Part IV]. This implies that for dipoles, e.g., the CH functional group in organic structures, the frequency of the oscillator will occur only over a narrow, fixed range. Thus, by measuring the characteristic frequency at which infrared absorption occurs, the identity of molecular species can be determined [see, e.g., Shimanouchi (1972, 1977)]. Frequently, however, other factors besides mass and bond strength must be considered when analyzing an infrared spectrum, and these are discussed in this text [see also Colthup *et al.* (1975)].

Infrared spectra of a-Si:H films (Fig. 1) show mainly the vibrational modes characteristic of various silicon–hydrogen bondings, i.e., $Si-H_x$ $(x = 1, 3)$. In Fig. 1, the absorption bands are due to three types of SiH_x modes. As discussed in Part III, the absorption at 2000 and 630 cm^{-1} is due to SiH_1 stretching and rocking modes, respectively. The weaker bands between 800 and 900 cm^{-1} are due to SiH_2 vibrational modes. The mass difference between Si and H is large and the vibrational spectrum of a-Si:H, e.g., Fig. 1, can be interpreted as largely due to the motion of the hydrogen atoms (Lucovsky, 1976). This assumption may not be valid when other elements such as fluorine are incorporated in the a-Si structure. In addition to the mass effect, near neighbors may influence the dipole oscillator by an inductive effect (Lucovsky, 1976). As discussed in Part IV, the SiH_x oscillator frequencies are affected by the presence of fluorine in a-Si:H,F films.

While the vibrational modes due to local, or short-range, bonding are evident in the a-Si:H spectrum (Fig. 1), vibrational modes due to the a-Si framework structure, i.e., the Si–Si bonding, also occur but are generally

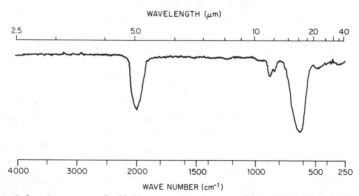

FIG. 1. Infrared spectrum of a-Si:H prepared by decomposition of silane in glow discharge. See Table II for identification of the absorption modes. The spectra of a-Si:H films are generally characterized by the presence of SiH_x $(x = 1, 2)$ vibrational modes.

weak (Part VII). These modes do not have first-order dipole moments, and infrared absorption occurs only by second-order processes.

In addition to characterizing species by either local or structural modes, the number of absorbing species can be determined from infrared data (Connell and Pawlik, 1976; Brodsky *et al.,* 1977; see also the Appendix). In the early development of a-Si:H material, calculations of this type, which are based on the magnitude of the integrated absorption for the species of interest, indicated that the amount of hydrogen in glow-discharge-prepared a-Si films was large, i.e., on the order of 5–20 at. %.

In the following sections, the types of instrumentation for infrared spectroscopy are described; the vibrational frequencies for the a-Si structure containing hydrogen, fluorine, carbon, boron, phosphorus, or nitrogen are summarized; and the relation between these data and hydrogen bonding in a-Si:H is reviewed.

II. Infrared Spectroscopy

1. INSTRUMENTATION

Infrared spectra are readily recorded by using either of two different types of instruments. These instruments separate the wavelengths of light by dispersion or interference. The latter technique requires use of the Fourier transform to obtain a spectrum. Commercial spectrometers using dispersion as the means of obtaining spectra have been available since the 1940s. Potts (1963) has reviewed the properties of that type of instrument. By contrast, commercial interferometric Fourier transform instruments for infrared spectroscopy were not practical until computers and software, e.g., the Cooley–Tukey algorithm, became widely available (Cooley and Tukey, 1965; Brigham, 1974, see pp. 7–9). These are essential for computing the Fourier transform in a reasonable amount of time. By the late 1960s commercial Fourier transform infrared spectrometers were widely available.

Nearly monochromatic infrared light is obtained with a dispersive instrument by refracting or diffracting and filtering incident polychromatic light. By this process the frequency of infrared radiation can be varied continuously, e.g., by rotating a diffraction grating (Potts, 1963; Colthup *et al.,* 1975), and infrared spectra can be recorded in real time. By contrast, monochromatic light is not obtained from an interferometer, which is shown schematically in Fig. 2, and spectra are not directly recorded as with a dispersive instrument.

In the interferometer, radiation from a source of infrared radiation (A) is directed to a 45° beam splitter (C). The split beams pass to a movable mirror

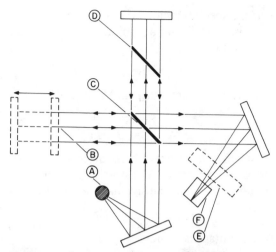

FIG. 2. Schematic diagram of an interferometer. For the Michelson interferometer shown, A is a source of infrared radiation, B a movable mirror whose position is precisely measured by a laser beam, C a 45° beam splitter, D a compensator plate, E a light filter, and F a detector.

(B) and a fixed mirror which is sometimes preceded by a compensator plate (D). This plate is designed to keep the optical path length of the fixed mirror arm the same as that of the movable mirror arm. The recombined beams eventually pass to a detector (F) after passing through a filter (E) which eliminates unwanted frequencies of light.

What the interferometer produces is a characteristic interference pattern as the mirror moves. The interference of a purely monochromatic beam of light is a cosine function of light intensity for each cycle of mirror displacement. Each wavelength, e.g., 2.5–40 μm, from a polychromatic source produces its own unique cosine function. The conventional spectrum representing light intensity as a function of wavelength is obtained by Fourier analysis of the complex interference pattern. Clearly, obtaining an infrared spectrum by this process is complicated. Compared to dispersion, interferometry has two distinct advantages in obtaining infrared spectra. These are known as Fellgett's and Jacquinot's advantages.

Based on Fellgett's analysis, the signal-to-noise ratio of a Fourier transform infrared spectrometer can be shown to be larger than that of a dispersive instrument for the same resolution and spectral range (Fellgett, 1958; see also Bell, 1972). This is due to the fundamental difference in the way the various frequencies of light are separated. Fellgett's analysis also assumes that the detector noise is independent of the signal, which is true for infrared detectors but is not a valid assumption for all light detectors. In a dispersive instrument, a relatively narrow band of frequencies is measured per unit time. If the spectral region is, in wave numbers, σ_1 to σ_2, and this region is scanned in time T with a resolution of R, the number of spectral elements M is given by

$$M = (\sigma_2 - \sigma_1)/R \quad \text{for} \quad \sigma_2 > \sigma_1. \quad (1)$$

The signal-to-noise ratio for a dispersive infrared spectrometer is then proportional to

$$(S/N)_{\text{disp}} = (T/M)^{1/2}. \quad (2)$$

In Fourier transform instruments, the combined interference patterns of many frequencies are measured. All infrared frequencies from σ_1 to σ_2 are measured simultaneously, and the signal-to-noise ratio is given by

$$(S/N)_{\text{interf}} = (T)^{1/2}. \quad (3)$$

Thus, Fellgett's advantage is expressed as

$$(S/N)_{\text{interf}}/(S/N)_{\text{disp}} = M^{1/2}, \quad (4)$$

and Eq. (4) is a measure of the improvement in signal-to-noise that can be obtained by interferometry.

One example of its significance is in the common practice of presenting spectra with expanded transmittance scales, i.e., for spectra of weak absorbers. As shown in Fig. 3, the spectrum (upper trace, Fig. 3) of an a-Si : H film is almost featureless due to weak absorbance, i.e., the film is too thin. The presence of dihydride, which is thought to be detrimental to

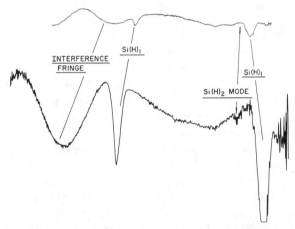

FIG. 3. Fellgett's advantage and transmittance scale expansion. Upper trace: an as-recorded spectrum of an a-Si : H film that is weakly absorbing; lower trace: transmittance scale expanded. Dihydride modes are evident (800–900 cm^{-1} region).

a-Si:H device performance (Jeffrey *et al.*, 1979), is not evident. Expansion of the transmittance scale (lower trace, Fig. 3) also increases the contribution of noise. Curve smoothing, if used to reduce the contribution of noise to the spectrum, will also reduce resolution. Thus, for infrared spectra with weak absorption bands, it is important to obtain the highest signal-to-noise ratio. Fellgett's advantage is significant in obtaining the quality of the expanded spectrum (lower trace, Fig. 3) which clearly shows the presence of dihydride modes (see Section 3).

In addition to measuring all frequencies simultaneously, interferometers used in Fourier transform spectroscopy have wide, usually circular, apertures. Thus, more light can pass through the interferometer than through a dispersive instrument. This is called Jacquinot's advantage. A more detailed discussion of these properties is given in the text by Bell (1972).

There are other properties of Fourier transform spectroscopy that can be used to improve the spectra of thin films. For example, spectra of thin films may have interference fringes superimposed over the absorption band (see Fig. 3). Since the amplitude of these fringes is proportional to the index of refraction for the film, amorphous silicon films can produce fringes of large amplitude. In Fourier transform spectroscopy these fringes can be removed from the spectrum by subtracting their frequency contribution from the interferogram, i.e., the measured, complex interference pattern that is to be Fourier transformed. A procedure for this has been reported by Hirschfeld and Mantz (1976).

Practical considerations, however, often dictate the type of instrument used for infrared analysis. For example, a dispersive instrument is more practical for measuring absorption data at fixed frequency in real time.

2. TECHNIQUES

For the study of a-Si:H films, special techniques can be used with either type of infrared spectrometer. For example, for infrared studies the use of thick 40–50 mil polished semiconductor wafers (e.g., crystalline Si) as substrates for the a-Si:H films is a common practice. Because the refractive indices of the film and of the substrate closely match, interference fringes within the a-Si:H film are greatly reduced. To enhance detection of weakly absorbing vibrational modes, germanium or gallium arsenide prisms can be used. These prisms are in the shape of long trapezoids and, although generally thicker, are similar to those used by Becker and Gobeli (1963) in their study of hydrogen on silicon. As discussed by Fahrenfort (1961), and in more detail by Harrick (1967), the way light reflects from an interface, such as the surface of a prism, depends on the angle of incidence. For light in a medium of higher refractive index n_1 entering a medium of lower refractive index n_2, there is a critical angle of incidence θ_c beyond which the light is

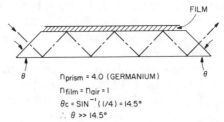

FIG. 4. Path of infrared light in a trapezoidal prism of high refractive index. The material for the prisms is, typically, germanium.

totally reflected at the interface, viz.,

$$\theta_c = \sin^{-1}(n_2/n_1). \tag{5}$$

Thus, when light enters a high-index semiconductor prism with an angle greater than θ_c, the light is totally reflected at each prism surface until it exits at the opposite end of the prism (Fig. 4). If the surfaces of the prism are coated with a film of high refractive index closely matching that of the prism, infrared light will pass through the film many times before it exits the prism. Whether the deposited layer is a simple film or a device structure such as a solar cell, the magnitude of the infrared absorption due to weak vibrational modes is greatly enhanced by multiple reflections through the film. The absorption by films that closely match the prism's refractive index follows Beer's law.

If absorbance is defined as $\log_{10}(1/T)$, where T is transmittance, then Beer's law states that absorbance by a homogeneous material is proportional to the concentration (or thickness) of the absorber.

When the difference in the indices is greater, the relation between the absorbance and the film thickness does not, in general, follow a simple relation, as is discussed in more detail by Harrick (1967).

By utilizing commercial infrared spectrometers and a wide variety of special techniques, the properties of hydrogen bonds in a-Si:H and related materials can readily be studied. These properties are discussed in Part III.

III. Vibrational Spectra of a-Si and a-Si:H

Following the report by Chittick et al. (1969) of the unusual electrical properties of a-Si deposited by glow discharge, it was thought that the structures of the a-Si films prepared by sputtering or evaporation and by glow discharge were different. Unhydrogenated a-Si exhibits modes related to the silicon–silicon bonds in the region below 800 cm^{-1}. However, information obtained from infrared spectroscopy and other techniques, such as Raman spectroscopy, indicated that the variously prepared amorphous structures were not substantially different from each other. In fact, the silicon–silicon bonding in these films was similar to that of crystalline silicon (Smith *et al.*, 1971). The properties of bonded hydrogen in hydrogenated amorphous germanium films (a-Ge:H) were reported in 1976 in a series of related papers. Lewis (1976) showed that hydrogenation altered the transport properties for a-Ge films. Connell and Pawlik (1976) reported on the variation in optical and infrared properties with hydrogen content in a-Ge:H. Finally, following the work of Spear and LeComber (1975), Paul *et al.* (1976) showed that the hydrogen bonds removed pseudogap states that are due to incomplete tetrahedral germanium or silicon bonding. Thus, infrared characterization of the types of silicon–hydrogen bonds formed became important. The infrared characterization of a-Si:H films is discussed in the following sections.

3. TYPES OF SILICON–HYDROGEN BONDS IN a-Si:H

Whereas the $Si-H_x$ and related $Ge-H_x$ ($x = 1-3$) vibrational modes for germanes and silanes (Shimanouchi, 1971, 1977), related organics (Colthup *et al.*, 1975), and related inorganics (Nakamoto, 1970) are known, some disagreement still exists concerning the assignment of the types of vibrational modes in the infrared spectrum of a-Si:H. The disagreement is noted mostly for samples having a high hydrogen content and concerns the multihydrogen modes (Paul, 1980). These are characteristic of a-Si:H films prepared at a substrate temperature lower than about 250°C. The spectra (Fig. 5) show the effect of substrate temperature on the hydrogen bonding in a-Si:H deposited by glow discharge. A listing of the types of vibrational modes for a-Si:H and their frequencies (in wave numbers) is given in Table II.

TABLE II

VIBRATIONAL FREQUENCIES FOR SiH_x
($x = 1-3$)

Group	Mode (cm⁻¹)		
	Stretching	Bending	Rocking
SiH	2000		630
SiH_2	2090	880	630
$(SiH_2)_n$	2090–2100	890, 845	630
SiH_3	2140	905, 860	630

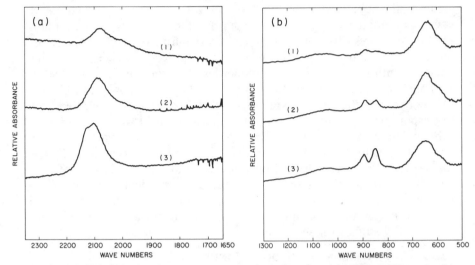

FIG. 5. Relative absorbance versus wave number for (a) 2300–1650 cm^{-1}, (b) 1300–500 cm^{-1}. Infrared spectra of a-Si:H films deposited on substrates at (1) 220°C, (2) 134°C, and (3) 46°C. The lower temperatures favor formation of multihydride Si–H$_x$ species. [From Catalano (1982).]

The assignments given in Table II are those most commonly reported in the literature. The assignments of SiH$_x$ ($x \neq 1$) functional groups for absorption bands near 2100 cm^{-1} are a topic of debate, and the actual assignments have yet to be fully resolved, as noted by Paul (1980) and more recently by Pollard and Lucovsky (1982).

The reasons for the continuing debate are related to the physical properties of a-Si:H. The infrared modes may be affected by (1) film morphology, i.e., the presence of columnar morphology (Fritzsche, 1980, in particular, pp. 449–454), (2) microstructure properties, i.e., the presence of voids with atomic dimensions (Shanks et al., 1980), and (3) the presence of polymer (SiH$_2$)$_n$ species (Lucovsky, 1976; Pollard and Lucovsky, 1982). The presence or absence of these physical properties has contributed to the debate concerning the SiH$_x$ ($x \neq 1$) assignments.

Consider, for example, the effect of the columnar film structure. Note that this structure is not a universal property of a-Si:H films but that it does occur under some deposition conditions. When this structure is present, the total surface area of the film is greatly increased. In general, the infrared spectra of surface material are different from those of bulk material, so from this aspect alone the film with a columnar structure will have unique infrared bands. Moreover, these surface regions may be sites for (SiH$_2$)$_n$, $n > 1$, polymer growth which may join adjacent columnar structures. If the surface structure is exposed to oxygen, silicon oxides sometime form at the exposed surfaces. Clearly, then, the film structure will have a significant effect on the infrared spectrum obtained for an a-Si:H film.

The presence of voids in a-Si:H films also is not a universal property of a-Si:H films but occurs with some deposition conditions, particularly when using the sputtering method. When present, voids are essentially small volumes with internal surfaces at which the hydrogen atoms can interact with each other. As will be discussed, the proton interactions at these sites can account for one type of absorption near 2100 cm^{-1}; see Shanks *et al.* (1980, 1981).

In general, however, from early work reporting the characteristic of the infrared bands (Zanzucchi *et al.,* 1977) and the mode assignments of Brodsky *et al.* (1977), there has been agreement on the assignment of the SiH$_x$ ($x = 1$) mode. The stretching mode occurs at 2000 cm^{-1} and a rocking mode occurs at 630 cm^{-1}.

The early assignments of the SiH$_2$ vibrational frequencies were first questioned by Freeman and Paul (1978). Their analysis of integrated absorption data showed that two types of modes existed in the 2090–2140 cm^{-1} region. The types of modes were not assigned although Paul (1980) later proposed that these were due to a monohydride species in two different environments. Following this, Lucovsky (1979, 1980), Lucovsky and Hayes (1979, particularly p. 228 and pp. 235–238), Lucovsky *et al.* (1979), Lucovsky and Rudder (1981), Knights (1979), Knights *et al.* (1978), John *et al.* (1980, 1981a,b), and Shanks *et al.* (1980, 1981) contributed to the present understanding of multihydride Si–H$_x$ ($x = 2, 3$) modes. Lucovsky (1976) demonstrated that the frequency v of the Si–H stretching modes in hydrogenated amorphous silicon can be deduced from an empirical relation

$$v_{\text{SiH}} = a + b \sum_{i=1}^{x} E(\text{R}_i), \tag{6}$$

where a and b are constants, $E(\text{R}_i)$ is a measure of the electronegativity of the atom or group R_i, and v_{SiH} is the calculated vibrational frequency. With this relation, mode assignments in a-Si:H and other related alloys could be compared to identical modes in substituted silane species.

Shanks *et al.* (1980), however, assigned the 2100 cm^{-1} mode, using a comparison of integrated absorption. This, and NMR data (Reimer *et al.,* 1981), led to the assignment of two types of Si–H modes, one due to a clustered phase and the other to a distributed phase (see Fig. 6). Finally, Pollard and Lucovsky (1982) have reported calculated data showing that the SiH$_3$ group does not have to be present, at least not in large amounts, to account for the 2140 cm^{-1} band. They propose that SiH$_2$ species can

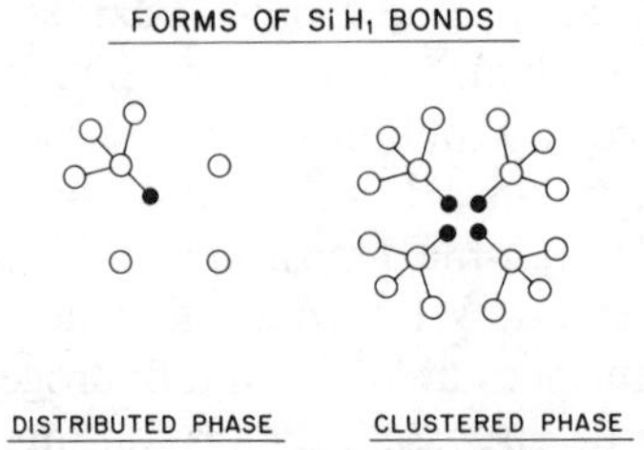

FORMS OF Si H₁ BONDS

DISTRIBUTED PHASE   CLUSTERED PHASE

FIG. 6. Schematic representation of two Si–H distributions in a-Si: H films, distributed or clustered phases. O, silicon; ●, hydrogen. [After Shanks *et al.* (1980, 1981).]

produce a 2140 cm^{-1} vibrational mode. The assignments given in Table II are for undoped a-Si: H films.

Similarly, the normal vibrations associated with each SiH$_x$ group and frequency (Fig. 7) represent undoped a-Si: H films. As noted by Cotton (1971), the normal vibrations are rarely pure bending, rocking, or stretching, and the motions given in Fig. 7 represent a best aproximation.

The SiH$_x$ ($x = 1 - 3$) vibrational modes are in the mid-infrared region. More information about fundamental modes is obtained in the low-energy, far-infrared region below about 400 cm^{-1}. Trodahl *et al.* (1980) have reported that broad or continuous absorption in this region can be explained by a charged void model. Note that in crystalline silicon, the donor or acceptor electronic transitions can be measured, at low temperature, in the far infrared (Kolbesen, 1975; Baber, 1980). Both the number and identity of the dopant species can be determined. Due to the distribution and type of states in a-Si: H, it is unlikely that donor or acceptor electronic states will be detected; e.g., boron is thought to be bonded to hydrogen rather than to silicon (Greenbaum *et al.,* 1982).

4. INTEGRATED ABSORPTION, SiH$_x$ MODES

In addition to the types of Si–H bonds, the number of each type and the total hydrogen content are significant and related to important properties such as the band-gap energy. Following the work of Brodsky *et al.* (1977) for each type of vibration, the number of oscillators N per unit volume is given by

$$N = \frac{cn\omega\mu}{2\pi^2 e^{*2}} \int \frac{\alpha(\omega)}{\omega} \, d\omega, \tag{7}$$

where c is the speed of light, n the refractive index, considered to be a constant over the mid-infrared region, ω the oscillator frequency, μ the reduced mass of the SiH species, and e^* effective charge for the oscillator in the matrix, which here is assumed to be a-Si.

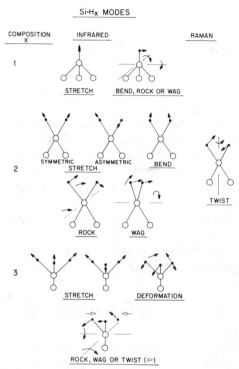

FIG. 7. Normal vibrations of Si–H$_x$ ($x = 1-3$) species. ●, hydrogen; ○, silicon. [Compare with Brodsky *et al.* (1977) and Lucovsky *et al.* (1979).]

The value of e^* depends on readily measurable parameters for a gas [see Wilson and Wells (1946)], but for a solid the calculated value must be corrected for the local environment, i.e., the local fields [see Brodsky *et al.* (1977)]. Note that such corrections depend on choosing a model for local bonding in a-Si:H. For example, one commonly used is the composite-type material model. Silicon–hydrogen oscillators are thought to be in an a-Si matrix, and a Maxwell–Garnett-type correction is used [see Heavens (1965)].

The derivation of Eq. (7) obtained by using this model is given in more detail in the Appendix. Note that how models for the local field correction actually represent a-Si:H determines the accuracy of the calculations of N. Moreover, it is evident that vibrational modes that are due to a mix of different oscillators, e.g., the 2100 cm^{-1} mode, cannot be used to calculate N. Finally, it is implicit in Eq. (7) that the density of the solid is a constant, which is not true for a-Si:H prepared under different experimental conditions. Thus, to obtain accurate values of N from Eq. (7), the density of the a-Si:H films must be measured and N corrected as necessary (Ross et al., 1982).

The total hydrogen content in a-Si:H has also been determined (Clark et al., 1977) by using secondary ion mass spectrometry and gamma-ray spectroscopy utilizing the nuclear reaction

$$^1H + {}^{15}N \rightarrow {}^{12}C + {}^4He + \delta(4.43 \text{ MeV}). \tag{8}$$

As described by Lanford et al. (1976) and, more recently, by Lanford and Burman (1982), the nuclear technique is an absolute method that provides good depth resolution with, typically, 50–100 Å depth resolution and with good sensitivity which is on the order of parts per thousand. Data obtained by this technique have been used to evaluate the accuracy of secondary ion mass spectrometry (SIMS).

For SIMS a primary beam of ions, typically oxygen ions (O_2^+ or O^-) or argon ions (Ar^+), impinges on a surface and surface material is sputtered away. In the course of this sputtering process, secondary ions representing the surface composition are created. In secondary ion mass spectrometry, these sputtered ions are mass analyzed to determine material composition.

Unlike many techniques for surface analysis, e.g., Auger spectroscopy, hydrogen can be detected by the SIMS technique. Thus, this technique can be used to obtain the profile, in depth, of hydrogen in a-Si:H films. Clark et al. (1977) find that the values of hydrogen content versus depth measured by the SIMS method agree to within 10% of the values determined by the nuclear method. In general, the SIMS method can be calibrated by standards prepared by hydrogen-ion implantation into crystal silicon. Fang et al. (1980) and Ross et al. (1982) have evaluated infrared, secondary ion, and the nuclear methods in detail and find good agreement if for the infrared the 630–640 cm^{-1} mode is used to obtain data for Eq. (7) and if the film density is known so that one may correct for density differences.

5. BONDING MODELS

Structural models have been proposed based on the extensive measurements of total hydrogen content in a-Si:H, as well as the extensive infrared, NMR, and other data.

A good example of these is the structural model proposed by Shanks et al. (1980) and shown in Fig. 8. Here the tetrahedral silicon units have a constant Si–Si bond length and varying bond angles. In the manner constructed, the structure exhibits hydrogen atom clusters consistent with the bonding shown in Fig. 6. In general, the characteristics of this model are consistent with the infrared spectra obtained. The model of Fig. 8 contains about 20 at. % H. Note that as the H concentration is increased, different models of the silicon structure are to be expected, e.g., the infrared data

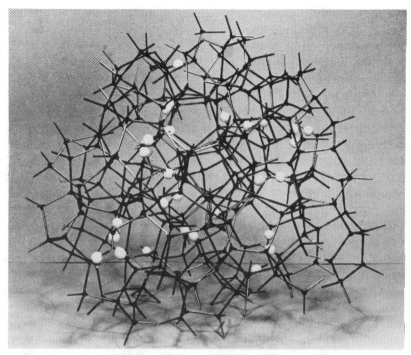

FIG. 8. Model of a-Si:H structure. [After Shanks *et al.* (1980).]

suggest that high hydrogen content films have extensive $(SiH_2)_n$ chainlike species not present in Fig. 8.

In addition to infrared, the ion implantation technique has also been used to develop models for bonding in a-Si:H. When high-energy ions in the keV to MeV range impinge on crystalline silicon, the lattice order is destroyed and an amorphous region is created. For example, after ion bombardment with Si^+ ions, the introduction of hydrogen ions (also by ion implantation) causes the incomplete silicon bonds to be hydrogenated. The uptake of hydrogen is generally no more than about 5 at. % and the dominant species formed is the silicon monohydride (Stein and Peercy, 1980). Note that in this process, both disorder and hydrogenation can be controlled independently (Peercy, 1981).

In an opposite approach, Oguz *et al.* (1980) report that He^+ bombardment of a-Si:H caused changes in the a-Si:H infrared spectrum. The 2000 and 850 cm^{-1} modes change but the multihydride modes at 2100, 890, and 650 cm^{-1} do not. This is explained as the effect of defects on the oscillator strengths of the various modes. Also of interest is the fact that a similar pattern of change was observed by Carlson *et al.* (1982). Thus the bonding in a-Si:H can be influenced by defects introduced by radiation. This is discussed in more detail in Part VI.

IV. Vibrational Spectra of a-Si:F and a-Si:H,F

The properties of a-Si:F and a-Si:H,F films were studied in detail as alternative, potentially more stable amorphous materials for solar cell fabrication, because of the absence of the Staebler–Wronski effect (Staebler and Wronski, 1977). This effect is thought to be due to light-induced structural changes in a-Si:H bonding. Ovshinsky and Madan (1978) and Madan *et al.* (1979) first reported on the properties of these films. The infrared spectra of films containing fluorine are considerably different from those of a-Si:H. This is partly a mass effect. The frequency of the Si–X oscillator depends on the mass of the atoms, viz.,

$$v \ (\text{cm}^{-1}) = \frac{1}{2\pi c} \left[k \left(\frac{1}{m_1} + \frac{1}{m_2} \right) \right]^{1/2}, \tag{9}$$

where v is the frequency of the oscillator in cm^{-1}, c the speed of light, k the force constant directly related to bond strength, and m_1 and m_2 are the masses of the vibrating atoms.

Since the mass of a fluorine atom is greater than that of a hydrogen atom, the frequency of the Si–F modes will be smaller. There should, however, be little difference in types of normal vibrations (as shown in Fig. 7), because the tetrahedral symmetries of a-Si:H, a-Si:F, and mixed alloys a-Si:H,F are the same. Identifying the frequency of the normal vibrations is, however, subject to the same general disagreements as exist for the assignments of a-Si:H normal modes and frequencies. The induction effect of fluorine may also be large and have an effect on the normal modes.

The Si–F_x ($x = 1, 2$) modes and frequencies have been reported by Ley *et al.* (1980), Shimada and Katayama (1980), Shimada *et al.* (1980), Lucovsky (1981), and Matsumura and Furukawa (1981). These are given in Table III. Of these assignments, the character of the SiF_2 species has been described by Lucovsky (1981) in terms of both isolated species and polymeric $(SiF_2)_n$, $n > 1$, species. The assignments of the modes and frequencies for these species are given in Table IV.

In addition to the presence of vibrational modes that can be assigned to isolated or polymeric SiF_x species, the Si–Si lattice modes are intensified by the presence of fluorine (Shimada and Katayama, 1980). Raman data obtained by Tsu *et al.* (1980) suggest that the addition of fluorine may increase the range of order in a-Si in a manner similar to that of the reported effect of high concentrations of P or As (Shen and Cardona, 1981).

Following from the interest in fluorinated a-Si:H films, the properties of

TABLE III

VIBRATIONAL FREQUENCIES FOR SiF$_x$ ($x = 1-3$)

	Mode (cm^{-1})		
Group	Stretching	Bending or wagging	Lattice
SiF$_4$	1010[a]	380[a]	
SiF$_3$	1015, 838[b]		
	920, 930[a]		
SiF$_2$	965, 920, 870, 827[b]		
	920–930[a]	300[a]	
SiF	850[b]		
	828[a]		
(SiF$_2$)$_n$	1015[c]		
F-induced			$\begin{cases} 650 \ (2 \ \text{LA}) \\ 515 \ (\text{TO})^b \\ 510 \ (\text{TO})^a \end{cases}$

[a]Ley et al., 1980.
[b]Shimada et al., 1980.
[c]Lucovsky, 1981.

chlorinated a-Si:H films have been studied. The films containing chlorine show enhancement of the Si–Si mode at 500 cm^{-1}. In addition, the Si–Cl mode is observed at 545 cm^{-1} (see Bruno et al., 1980; Fortunato et al., 1981; and Chevallier et al., 1982). It is thought that SiCl$_4$ forms when these films are annealed, and a new band appears at 615 cm^{-1} (Kalem et al., 1981). As with hydrogen and fluorine, the introduction of chlorine into a-Si films has the effect of reducing the number of dangling bonds.

TABLE IV

STRETCHING MODES FOR SiF SPECIES[a]

	Stretching modes (cm^{-1})	
Species	Symmetric	Antisymmetric
SiF (isolated)	827	930
SiF$_2$:(SiF$_2$)$_n$ (terminal SiF$_2$)	870	965
SiF$_2$:(SiF$_2$)$_n$ (interior SiF$_2$)	925	1012

[a]After Lucovsky (1981).

V. Effect of Dopant and Impurities on a-Si:H Vibrational Spectra

The infrared spectra of dopants, typically boron or phosphorus, and of impurities, such as carbon, oxygen, or nitrogen, in a-Si:H (or a-Si:F) are often characterized by modes of hydrogen (or fluorine) bonded to the impurity. For example, with a high concentration of boron, a-Si:H will exhibit various B–H modes. Phosphorus and oxygen, however, appear to be exceptions. When incorporated into a-Si:H, detectable numbers of P–H or O–H species are not found (Pontuschka *et al.*, 1982). Furthermore, in a-Si:H the position in and bonding to the Si–Si structure by dopants or impurities are not well understood. It is known that dopants can be clustered and not tetrahedrally bonded (Tsai, 1979; Greenbaum *et al.*, 1982). Overall, the vibrational modes associated with dopants and impurities in an amorphous structure of a-Si:H are generally broad and are frequently mixed modes representing more than one type of oscillator.

6. BORON- AND PHOSPHORUS-RELATED VIBRATIONAL MODES

Whereas Blum *et al.* (1977) reported on the infrared absorption of amorphous boron films containing carbon and hydrogen, the first extensive study of amorphous silicon–boron alloys was reported by Tsai (1979). From this latter study, the boron-related vibrational modes given in Table V are assigned.

Modes of B–O or Si–O were not observed, indicating that the films of the study were stable in air. The broadness of the B–B mode and the characteristics of the 1900–2000 cm^{-1} band suggest the presence of three center B–H–B bonds in a-Si:H,B films. This is in agreement with the NMR study by Greenbaum *et al.* (1982). The mode associated with hydrogen in amorphous B films was also reported by Blum *et al.* (1977).

These data suggest that boron is incorporated in a-Si:H in a complex mix of bonds. Probably only a small percentage of the boron is electrically active.

TABLE V

VIBRATIONAL MODES FOR BH AND RELATED SPECIES[a]

Vibrational mode	Frequency (cm^{-1})
B–H stretching vibration	2560
B–B vibrations	~1100 (broad)
B–H–B bridge bonds	~1900–2000
B amorphous cluster (H containing)	~2000–1200 (broad, featureless)
Si–B	~1400

[a]After Tsai (1979).

In addition, boron and other dopants are known to form complexes with carbon or transition metals such as iron in crystalline silicon (Wünstel and Wagner, 1982). The presence of these impurities in doped a-Si:H may further reduce the number of active dopants.

For phosphorus, Shen and Cardona (1981) report that a-Si:H films with up to 20 at. % phosphorus did not show P–H absorption modes. The authors note, however, that the weakly allowed infrared absorption by the Si–Si network is greatly enhanced by the presence of the dopant. Fluorine was also found to enhance the a-Si modes. The basis for this effect is unknown but may relate to a change in the characteristics of long-range order in these materials, as suggested by Tsu et al. (1980).

7. CARBON-, OXYGEN-, AND NITROGEN-RELATED VIBRATIONAL MODES

The infrared spectra of a-Si:H,C show Si–C absorption in the same region in which the lattice mode for hexagonal or cubic silicon carbide occurs; these modes are listed in Table VI. Whereas the crystalline Si–C forms have definite stoichiometry, Anderson and Spear (1977) note that amorphous carbides and nitrides can be prepared with a wide range of compositions. This is a characteristic of glow-discharge deposition.

In addition to SiC bonds, glow-discharge decomposition of, e.g., a silane–methane mixture, produces SiC films that show Si–H and C–H modes. The films are thus a-Si:C,H. Additional modes reported by Wieder et al. (1979) and Catherine and Turban (1980) are given in Table VII. Wieder et al. (1979) report that one hydrogen is generally bonded to the Si atom while two or three hydrogens are bonded to the C atom. Catherine and Turban (1980) report that the CH_x modes are detectable only in films with a high

TABLE VI

COMPARISON OF VIBRATIONAL MODES FOR
CRYSTALLINE AND AMORPHOUS SiC

Carbide structure	Mode frequency (cm^{-1})
hexagonal[a]	794, 786, 883 (weak)
cubic[b]	794
amorphous, from CH_x[c]	760–720
amorphous, from CF_x[d]	750

[a]Spitzer et al., 1959a.
[b]Spitzer et al., 1959b.
[c]Wieder et al., 1979, and Catherine and Turban, 1980.
[d]Saito and Yamaguchi, 1982.

TABLE VII

CHARACTERISTIC VIBRATIONAL MODES FOR
a-Si : H,C

a-Si : H,C modes	Mode frequency (cm⁻¹)
SiC stretching vibration	760–720, 1000 (shoulder)
Si–CH₃, CH₃ wag	960, also a
CH$_x$	a
SiH$_x$	2000, 630–650^b

^{a}2972–2952 (asymmetric stretch), 2882–2862 (symmetric stretch), 1475–1450 (asymmetric deformation), and 1383–1377 (symmetric deformation) for CH₃ modes; 2936–2916 (asymmetric stretch) and 2863–2843 (symmetric stretch) for CH₂ modes. [After Colthup et al. (1975).]

^{b}Generally monohydride species are observed with modes at 2000 and 650–630. [After Wieder et al. (1979).]

carbon content. Data for vibrational modes of a-Ge : H,C are also given by these authors.

In addition to a-Si : H,C films, vibrational spectra of the fluorinated form, i.e., a-Si : F,C, have been analyzed by Saito and Yamaguchi (1982), whereas optical data for a-Si : C films have been reported by Anderson (1977), Jones and Stewart (1981), and McKenzie et al. (1982). The fluorinated films are produced from decomposition of CF₄, and the modes given in Table VIII are observed.

Although the addition of carbon in a-Si:H films may be desirable in device materials, e.g., to increase the band gap of the film, oxygen and nitrogen in the films may lead to poor or unpredictable device performance. Oxygen in a-Si:H is, however, easily detected by the presence of the relatively strongly absorbing SiO$_x$ mode in the 1100–1000 cm⁻¹ region [see, e.g., Freeman and Paul (1978)].

The optical properties of both SiO₂ (Philipp, 1971, 1979; Pliskin and Lehman, 1965) and SiO (Hass and Salzberg, 1954) are well known. In the a-Si:H structure the SiO$_x$ mode generally appears at about 1000 cm⁻¹ but may absorb at a higher wave number after annealing (see Part VI). The basis for this change is not well understood. John et al. (1981b) and Knights et al. (1980) have reported on the properties of oxidized a-Si:H, whereas Yacobi et al. (1981) describe the effect of oxygen on the optoelectronic properties of a-Si:H.

As noted earlier, the oxygen in a-Si:H does not react with hydrogen to form hydroxyl groups (Knights et al., 1980). Finally, the effect of nitrogen in

TABLE VIII

CHARACTERISTIC VIBRATIONAL MODES FOR
a-Si:F,C

a-Si:F,C modes	Mode frequency (cm^{-1})
CF_2	~1100
SiF_2, SiF stretching	870, 820
SiF_4	1015

crystalline silicon has only recently been studied by Tajima *et al.* (1981), using photoluminescence to measure its properties. Similarly, the effects of incorporating nitrogen in amorphous silicon have not been widely studied.

Watanabe *et al.* (1982) and Sasaki *et al.* (1982) have reported on the properties of a-Si:N,H films. In addition to the SiH_x mode ($x = 1, 2$) in the 2100–2000 cm^{-1} region, a broad SiN mode is observed at 840 cm^{-1}. The frequency of the SiN mode is a function of nitrogen content in the glow discharge. Electrical properties of a-Si:N,H are reported by Sasaki *et al.* (1982).

VI. Infrared Measurements of Films on Prisms, Studies of a-Si:H Stability

As previously described, the infrared spectra of thin films can be enhanced by coating a trapezoidal prism with the film material and using the prism to obtain multiple passes of light through the film (see Fig. 4). In this way the absorption of weakly absorbing modes can be enhanced. This technique has been used at RCA Laboratories to measure small changes in Si–H absorption at 2000 cm^{-1} after exposure to the films to light, to electron-beam bombardment, or to heat (Carlson *et al.,* 1982).

As reported by Carlson *et al.* (1982), depositions contaminated with air or water vapor produce films that exhibit reversible light-induced changes in Si–H content. The films were contaminated because the presence of contaminants enhances the magnitude of the change in Si–H content.

Typical difference spectra are shown in Fig. 9. Spectrum A is the difference between spectra of the a-Si:H film prior to and after light exposure. Similarly, spectrum B is the difference between spectra prior to and after an anneal. The spectra C and D are difference spectra due to a second light soaking and subsequent anneal. Changes occur near 2000 cm^{-1} due to changes in SiH content and near 1100 cm^{-1} due to changes in SiO_x composition and structure. After the second anneal, the SiO_x absorption is characteristic of a crystalline material.

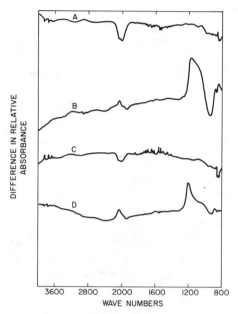

FIG. 9. Difference spectra for a-Si:H,O films after exposure to light or heat. A: difference between spectra prior to and after light exposure; B: difference between spectra prior to and after anneal at ~160°C for ~1 hr (after A); C: difference between spectra prior to and after second light exposure (after B); D: difference between spectra prior to and after second anneal (after C). [After Carlson et al. (1983).]

Crystalline aggregates of SiO_x are known to reduce transport of carriers in crystalline silicon. The change in SiH is about 1–2%, based on the magnitude of the change in absorption. Note that this is about a factor of 10^3 greater in magnitude than the number of metastable (gap states) centers reported by Crandall (1981). Apparently the light-induced effects observed can occur without creating gap states, e.g., by hydrogen diffusion and rebonding (Carlson et al., 1982). Similar results for changes in oxygen-contaminated a-Si:H induced by electron-beam bombardment are obtained (see Chapter 11 by Schade). From these studies, the stability of Si–H bonds is clearly affected by light, electron-beam bombardment, and heat for oxygen-contaminated films containing ~1 at. % oxygen. This demonstrates how the prism technique can be used to enhance absorption and thus to obtain information about the stability of Si–H bonding in a-Si:H.

VII. Conclusions

Vibrational spectra are, in general, used to develop an understanding of chemical bonding in solids. More specifically, the frequency of the various atomic (or lattice) motions is a basic property of materials. For example, the presence of quantized vibrational modes is directly related to heat capacity.

For crystalline silicon, our present understanding of silicon–silicon and silicon–impurity bonds has evolved from studies of vibrational spectra. As a result of these experimental studies, many methods of calculating vibrational parameters have been studied and this theoretical work is still in progress [e.g., see Barker and Sievers (1975, particularly pp. S161–S171) and Newman, (1969)].

In contrast, studies of the vibrational properties of a-Si:H are at an experimental stage. For example, the role of phonon modes in a-Si:H materials is of importance since many energy transfer processes depend on these modes. Street (1981) (see in particular pp. 602–607) has discussed the role of phonons in the photoluminescence of a-Si:H. Their role in Raman scattering will be discussed in Chapter 6. Phonon spectra will be the subject of much further work.

In this chapter, experimental data concerning the infrared vibrational spectra of a-Si:H, and related materials, have been reviewed. From these data, the general types of bonds present in a-Si:H are now well known. These are of fundamental importance in understanding the properties of hydrogenated amorphous silicon.

Appendix: Calculation of the Number of SiH$_x$ Oscillators in a-Si:H

From measured optical data, the concentration of electronic or vibrational states can be calculated by the use of classical dispersion relations. This method was used in the late 1920s by Smakula (1930) [and was reviewed by Markham (1966)] to determine the number of color centers in alkali halide single crystals. Connell and Pawlik (1976) first attempted to use the dispersion relations to calculate the number of GeH$_x$ species in a-Ge:H. Basically, an effective oscillator-strength parameter that includes a so-called local field correction must be calculated. The local field correction is well known for crystalline materials [see, e.g., Seitz (1940) and Stern (1963)]. Determining the local field correction for a species in an amorphous solid is a unique, new problem. Brodsky et al. (1977) have evaluated the problem in detail and their calculation is described as follows. To calculate the number of Si–H bonds N from the absorption coefficient α at discrete frequencies ω, we use

$$N \propto \int \frac{\alpha(\omega)}{\omega} \, d\omega \qquad (10)$$

and, following Brodsky et al. (1977) and Connell and Pawlik (1976),

$$N = \frac{cn\omega\mu}{2\pi^2 e_s^{*2}} \int \frac{\alpha(\omega)}{\omega} \, d\omega, \tag{11}$$

where μ is the reduced mass for the Si–H oscillator, c the speed of light, n the refractive index assumed to be a constant, α the absorption coefficient, and e_s^* a term for the effective charge of the vibrational mode in the a-Si:H structure. The value for this parameter is estimated from infrared data for gaseous silanes. The gas-phase parameter e_g^* is given by

$$e_g^{*2}/\mu = c\omega\Gamma/2\pi^2 N_A \zeta. \tag{12}$$

Here N_A is Avogadro's number, ζ the number of the Si–H bonds per molecule, and Γ is given by

$$\Gamma = \frac{1}{N} \int \frac{\alpha}{\omega} \, d\omega. \tag{13}$$

For N moles of gas, e.g., silane, per unit volume, Γ is given in units of square centimeters per millimole and represents the oscillator strength. To calculate e_s^*, a model of the local field must be assumed. For a homogeneous medium

$$e_s^* = \tfrac{1}{3}(\varepsilon_m + 2)e_g^*, \tag{14}$$

where ε_m is the electronic dielectric constant of the a-Si:H. However, for a-Si:H, the Si–H oscillator is in an a-Si matrix and a type of Maxwell–Garnett local field correction is favored. This is given by

$$e_s^{*2} \sim \frac{9\varepsilon_m^2}{(1 + 2\varepsilon_m)^2} e_g^{*2}, \tag{15}$$

where ε_m is the dielectric constant for the matrix, i.e., a-Si. This expression gives the local field correction for noninteracting spheres in the a-Si medium. Equations (11), (12), and (15) give

$$N = \frac{(1 + 2\varepsilon_m)^2}{9\varepsilon_m^2} \frac{N_A n}{(\Gamma/\zeta)} \int \frac{\alpha(\omega)}{\omega} \, d\omega. \tag{16}$$

For the Si–H stretching mode 2000 cm^{-1}, Brodsky et al. (1977) reduce the expression to

$$N_s = \frac{(1 + 2\varepsilon_m)^2 (\varepsilon_m)^{1/2}}{9\varepsilon_m^2} (1.72 \times 10^{20} \text{ cm}^{-2}) \int_{\omega_s} \frac{\alpha(\omega)}{\omega} \, d\omega. \tag{17}$$

Typically, the values of N_s calculated are on the order of 10^{21}–10^{22} bonds cm^{-3}. Considering the uncertainties in the evaluation of the local field correction, the calculated values are reasonably accurate. These calculated values were particularly important in the early characterization of a-Ge:H and a-Si:H. These results indicated that the amount of hydrogen incorporated in the a-Si:H structure was very large, i.e., on the order of 5–20 at. %.

ACKNOWLEDGMENTS

Permission to use figures and data from H. R. Shanks (Ames Laboratory, USDOE, Ames, Iowa) and A. Catalano (RCA Laboratory, Princeton, N.J.) is appreciated. The review of this manuscript by R. E. Honig is gratefully acknowledged.

REFERENCES

Alben, R., Weaire, D., Smith, J. E., Jr., and Brodsky, M. H. (1975). *Phys. Rev. B* **11**, 2271–2296.
Anderson, D. A. (1977). *Philos. Mag.* **35**, 17–26.
Anderson, D. A., and Spear, W. E. (1977). *Philos. Mag.* **35**, 1–16.
Baber, S. C. (1980). *Thin Solid Films* **72**, 201.
Barker, A. S., Jr., and Sievers, A. J. (1975). *Rev. Mod. Phys.* **47** Suppl. 2, S1–S180.
Becker, G. E., and Gobeli, G. W. (1963). *J. Chem. Phys.* **38**, 2942–2945.
Bell, R. J. (1972). "Introductory Fourier Transform Spectroscopy," pp. 19–25. Academic Press, New York.
Blum, N. A., Feldman, C., and Satkiewicz, F. G. (1977). *Phys. Status Solidi A* **41**, 481–486.
Brigham, Oran E. (1974). "The Fast Fourier Transform," pp. 7–9. Prentice-Hall, Englewood Cliffs, New Jersey.
Brodsky, M. H., Cardona, M., and Cuomo, J. J. (1977). *Phys. Rev. B* **16**, 3556–3571.
Brodsky, M. H., Title, R. S., Weiser, K., and Pettit, G. D. (1970). *Phys. Rev. B* **1**, 2632–2641.
Bruno, G., Capezzuto, P., Cramarossa, F., and D'Agostino, R. (1980). *Thin Solid Films* **67**, 103.
Carlson, D. E. (1977). *IEEE Trans. Electron Devices* **ED-24**, 449–453.
Carlson, D. E., and Wronski, C. R. (1976). *Appl. Phys. Lett.* **28**, 671–673.
Carlson, D. E., Moore, A. R., Szostak, D. J., Goldstein, B., Smith, R. W., Zanzucchi, P. J., and Frenchu, W. R. (1983). *Solar Cells* **9**, 19–23.
Catalano, A. (1982). Unpublished work, RCA Laboratory, Princeton, New Jersey.
Catherine, Y., and Turban, G. (1980). *Thin Solid Films* **70**, 101–104.
Chevallier, J., Kalem, S., Dallal, S. Al., and Bourneix, J. (1982). *J. Non-Cryst. Solids* **51**, 277.
Chittick, R. C., Alexander, J. H., and Sterling, H. F. (1969). *J. Electrochem. Soc.* **116**, 77–81.
Clark, G. J., White, C. W., Allred, D. D., Appleton, B. R., Magee, C. W., and Carlson, D. E. (1977). *Appl. Phys. Lett.* **31**, 582–585.
Colthup, N. B., Daly, L. H., and Wiberly, S. E. (1975). "Introduction to Infrared and Raman Spectroscopy," 2nd ed., pp. 329–342. Academic Press, New York.
Connell, G. A. N., and Pawlik, J. R. (1976). *Phys. Rev. B* **13**, 787–804.
Cooley, J. W., and Tukey, J. W. (1965). *Math. Comp.* **19**, 297–301.
Cotton, F. A. (1971). "Chemical Applications of Group Theory," 2nd ed., pp. 325–326. Wiley (Interscience), New York.
Crandall, R. S. (1981). *Phys. Rev. B* **24**, 7457–7459.
Fahrenfort, J. (1961). *Spectrochim. Acta* **17**, 698–709.

Fang, C. J., Gruntz, K. J., Ley, L., Cardona, M., Demond, F. J., Müller, G., and Kalbitzer, S. (1980). *J. Non-Cryst. Solids* **35/36**, 255–260.

Fellgett, P. (1958). *J. Phys. Radium* **19**, 187.

Fortunato, G., Evangelisti, F., Bruno, G., Capezzuto, P., Cramarossa, F., Augelli, V., and Murri, R. (1981). *J. Non-Cryst. Solids* **46**, 95–104.

Freeman, E. C., and Paul, W. (1978). *Phys. Rev. B* **18**, 4288–4300.

Fritzsche, H. (1980). *Solar Energy Mat.* **3**, 447–501. (See also pp. 449–454.)

Greenbaum, S. G., Carlos, W. E., and Taylor, P. C. (1982). *Solid State Commun.* **43**, 663–666.

Hamakawa, Y. (Ed.) (1981). "Amorphous Semiconductor, Technologies and Devices 1982," Chap. 4, pp. 133–210. Ohmsha, Tokyo, and North-Holland Publ., Amsterdam.

Harrick, N. J. (1967). "Internal Reflection Spectroscopy," pp. 22–23. Wiley (Interscience), New York.

Hass, G., and Salzberg, C. D. (1954). *J. Opt. Soc. Am.* **44**, 181–187.

Heavens, O. S. (1965). "Optical Properties of Thin Solid Films," pp. 177–179. Dover, New York.

Hirschfeld, T., and Mantz, A. W. (1976). *Appl. Spectrosc.* **30**, 552–553.

Jeffrey, F. R., Shanks, H. R., and Danielson, G. C. (1979). *J. Appl. Phys.* **50**, 7034–7038.

John, P., Odeh, I. M., Thomas, M. J. K., Tricker, M. J., Riddoch, F., and Wilson, J. I. B. (1980). *Philos. Mag. B* **42**, 671–681.

John, P., Odeh, I. M., Thomas, M. J. K., Tricker, M. J., and Wilson, J. I. B. (1981a). *Phys. Status Solidi B* **103**, K141–K146.

John, P., Odeh, I. M., Thomas, M. J. K., Tricker, M. J., and Wilson, J. I. B. (1981b). *Phys. Status Solidi B* **105**, 499–505.

Jones, D. I., and Stewart, A. D. (1981). *J. Phys. Colloq. Orsay, Fr.* **42** Suppl. 10, C4-1085–1088.

Kalem, S., Chevallier, J., Dallal, S. A., and Bourneix, J. (1981). *J. Phys. Colloq. Orsay, Fr.* **42**, Suppl. 10, C4-361.

Knights, J. C. (1979). *Jpn. J. Appl. Phys.* **18**, Suppl. 18-1, 101–108.

Knights, J. C., Lucovsky, G., and Nemanich, R. J. (1978). *Philos. Mag. B* **37**, 467–475.

Knights, J. C., Street, R. A., and Lucovsky, G. (1980). *J. Non-Cryst. Solids* **35/36**, 279–284.

Kolbesen, B. O. (1975). *Appl. Phys. Lett.* **27**, 353.

Lanford, W. A., and Burman, C. (1982). *Appl. Phys. Lett.* **41**, 473–475.

Lanford, W. A., Trautvetter, H. P., Ziegler, J. F., and Keller, J. (1976). *Appl. Phys. Lett.* **28**, 566–568.

Lanford, W. A., and Burman, C. (1982). *Appl. Phys. Lett.* **41**, 473–475.

Lewis, A. J. (1976). *Phys. Rev. B* **14**, 658–668.

Ley, L., Shanks, H. R., Fang, C. J., Gruntz, K. J., and Cardona, M. (1980). *J. Phys. Soc. Jpn.* **49**, Suppl. A, 1241–1244.

Lucovsky, G. (1979). *Solid State Commun.* **29**, 571–576.

Lucovsky, G. (1980). *Solar Cells* **2**, 431–442.

Lucovsky, G. (1981). *In* "Tetrahedrally Bonded Amorphous Semiconductors" (R. A. Street, D. K. Biegelsen, and J. C. Knights, eds.), pp. 100–105. Amer. Inst. Phys., New York.

Lucovsky, G., and Hayes, T. M. (1979). *In* "Amorphous Semiconductors" (M. H. Brodsky, ed.), pp. 215–250. Springer-Verlag, Berlin and New York.

Lucovsky, G., and Rudder, R. A. (1981). *In* "Tetrahedrally Bonded Amorphous Semiconductors" (R. A. Street, D. K. Biegelsen, and J. C. Knights, eds.), pp. 95–99. Amer. Inst. Phys., New York.

Lucovsky, G., Nemanich, R. J., and Knights, J. C. (1979). *Phys. Rev. B* **19**, 2064–2073.

McKenzie, D. R., McPhedran, R. C., Botten, L. C., Savvides, N., and Netterfield, R. P. (1982). *Appl. Opt.* **21**, 3615–3617.

Madan, A., Ovshinsky, S. R., and Benn, E. (1979). *Philos. Mag. B* **40**, 259–277.

Markham, J. J. (1966). "F-Centers in Alkali Halides." *Solid State Phys. Suppl.* **8**, pp. 26–35.
Matsumura, H., and Furukawa, S. (1981). *In* "Amorphous Semiconductor, Technologies and Devices 1982" (Y. Hamakawa, ed.), pp. 88–108. Ohmsha, Tokyo, and North-Holland Publ., Amsterdam.
Nakamoto, K. (1970). "Infrared Spectra of Inorganic and Coordination Compounds," 2nd ed., pp. 106–112. Wiley (Interscience), New York.
Newman, R. C. (1969). *Adv. Phys.* **18**, 545–663.
Oguz, S., Anderson, D. A., Paul, W., and Stein, H. J. (1980). *Phys. Rev. B* **22**, 880–885.
Ovshinsky, S. R., and Madan, A. (1978). *Nature (London)* **276**, 482–484.
Paul, W. (1980). *Solid State Commun.* **34**, 283–285.
Paul, W., Lewis, A. J., Connell, G. A. N., and Moustakas, T. D. (1976). *Solid State Commun.* **20**, 969–972.
Peercy, P. S. (1981). *Nucl. Instrum. Methods* **182/183**, 337–349.
Philipp, H. R. (1971). *J. Phys. Chem. Solids* **32**, 1935–1945.
Philipp, H. R. (1979). *J. Appl. Phys.* **50**, 1503–1507.
Pliskin, W. A., and Lehman, H. S. (1965). *J. Electrochem. Soc.* **112**, 1013–1019.
Pollard, W. B., and Joannopoulos, J. D. (1981). *Phys. Rev. B* **23**, 5263–5268.
Pollard, W. B., and Lucovsky, G. (1982). *Phys. Rev. B* **26**, 3172–3180.
Pontuschka, W. M., Carlos, W. W., Taylor, P. C., and Griffith, R. W. (1982). *Phys. Rev. B* **25**, 4362–4376.
Potts, W. R., Jr. (1963). "Chemical Infrared Spectroscopy," Vol. 1. Wiley, New York.
Reimer, J. A., Vaughan, R. W., and Knights, J. C. (1981). *Phys. Rev. B* **24**, 3360–3370.
Ross, R. C., Tsong, I. S. T., Messier, R., Lanford, W. A., and Burman, C. (1982). *J. Vac. Sci. Technol.* **20**, 406–409.
Saito, N., and Yamaguchi, T. (1982). *Phys. Status Solidi A* **69**, 133–138.
Sasaki, G., Kondo, M., Fujita, S., and Sasaki, A. (1982). *Jpn. J. Appl. Phys.* **21**, L377–L378.
Seitz, F. (1940). "The Modern Theory of Solids," Chap. 17, p. 603. McGraw-Hill, New York.
Shanks, H., Fang, C. J., Ley, L., Cardona, M. C., Demond, F. J., and Kalbitzer, S. (1980). *Phys. Status Solidi B* **100**, 43–56.
Shanks, H. R., Jeffrey, F. R., and Lowry, M. E. (1981). *J. Phys. Colloq. Orsay, Fr.* **42** *Suppl. 10*, C4-773–777.
Shen, S. C., and Cardona, M. (1981). *Phys. Rev. B* **23**, 5322–5328.
Shimada, T., and Katayama, Y. (1980). *J. Phys. Soc. Jpn.* **49**, Suppl. A., 1245–1248.
Shimada, T., Katayama, Y., and Horigome, S. (1980). *Jpn. J. Appl. Phys.* **19**, L265–L268.
Shimanouchi, T. (1971, 1977). "Tables of Molecular Vibrational Frequencies," Consolidated Vols. I and II. *Nat. Std. Ref. Data Ser. (U.S. Nat. Bur. Stand.)*
Smakula, A. (1930). *Z. Phys.* **59**, 603.
Smith, J. E., Jr., Brodsky, M. H., Crowder, B. L., Nathan, M. I., and Pinczuk, A. (1971). *Phys. Rev. Lett.* **26**, 642–646.
Spear, W. E., and LeComber, P. G. (1975). *Solid State Commun.* **17**, 1193–1196.
Spitzer, W. G., Kleinman, D., and Walsh, D. (1959a). *Phys. Rev.* **113**, 127–132.
Spitzer, W. G., Kleinman, D. A., and Frosch, C. J. (1959b). *Phys. Rev.* **113**, 133–136.
Staebler, D. L., and Wronski, C. R. (1977). *Appl. Phys. Lett.* **31**, 292–294.
Stein, H. J., and Peercy, P. S. (1980). *Phys. Rev. B* **22**, 6233–6239.
Stern, F. (1963). *Solid State Phys.* **15**, 342–344.
Street, R. A. (1981). *Adv. Phys.* **30**, 593–676.
Taft, E. A. (1971). *J. Electrochem. Soc.* **118**, 1341–1346.
Tajima, M., Masui, T., Abe, T., and Nozaki, T. (1981). *Jpn. J. Appl. Phys.* **20**, L423–L425.
Trodahl, H. J., Fee, M., Livick, N., and Buckley, R. G. (1980). *Solid State Commun.* **35**, 551–552.
Tsai, C. C. (1979). *Phys. Rev. B* **19**, 2041–2055.

Tsu, R., Izu, M., Cannella, V., Ovshinsky, S. R., Jan, G-J., and Pollak, F. H. (1980). *J. Phys. Soc. Jpn.* **49,** *Suppl. A,* 1249–1252.

Watanabe, H., Katoh, K., and Yasui, M. (1982). *Jpn. J. Appl. Phys.* **21,** L341–L343.

Wieder, H., Cardona, M., and Guarnieri, C. R. (1979). *Phys. Status Solidi B* **92,** 99–112.

Wilson, E. B., Jr., and Wells, A. J. (1946). *J. Chem. Phys.* **14,** 578–580.

Wronski, C. R. (1979). *IEEE Trans. Electron Devices* **ED-24,** 351–357.

Wünstel, K., and Wagner, P. (1982). *Appl. Phys. A* **27,** 207–212.

Yacobi, B. G., Collins, R. W., Moddel, G., Viktorovitch, P., and Paul, W. (1981). *Phys. Rev. B* **24,** 5907–5912.

Zanzucchi, P. J., Wronski, C. R., and Carlson, D. E. (1977). *J. Appl. Phys.* **48,** 5227–5236.

SEMICONDUCTORS AND SEMIMETALS, VOL. 21, PART B

CHAPTER 5

Electroreflectance and Electroabsorption

Yoshihiro Hamakawa

FACULTY OF ENGINEERING SCIENCE
OSAKA UNIVERSITY
TOYONAKA, OSAKA, JAPAN

I. Introduction

In this chapter we review recent aspects of electroreflectance and electroabsorption studies in hydrogenated tetrahedrally bonded amorphous semiconductors. In Part II, a historical background of modulation spectroscopy and its availability to amorphous material characterizations are briefly discussed, together with its difficulties and usefulness. Then in Part III some recent results on electroreflectance investigations of a-Si:H, a-SiC:H, and μc-Si are introduced. New topics on back-surface-reflected electroabsorption and its efficient application to a-Si device physics are described in Part IV. Finally in Part V, the knowledge obtained is summarized, and some other possibilities for this technique as a tool for exploring the vast field of amorphous materials science are forecast.

II. Usefulness and Difficulties of the Modulation Spectroscopic Characterization of Amorphous Solids

Modulation spectroscopy has been very useful in the study of band structure of crystalline solids. A great advantage of this characterization method is that a measurement of the differential spectrum produced by an externally applied energy perturbation makes it possible to enhance optical structures with high resolution by using phase sensitive detection techniques

141

(Cardona, 1969). Among various energy modulating parameters in differential spectroscopy, electric field modulations, also so-called *electroabsorption* (EA) and *electroreflectance* ER is the most advanced one both theoretically and technologically. Refined theories are available; for example, a low-field third-derivative ER (Aspnes and Bottka, 1972) has achieved a quantitative treatment of the modulated optical spectrum.

In first-derivative modulation spectroscopy, in which lattice periodicity is retained, momentum is still a good quantum number to within a reciprocal lattice vector, and optical transitions remain vertical in the one-electron band picture, although the threshold energy changes due to the perturbation. Since the change in threshold energy is generally small on the scale of the energy gap, the perturbation-induced changes in the dielectric function are of first order and can be approximated by the first-derivative of the dielectric function. On the other hand, in electromodulation the perturbation $e\mathbf{F} \cdot \mathbf{r}$ destroys the translational invariance of the Hamiltonian in the direction of the field, in contrast to other first-derivative modulation spectroscopy methods. The loss of translational invariance under the electric field is the origin of the complexity of ER. A comparison of major differences between first-derivative and electromodulation spectroscopy is schematically illustrated in Fig. 1 for the imaginary part of the dielectric function at a three-dimensional M_0-type edge.

If an electric perturbation is applied to the crystal, translational invariance is lost in the field direction. The momentum of accelerated electrons in the field direction is no longer a good quantum number. Consequently, the one-electron Bloch functions of the unperturbed crystal become mixed. This effect is equivalent to spreading the formerly mentioned sharp vertical transitions over a finite range of initial and final momenta, as shown at the bottom of Fig. 1. In the case that the field is not too large, the mixing of unperturbed wavefunctions will be restricted to those wavefunctions near the originally vertical allowed transitions. Eventually, smearing of the optical constants yields a more complicated difference spectrum having changes in sign, and the difference can be approximated only by higher derivatives of the unperturbed dielectric function.

Another explanation for the origin of the third-derivative nature of low-field ER is given by Aspnes and Bottka (1972) in terms of an averaging effect of the momentum-space mixing of wavefunctions. Considering the time dependence of the average energy of a band electron in a perturbation turned on instantaneously at $t = 0$, the perturbation simply produces, if periodicity is preserved, a discontinuous change in $E_g (= \hbar\omega_g)$ or Γ, the zero-order (constant) terms in a Taylor series expansion of the average energy as a function of time (Aspnes, 1973). With electric field perturbation, the discontinuity occurs in the acceleration of a band electron, and appears in the second-order or quadratic term in the time-dependent average energy.

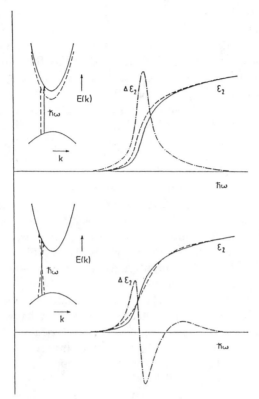

FIG. 1. Schematic diagrams of the change in the imaginary part of the dielectric function at the three-dimensional M_0-type edge for first-derivative (top) and third-derivative (bottom) modulation spectroscopies. ———, unperturbed; — — —, perturbed; —·—·—, difference. The effect of the perturbation on the energy band diagram and on the optical transition is shown in the inset. [After Aspnes (1973). Copyright North-Holland Publ. Co., Amsterdam, 1973.]

When transforming into the frequency domain to examine the spectral dependence of the dielectric function $\varepsilon(\omega)$, it is found that a small discontinuity in the zero-order term of E_g and Γ leads to a first-order change in $\varepsilon(\omega)$ (first-derivative spectrum). Similarly, a small discontinuity in the second-order term of the acceleration leads to a third-order change in $\varepsilon(\omega)$ (third-derivative spectrum). This argument applies as long as the average change in energy produced by field acceleration prior to a scattering event is small compared to an uncertainty associated with the lifetime broadening or disorder inherent in the structure.

From the analysis of the ER and EA signals with this theoretical background, much information on the electronic band structure, such as the critical point of the joint density of states, effective mass near the band edge, and even the type of band edges, has been most precisely obtained (Hamakawa and Nishino, 1976). Thus this type of modulation spectroscopy becomes a useful tool for high resolution characterization of the electronic band structure of crystals, and is still spreading into the study of various new mixed-compound semiconductors such as InGaAsP [e.g. Yamazoe et al. (1981)] and CuInSe$_2$ [e.g. Shay (1972)].

In the amorphous semiconductor area, on the other hand, relatively little had been done during the 1970s. There were some trials of ER studies on a-Se by Weiser and Stuke (1969), on a-Ge by several institutes, including Seraphin's group (Piller et al., 1969), and on a-Si:H by Okamoto et al. (1979) and Freeman et al. (1979).

Because there is an internal localized electric field due to the disorder inherent in the amorphous structure, no sharp band edge in the density of electronic states has been observed. Moreover, there is another difficulty in getting the fine electro-optical modulation signal because of a superposition of interference effects in a thin amorphous film sample. An example of an ER signal afflicted with these interference fringes is shown in Fig. 2, together with its real absorption spectrum near the optical band edge (Okamoto et al., 1979). Freeman et al. (1979) have discussed these problems with respect to ER measurements in a thin film of a-Si:H and have postulated a limit to the usefulness of ER as a tool for the characterization of a-Si. Another technical difficulty is that ER and EA signals involve a considerable thermoreflectance component, as was pointed out by Brodsky and Leady

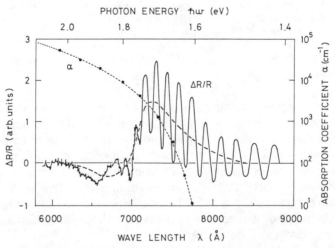

FIG. 2. Electroreflectance spectrum near the optical energy gap in a-Si:H with Schottky-barrier electromodulation at RT. [After Okamoto et al. (1979). Copyright North-Holland Publ. Co., Amsterdam, 1979.]

(1980), due to the Joule heating by the applied modulation field. Thus to eliminate this thermal dark current, the sample must have a good blocking barrier such as an $n-i-p$ or a Schottky junction for the longitudinal (surface-barrier) mode of field modulation.

Apart from these difficulties in the electro-optical signal probing of amorphous materials, there is no method thus far for a definitive band edge energy assignment, except for this spectroscopic identification by electric field modulation. In fact, the optical energy gap reported for most amorphous materials has been assigned by an extrapolation of $\sqrt{\alpha\hbar\omega}$ versus $\hbar\omega$ plots using an empirical rule, since there is no exact theoretical background for the $\alpha \propto |\hbar\omega - \varepsilon_{g_{opt}}|^2/\hbar\omega$ relationship. On the other hand, a sharp dominant structure in both EA and ER signals is seen at energies at which the critical point occurs in the interband joint density-of-states function. In other words, ER and EA signals indicate more meaningful band edge assignment on the basis of electronic band structure. According to some recent work in this field, moreover, the clarity of the signal and the line shape of the electro-optical spectra indicate the quality of the a-Si: H film which depends on the degree to which hydrogen passivates the electronic activity of the dangling bonds in the amorphous network. Recently, another utilization of EA spectroscopy for the determination of the internal electric field and built-in potential in the $p-i-n$ junction has been discovered. This technique is readily available for the characterization of a-Si photovoltaic devices. These items will be introduced and discussed in the following sections.

III. Electroreflectance

It has been shown elsewhere that the introduction of hydrogen into tetrahedrally bonded amorphous semiconductors is effective in reducing the density of midgap states by several orders of magnitude. This effect is called hydrogen passivation, or termination, of dangling bonds, and the experimental verifications of this effect were done by many workers (Madan et al., 1976; Tsai and Fritzsche, 1979). Freeman and Paul (1979) have recently studied the effect of hydrogenation on the optical properties in a-Si: H, and suggested that the dangling bond passivation effect makes the optical band edge steeper and shifts it to higher energies.

With the progress made in improving film quality after 1980, much clearer ER signals have been observed from both a-Si: H and microcrystalline silicon (μc-Si) (Hamakawa, 1982). Figure 3 shows an example of data taken by Tsu et al. (1980) on fluorinated a-Si (a-Si: F, H) using ER with an electrolyte interface. As can be seen from these data, not only the signal due to the fundamental edge but also those of higher interband transitions are observed.

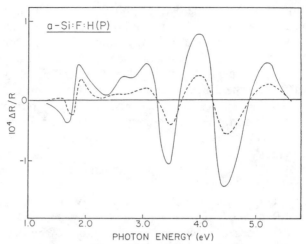

FIG. 3. Electrolyte electroreflectance spectra for a-Si:F:H doped with P (0.1% of PH₃ in SiF₄). ———, sample 35–A–B with 0.3-μm-thick film; ————, sample 35–B with 0.15-μm-thick film.

However, the features of the $\Delta R/R$ spectrum are greatly broadened compared with those obtained from single crystalline silicon (Frova *et al.,* 1967), and are more like those of a-Si:H (Nonamura *et al.,* 1981). Using the ER line shape to assess the film quality, Hamakawa *et al.* (1981) of Osaka University have worked out a systematic investigation of the relationship between ER signals and glow-discharge conditions.

Figure 4 shows a set of experimental data for rf powers of $P_{rf} = 35$, 100, and 140 W, respectively. Electroreflectance measurements were carried out at room temperature with a constant modulation voltage of 3 V in the Schottky-barrier geometry. As seen in the figure, the ER spectrum with $P_{rf} = 35$ W has a broad single peak and clearer line shape than that obtained with sputtered a-Si:H films (Freeman *et al.,* 1979). Although the signal amplitude decreases with increasing rf power, the observed line shape gets clearer and becomes similar to that of the M_0 critical point in crystalline semiconductors. Optical transitions between extended states in the completely disordered system would be altered only slightly by the presence of the external electric field modulation because the lifetime of an electron in the hypothetical Bloch states is so short; that is, the lifetime broadening prevails over the influence of the external electric field to diminish the modulated reflectance signal associated with transitions between the extended states. This is the case seen in a-Si:H prepared by using low-power plasma deposition or by sputtering, in which the field-induced change in the dielectric constant might be dominated by transitions between tail states

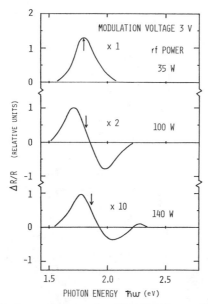

FIG. 4. Change in the electroreflectance signal $|\Delta R/R|$ with the rf power of plasma decomposition of a $SiH_{4(10)} + H_{2(90)}$ gas mixture. The arrows indicate the optical energy gap identified by $\sqrt{\alpha} \hbar \omega$ versus $\hbar \omega$ plots of absorption spectra. [After Nonomura *et al.* (1981).]

and extented states, as suggested by Jalali and Weiser (1980). On the other hand, ER spectra in plasma-deposited a-Si : H with higher rf power exhibit a broad but distinct structure resulting from the field-induced change in optical transitions between parabolic extended bands. This implies that the higher power deposition promotes structural order and thus weakens the lifetime broadening.

The dependence of peak $\Delta R/R$ on applied modulation voltage V_{app} is plotted in Fig. 5. The slope in log $\Delta R/R$ versus log V_{app} plots is very close to 2 for all the peaks in the ER spectra. The electric field is applied almost uniformly through the a-Si : H *i*-layer when the film thickness is very thin (about 2000 Å) (Okamoto *et al.,* 1981), in contrast to the case of bulk crystalline semiconductors in which the applied electric field extends only in the surface barrier region. This result indicates the square power dependence of $\Delta R/R$ on the modulating electric field. Thus it could be suggested from this result that the ER theory in the low-field regime developed by Aspnes and Bottka (1972) for crystalline solids holds good to a certain extent in the case of a-Si : H deposited at higher power. Then one could determine the position of the energy gap, i.e., the fundamental edge (indicated by arrows in Fig. 4), by the three-point adjusting method. For a-Si : H deposited at lower power, the electric field dependence of ER spectra coincides with

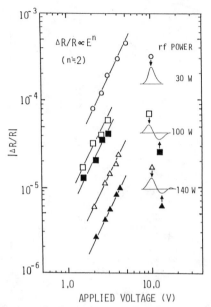

FIG. 5. Dependence of peak $\Delta R/R$ in a-Si:H on applied modulation voltage V_{app}. The rf power for plasma decomposition of $SiH_{4(10)} + H_{2(90)}$ is shown for each curve.

that theoretically verified by Esser (1972) for the electro-optical signal associated with localized states.

Figure 6 shows the absorption coefficient spectra near the fundamental edge of the films used for ER measurements. As can be seen from the figure, a change in the substrate temperature causes a parallel shift of the absorption spectra and also a change of slope in log α versus $\hbar\omega$ plots in the lower-photon-energy region. The former corresponds to the change in the optical energy gap with hydrogen content in the a-Si network and the latter indicates the spreading of the tail of localized states with decreasing substrate temperature. Deposition at the higher power gives rise to a drastic change in absorption spectra, as was found in the ER spectra shown in Fig. 4. Probably, because of the structural order brought about in the a-Si:H network by deposition at higher power, the absorption coefficient can no longer be fitted by the $\sqrt{\alpha\hbar\omega}$ versus $\hbar\omega$ plot that characterizes the nondirect transition that is unique to disordered systems. These observations also support the interpretation of the dependence of ER spectra on rf power or substrate temperature.

Figure 7 shows the substrate temperature T_s dependence of ER spectra in a-Si:H films prepared by using horizontal-mode rf power deposition for photovoltaic device applications (Okamoto *et al.*, 1980). The arrows in the

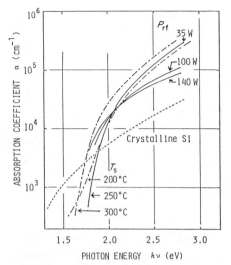

FIG. 6. Spectra of the absorption coefficient $\alpha(\hbar\omega)$ near the optical band edge of a-Si : H and μc-Si : H films prepared at various plasma rf powers. Dashed line: $\alpha(\hbar\omega)$ for crystalline Si; solid line: $T_s = 250°C$; dot-dash line: $P_{rf} = 35$ W.

figure indicate the positions of the optical gap. It is noticeable that the increase of substrate temperature shifts the ER peak toward lower energy, in accord with the change in the optical gap ($E_{g_{opt}}$) and also decreases the half-width of the ER spectra. The narrowing of the half-width might indicate a shrinking of tail states with increasing substrate temperature. This fact implies that local disorder in the amorphous network might be smeared out

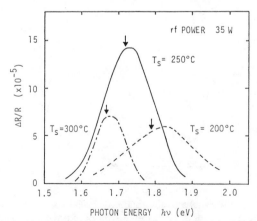

FIG. 7. Changes in the ER signal line shape with the substrate temperature, fabricated at constant rf power. The sample prepared at $T_s = 250°C$ has the best film quality.

by increasing the substrate temperature, as has been suggested by Freeman and Paul (1979).

Another efficient utilization of ER measurements is that for the study of compositional dependence of electronic band parameters in amorphous semiconductor alloys. Recently, Nonomura *et al.* (1983a) have studied the ER of hydrogenated silicon carbide (a-SiC:H) fabricated by the horizontal-mode plasma decomposition of $(SiH_{4(1-x)} + CH_{4(x)})$ mixed gases. Figure 8 shows the raw data obtained in this measurement. As can be seen from these data, a beautiful compositional dependence on the carbon fraction in a-SiC:H is observed. Although we have no good insight to make a more detailed analysis on broadened line spectra such as are obtained here, a relatively good coincidence is obtained between energies of $\Delta R/R$ peaks and extrapolated thresholds of $\sqrt{\alpha\hbar\omega}$ versus $\hbar\omega$ plots, as denoted by the arrows in the figure.

The author could not discuss further the diagnostics of ER signals at the present preliminary stage of data on a-Si:H and a-SiC:H. However, one might expect that very important information could be obtained in the near future from advanced measurements of various parameters, such as those for the effects of hydrogen content and impurity interactions in a-Si:H (P, B) and a-SiC:H (P, B).

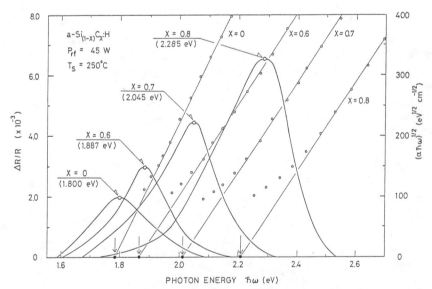

FIG. 8. Variations of ER signal and absorption coefficient spectra with carbon content in a-SiC:H alloys. Circles show peak energy of $\Delta R/R$ spectra and dots with vertical arrows indicate the threshold energy of $\sqrt{\alpha\hbar\omega}$ versus $\hbar\omega$ extrapolation. [After Nonomura *et al.* (to be published)].

IV. Electroabsorption

The history of the EA study of amorphous semiconductors seems very superficial compared to the ER work. Although several groups are working in this field [e.g., on SiAsTe, Taneki *et al.* (1974)], no published data have been seen. Recently, in the course of studying the optical confinement of near band-edge photons in $p-i-n$ a-Si:H solar cells, an EA signal about two orders of magnitude larger than the ER signal has been found. This was named the *back surface reflected electroabsorption* effect (BASREA). Extensive work on this effect was pursued both theoretically and experimentally (Nonomura *et al.,* 1983b). Figure 9 shows a schematic illustration of the BASREA measurement system and the light path in the transparent electrode $(\text{ITO/SnO}_2)/p-i-n$ a-Si:H/metal sample. We shall call ΔS the BASREA signal.

A typical ΔS spectrum observed in an a-Si:H $p-i-n$ junction and the absorption coefficient α of the i-layer used for the cell are shown in Fig. 10. From this figure, it is seen that the ΔS signal appears at the low absorption region in the vicinity of the optical absorption edge, which is defined by the $\sqrt{\alpha}\hbar\omega$ versus $\hbar\omega$ plot. In actual a-Si solar cells, the total thickness of the a-Si:H layer is around 6000 Å, and the light penetration depth ($\sim 1/\alpha$) for the wavelength at which the ΔS signal appears is much greater than the total thickness, as is shown in Fig. 11 (Hamakawa, 1982). In such a situation, it is reasonable to consider that the observed ΔS signal comes mostly from the

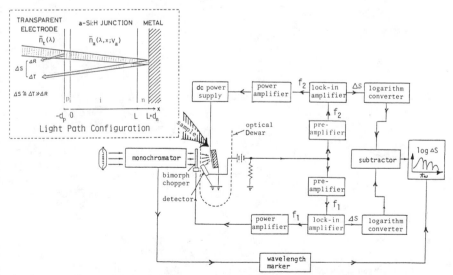

FIG. 9. Schematic diagram of the back surface reflected EA (BASREA) measurement system and the light path in the sample.

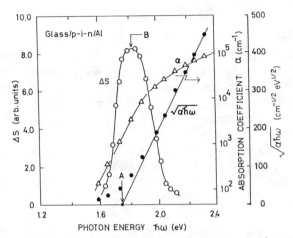

FIG. 10. Typical BASREA signal $\Delta S(\hbar\omega)$ and absorption spectra near the optical band edge of a-Si: H at RT.

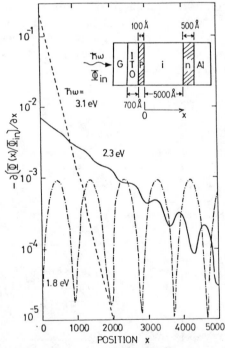

FIG. 11. Results of computer calculations of light penetration into an a-Si:H $p-i-n$ junction. Most of the 1.8 eV photons which are just above the band edge are reflected from the back surface and come out again through the i-layer. [After Hamakawa (1982).]

field modulated component of the light reflected at the back surface passing through the high electric field region within the i-layer.

In the geometrical arrangement shown in the insert of Fig. 9, the ΔS measurement is carried out by directing a light of wavelength λ in the positive direction of x and detecting the change in the reflection induced by the applied voltage V_a. According to the BASREA theory (Nonomura et al., 1983b), the signal ΔS has been calculated as

$$\Delta S(\lambda; V_{dc}, \Delta V_{pp}) = D(\lambda) \int_0^L \Delta E^2(x; V_{dc}, \Delta V_{pp}) \, dx, \tag{1}$$

where $V_a = V_{dc} + \Delta V_{pp}$, ΔV_{pp} is the modulation, ΔE is the local field increment, and

$$D(\lambda) = 2 \, \mathrm{Re} \left[\frac{(1 - \rho_f^2)\rho_b \gamma^2(\lambda; 0)}{(1 + \gamma^2 \rho_f \rho_b)^2} \left\{ C(\lambda) - i \frac{4\pi}{\lambda} B(\lambda) \right\} \right]. \tag{2}$$

Here $B(\lambda)$ and $C(\lambda)$ are the field independent optical constants for determination of the line shape of ΔS. When deriving Eq. (1), the reflectance R (under the applied voltage V_a) is formulated by taking into account the multiple reflection effect within the a-Si : H layer as

$$R = |\hat{\rho}|^2, \tag{3}$$

where

$$\hat{\rho} = (\rho_f + \rho_b \gamma^2)/(1 + \gamma^2 \rho_f \rho_b), \tag{4}$$

where ρ_f and ρ_b are reflection coefficients at the front surface (transparent electrode/a-Si : H) and back surface (a-Si : H/metal) interfaces, respectively. The function $\gamma(\lambda, 0)$ in Eq. (2) is defined as

$$\gamma(\lambda; 0) = \exp \left\{ i \frac{2\pi}{\lambda} \int_{-d_p}^{L+d_n} \tilde{n}_a(\lambda, x; 0) \, dx \right\}, \tag{5}$$

where $\tilde{n}_a$ is the complex refractive index of a-Si : H. It is worth noting that the signal detected by BASREA measurement is described by a combination of two separate functions, one of which is dependent only on the light wavelength and the other only on the electric field.

Figure 12(a) shows the peak intensities of BASREA spectra ΔS plotted as functions of the dc bias voltage V_{dc} for various ac modulation voltages ΔV_{pp} (Nonomura et al., 1982). The result in Fig. 12(a) is for a $SnO_2/p-i$(a-Si : H)$-n(\mu c$-Si)/Al type solar cell. The light wavelength used for the measurement was 6700 Å. It can be seen that ΔS has a linear dependence on V_{dc} and that all the extrapolated straight lines for various ΔV_{pp} intercept the dc bias voltage axis at the same point. Figure 12(b) demonstrates the dc bias

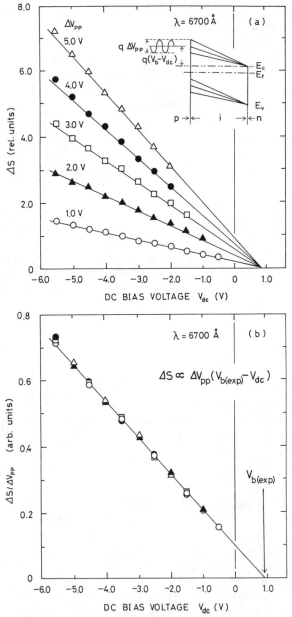

FIG. 12. Variation of (a) BASREA signal peak ΔS and (b) the normalized signal $\Delta S/\Delta V_{pp}$ with dc bias voltage V_{dc} with constant ac modulation voltage ΔV_{pp}. Note the beautiful relationship shown in (b). $\triangle$, 5.0 V; $\bullet$, 4.0 V; $\square$, 3.0 V; $\blacktriangle$, 2.0 V; $\bigcirc$, 1.0 V. [After Nonomura et al. (to be published).]

dependence of ΔS divided by ΔV_{pp}. Note that all data points fall on a single straight line. From this behavior, the dc and ac bias dependences of ΔS are expressed empirically by

$$\Delta S \propto \Delta V_{pp}(V_{b(exp)} - V_{dc}), \qquad (6)$$

where the crossing point on the dc bias axis is denoted by $V_{b(exp)}$. This characteristic relation can be easily obtained in two simple cases by using Eq. (1). These two cases are uniform and exponential-like distribution of the internal electric field $E_i(x; V_a)$ in the i-layer. In the case of the uniform field distribution, the internal electric field with externally applied voltage V_a is described as $E_i(x; V_a) = (V_b - V_a)/L$, where we define V_b as the electric potential within the i-layer of an a-Si:H solar cell in the case of no applied bias;

$$V_b = \int_0^L E_i(x; 0) \, dx. \qquad (7)$$

Figure 13 shows a correlation between the V_b determined from BASREA, and the open circuit photovoltage V_{oc} under AM 1 (100 mW/cm²) illumination in various types of a-Si:H solar cells. In a Schottky-barrier cell utilizing a gold contact, V_b is about 0.7 V and V_{oc} is 0.55 V. The V_b of the inverted-type cell ITO/$n(\mu$c-Si)–i(a-Si:H)–p(a-SiC:H)/SUS is about 1.2 V and its

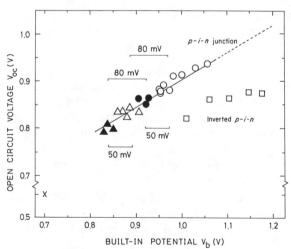

FIG. 13. Relation between built-in potential V_b determined by BASREA and the open circuit voltage V_{oc} measured from the photovoltaic performance of various types of a-Si solar cells having different junction structures. ▲, SnO₂/p–i–n(a-Si)/Al; △, SnO₂/p–i(a-Si)–$n(\mu$c-Si)/Al; ●, SnO₂/p(a-SiC)–i–n(a-Si)/Al; ○, SnO₂/p(a-SiC)–i(a-Si)–$n(\mu$c-Si)/Al; □, ITO/$n(\mu$c-Si)–i(a-Si)–p(a-SiC)/S.S.; ×, Au/i(a-Si)–n(a-Si)/S.S.

V_{oc} is 0.88 V. The V_b of the $p-i-n$ type homojunction cell constructed on the transparent electrode is around 0.84 V and the V_{oc} is about 0.79 V. The value of V_b is enhanced by about 50 mV when n-type μc-Si is used instead of n-type a-Si:H. In addition, an 80 mV increase of V_b was found in the case of p-type a-SiC:H. It is seen that V_{oc} becomes larger with increasing V_b.

The observed characteristics can be interpreted by using the relation

$$V_{oc} = \int_0^L E_i(x; 0) \left[\frac{\Delta\sigma(x)}{\sigma_0(x) + \Delta\sigma(x)} \right] dx, \tag{8}$$

in which $\sigma_0(x)$ and $\Delta\sigma(x)$ are dark conductivity and photoinduced change of conductivity, respectively. The inverted-type cell has a V_{oc} of 0.88 V in spite of its large built-in potential. This lower V_{oc} compared with that of cells constructed on a transparent electrode might be due to the space charge smearing effect caused by photogenerated carriers. The maximum V_{oc} achieved so far is 936.6 mV, in a SnO_2/p(a-SiC:H)$-i$(a-Si:H)$-n$(μc-Si)/Al type cell. The V_b of this cell is 1.06 V. If the built-in potential could be increased to 1.2 V by improving the transparent electrode, it would be possible to obtain a V_{oc} close to 1.0 V, according to the extrapolation of the solid line in Fig. 13.

V. Conclusion

In concluding this review on recent aspects of ER and EA studies on amorphous semiconductors, the author wishes to give a few summarizing comments.

(a)   Clear signals of electric-field-modulated reflectance and absorption are observed in hydrogen-passivated, tetrahedrally bonded amorphous semiconductors such as a-Si:H, μc-Si:F:H, and a-SiC:H.

(b)   The line shape and structure of the modulated signals are clearer in the better quality samples. This means that a measurement of ER and EA becomes a good measure of film quality.

(c)   The structure in observed ER and EA spectra is still too poor to diagnose detailed band parameters, as has been done in crystalline semiconductors. However, the spectral peak always lies around the absorption edge. Therefore, electric field modulation spectroscopy is also a useful tool for determining the effective optical energy gap of the amorphous semiconductors.

(d)   Temperature dependences and low temperature measurements of ER and EA at electromodulated interfacial barriers in a-Si and $p-i-n$ junctions and heterojunctions might give more detailed information for the analysis.

(e)   This technique would be useful in studying the compositional dependence of band parameters in mixed alloys of amorphous semiconductors such as a-SiC, a-SiN, and a-SiGe.

(f)   Surface barrier and/or interface barrier modulation in Schottky-type junctions, $p-i-n$ a-Si/a-SiC heterojunctions, etc., might also be a very good probe for finding the built-in potential and its distribution in various devices.

(g)   A wide variety of needed information such as the effect of hydrogen content on the band parameters, various impurities, and defect interactions in amorphous material could be obtained with this modulation spectroscopy.

(h)   With further progress in improvement of film quality it might be possible to make semiquantitative line spectrum analyses using electro-optical theory.

REFERENCES

Aspnes, D. E. (1973). *Surf. Sci.* **37,** 418.
Aspnes, D. E., and Bottka, N. (1972). Modulation Techniques. *In* "Semiconductors and Semimetals" (R. K. Willardson and A. C. Beer, eds.), Vol. 9, Chap. 6. Academic Press, New York.
Brodsky, M., and Leady, P. (1980). *J. Non-Cryst. Solids* **35/36,** 487.
Cardona, M. (1969). "Modulation Spectroscopy." Academic Press, New York and London.
Esser, B. (1972). *Phys. Status Solidi B* **51,** 735.
Freeman, E. C., Anderson, D. A., and Paul, W. (1979). *Bull. Am. Phys. Soc.* **24,** 339.
Freeman, E. C., and Paul, W. (1979). *Phys. Rev. B* **20,** 716.
Frova, A., Handler, P., Germano, F. A., and Aspnes, D. E. (1967). *Phys. Rev.* **145,** 575.
Hamakawa, Y. (1982). *In* "Amorphous Semiconductor Technologies and Devices" (Y. Hamakawa, ed.), Chap. 4. Ohm, Tokyo, and North-Holland, Amsterdam and New York.
Hamakawa, Y., and Nishino, T. (1976). *In* "Optical Properties of Solids, New Developments" (B. O. Seraphin, ed.), Chap. 6. North-Holland, Amsterdam, and Oxford and American Elsevier, New York.
Hamakawa, Y., Okamoto, H., and Nishino, T. (1981). Electronic States of a-Si. Fact Rep. p. 36. Sunshine Project Headquarters Office, Agency Ind. Sci. and Technol., Minist-Int. Trade and Ind., Japan.
Jalali, S., and Weiser, G. (1980). *J. Non-Cryst. Solids* **41,** 1.
Madan, A., LeComber, P. G., and Spear, W. E. (1976). *J. Non-Cryst. Solids* **20,** 239.
Nonomura, S., Okamoto, H., Nishino, T., and Hamakawa, Y. (1981). *J. Phys. Colloq. Orsay, Fr.* **42,** C4-761.
Nonomura, S., Okamoto, H., and Hamakawa, Y. (1982). *Jpn. J. Appl. Phys.* **21,** L464.
Nonomura, S., Okamoto, H., Fukumoto, K., and Hamakawa, Y. (1983a). *J. Non-Cryst. Solids* **59/60,** 1099.
Nonomura, S., Okamoto, H., and Hamakawa, Y. (1983b). *Appl. Phys. A* **32.**
Okamoto, H., Nitta, Y., Adachi, T., and Hamakawa, Y. (1979). *Surf. Sci.* **86,** 486.
Okamoto, H., Nitta, Y., Yamaguchi, T., and Hamakawa, Y. (1980). *Solar Energy Mat.* **2,** 313.
Okamoto, H., Yamaguchi, T., Nonomura, S., and Hamakawa, Y. (1981). *J. Phys. Colloq. Orsay, Fr.* **42,** C4-507.

Pille, H., Seraphin, B. O., Markel, K., and Fischer, J. E. (1969). *Phys. Rev. Lett.* **23,** 775.
Shay, J. L. (1972). *Proc. Int. Conf. Phys. Semicond., 11, Warsaw,* p. 787.
Taneki, Y., Nunoshita, H., Okamoto, H., and Hamakawa, Y. (1974). Semicond. Lab. Rep., Osaka University.
Tsai, C. C., and Fritzsche, H. (1979). *Solar Energy Mat.* **1,** 29.
Tsu, R., Izu, M., Cannela, V., and Ovshinsky, S. R. (1980). *J. Phys. Soc. Jpn., Suppl. A* **44,** 1249.
Weiser, G., and Stuke, J. (1969). *Phys. Status Solidi* **35,** 747.
Yamozoe, Y., Nishino, T., and Hamakawa, Y. (1981). *IEEE J. Quantum Electron.* **17,** 139.

SEMICONDUCTORS AND SEMIMETALS, VOL. 21, PART B

CHAPTER 6

Raman Scattering of Amorphous Si, Ge, and Their Alloys

Jeffrey S. Lannin†

DEPARTMENT OF PHYSICS
PENNSYLVANIA STATE UNIVERSITY
UNIVERSITY PARK, PENNSYLVANIA

I. Introduction

Raman scattering measurements provide a means of studying the lattice dynamics of disordered solids and liquids. In addition, these measurements give insight into structure, bonding, and the nature of disorder. The study of the intensity distribution of inelastically scattered light thus provides valuable dynamical and structural information about amorphous solids and their variations with preparation conditions or on alloying.

In ordered crystalline semiconductors conservation of crystal momentum results in narrow $q \simeq 0$ phonon wave-vector-associated first-order Raman scattering processes. Disorder, which arises, for example, from impurities, defects, grain size, or anharmonic effects, may modify this condition. The initial effect of increasing disorder is generally a relatively symmetric broad-

† This research was supported by National Science Foundation grant 81-09033.

159

ening of the first-order response. In contrast, it is observed that the loss of periodicity, as it arises in noncrystalline solids or liquids, results in rather broad Raman spectra in which the absence of crystal momentum conservation implies the participation of all the vibrational eigenstates (Shuker and Gammon, 1970). Between these two extremes there lie, however, intermediate situations in which the crystalline response in highly disordered alloys or small crystals may have additional broad features associated with an admixture of $q \neq 0$ phonons into the scattering process. Under special conditions associated with static or dynamical correlation effects (Martin and Galeener, 1981; Shanabrook and Lannin, 1981) or narrow phonon bands, amorphous solids may also yield Raman scattering spectra that may have sharper features superposed on a broad response.

In this review emphasis is placed on the role of local order in Raman scattering in a-Si, a-Ge, and their alloys, including those with hydrogen. In these systems the Raman spectra are predominantly influenced by changes in short-range order due to modifications of structural and compositional disorder. The dependence of the Raman scattering processes on short-range order is a consequence of predominant short-range interactions dictating the vibrational eigenstates as well as the modulations of the electronic susceptibility by these phonons. In addition, the "three-dimensional" character of the bonding in these systems, as well as the substantial fluctuations in short-range order, implies that intermediate-range order effects are expected to be reduced. This contrasts with group V and group VI amorphous semiconductors in which smaller bond-angle variations allow correlation effects of intermediate range to be observed (Shanabrook and Lannin, 1981). This emphasis on short-range order does not, however, imply that structural configurations of intermediate range do not contribute to the Raman spectra. For example, as discussed in Section 7, the possible influence of ring configurations may be of importance for a selected spectral range.

In contrast to diffraction methods, vibrational probes are an indirect means of obtaining information about structural changes. Raman scattering and infrared absorption involve matrix-element or coupling-parameter effects that require in principle a detailed theoretical analysis for disordered solids. In practice, a phenomenological analysis that attempts to separate matrix-element effects from the density of states appears to be a useful starting point. This separation, which assumes relatively smooth, though significant, frequency-dependent matrix elements, is employed to obtain structural and dynamical information about a-Si and a-Ge. Emphasis is placed on the considerable sensitivity of the Raman spectra to small changes in short-range order in these systems relative to diffraction methods. In addition, these spectra in both a-Si and a-Ge indicate the variability with preparation conditions of the amorphous state as well as provide a means for quantitatively measuring the degree of disorder. Section 6 also indicates that changes in the Raman spectra with disorder exhibit a correspondence to variations in the optical gap, thus emphasizing the interrelation of phonon and electron states.

Although the addition of H to a-Si and a-Ge is of considerable current interest, limited structural information is available about changes in host short-range order (see Chapter 11 by Guttman in Vol. 21A). Such information, as well as the extent to which such alloys are microscopically heterogeneous, is of considerable importance in attempting to understand the physical properties of these materials. In Section 7 the influence of H on the Raman spectra is discussed in terms of possible short-range order modifications. Recent studies of a-Si$_{1-x}$H$_x$ alloys have demonstrated that H, depending on its concentration, may have more than one function to the extent that it may form new alloy electronic states as well as modify the local Si host order.

Recent interest also exists in microcrystalline Si and Si$_{1-x}$H$_x$ alloys. The nature of the disorder here is qualitatively quite different from that of amorphous solids, except possibly in the limit of very small crystallites that are a few unit cells in diameter. Even in such presently hypothetical materials, theoretical studies suggest that unless large grain boundary strains are present the Raman scattering response will differ qualitatively from that of true amorphous materials. In the case of models of a-Si and a-Ge, this difference may be noted by substantial fluctuations in bond angle at each atom in the network.

Shown in Table I is a qualitative comparison of some of the types of disorder in tetrahedrally based amorphous and crystalline solids, their origin, and the qualitative first-order Raman response. In contrast to amorphous solids, in which the loss of crystal momentum conservation results in substantial scattering intensity from all phonon modes, both microcrystalline and concentrated crystalline alloys are observed to yield a strong scattering peak or peaks that are derived from the allowed elemental crystalline $q \simeq 0$ modes. In alloys, the low-frequency disorder-induced scattering appears to be relatively weak in intensity.† Thus, within group IV materials, a qualitative distinction between the scattering from amorphous solids and disordered crystalline systems is suggested. The special case of very small crystallites is discussed in Section 8. Whereas the vertical direction in Table I indicates the general trend of increasing disorder, additional disorder mechanisms, e.g., those due to vacancies (Suzuki and Maradudin,

---

† The actual low-frequency scattering intensity may, however, be comparable to that in the corresponding amorphous solid.

TABLE I

QUALITATIVE EFFECTS OF DISORDER ON FIRST-ORDER RAMAN SCATTERING IN
DISORDERED TETRAHEDRALLY BASED SOLIDS

Type of disorder	Disorder mechanisms	First-order Raman response
Thermal	Anharmonicity	Broadened $q \simeq 0$ peak; very weak asymmetry[a]
Crystal size	Boundary scattering; lattice distortions	Broadened $q \simeq 0$ peak; admixture of $q \neq 0$ optic modes yields asymmetric peak; very weak low frequency scattering
Alloy	Local mass/force constant variations	Broadened $q \simeq 0$ peak; admixture of $q \neq 0$ optic modes; local, resonant mode scattering; weak low frequency disorder-induced scattering[b]
Amorphous	Local bond angle and distance fluctuations; bonding defects	No $q = 0$ scattering; broad disorder-induced scattering at all frequencies
Amorphous alloys	Compositional fluctuations; mass/force constant variations	No $q = 0$ scattering; broadened or narrowed host band in solid solutions; local, resonant modes

[a] Cowley (1971).
[b] Xinh (1968).

1983) or defect complexes, have been omitted. Although alloying often increases both compositional and structural disorder, in the case of a-$Si_{1-x}H_x$ alloys, improved short-range bond-angle order appears improved as suggested below.

In the following sections first- and second-order Raman scattering processes in noncrystalline solids are considered. This is followed by a review of relevant theoretical results in a-Si and a-Ge. An overview of experiment is then presented prior to a discussion of experimental results in these systems and their alloys with and without hydrogen. This is followed by theoretical and experimental results concerning Raman scattering in microcrystalline systems and concluding comments on future areas of possible study.

II. Raman Processes and Phenomenology

1. FIRST-ORDER RAMAN SCATTERING

The first-order Raman scattering process associated with spontaneous phonon emission for Stokes scattering is shown schematically in Fig. 1 for

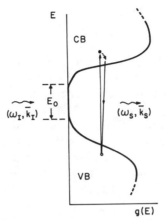

FIG. 1. Schematic of first-order Stokes Raman process in an amorphous semiconductor for $\hbar\omega_I > E_0$. (In the figures, vectors are indicated by a bar over the vector symbol. In the text, vectors are indicated by bold type.) VB and CB denote valence and conduction bands.

an amorphous semiconductor whose electronic density of states is $g(E)$. The example shown is for $\hbar\omega_I > E_0$, where ω_I and ω_S are the incident and scattered frequencies, respectively. Alternative processes occur in which $\hbar\omega_I < E_0$ as well as with phonon emission by the hole rather than the electron. The optical gap indicated here by E_0 represents that obtained from absorption measurements at values of the absorption coefficient α for which $(\alpha\hbar\omega_I)^{1/2}$ varies as $\hbar\omega_I - E_0$.

An alternative description of the Raman process is one in which the vibrational excitations weakly modulate the electronic polarizability tensor P_{ij} proportional to the displacement amplitudes $u_{\sigma,q}$ from static equilibrium, where σ labels the phonon branch. For an opaque crystalline solid the differential photon-counting cross section for first-order Stokes processes may be represented by the relation (Klein *et al.*, 1978)

$$\frac{1}{I_I} \frac{d^2 I^{(1)}(\omega)}{d\omega_S \, d\Omega}\bigg|_{(\alpha)} = \frac{\omega_S^3 \omega_I A}{\cos\theta_I c^4} T_{(\alpha)}(\omega_I, \omega_S)$$

$$\times \sum_{\sigma,q} \frac{n(\omega_\sigma) + 1}{\omega_\sigma} \left| \left\langle \frac{\partial P_{ij}}{\partial u_{\sigma,q}^*} \right\rangle \right|^2_{(\alpha)} \delta[\omega - \omega_\sigma(\mathbf{q})] \, \delta(\mathbf{k} - \mathbf{q}),$$

$$(1)$$

where $\omega = \omega_I - \omega_S$ is the phonon frequency, $\mathbf{k} = \mathbf{k}_I - \mathbf{k}_S$ the wave-vector difference of the incident and scattered fields, θ_I the angle of incidence, Ω the solid angle, and A the illumination area. The term $T_{(\alpha)}(\omega_I, \omega_S)$ is a transmission plus attenuation factor arising from the opaque nature of the sample and includes a solid angle collection factor. It has been shown (Mills *et al.*,

1970; Klein *et al.*, 1978) that

$$T_{\{\alpha\}}(\omega_I, \omega_S) = |\tau_I \tau_S / \sqrt{\tilde{\varepsilon}}|^2_{\{\alpha\}} \delta/2, \tag{2}$$

where τ_I and τ_S are the incident and scattered transmission coefficients of the vacuum–film interface discussed in Part IV [Eqs. (17)–(19)], $\tilde{\varepsilon}$ is the complex dielectric constant, and δ the penetration depth. The subscript $\{\alpha\}$ is a condensed tensorial label for the scattering configuration, specified by the unit polarization directions e_I and e_S of the incident and scattered electric fields. Equation (1) neglects spatial dispersion (Zeyher *et al.*, 1974).

For an amorphous solid the absence of periodicity implies the loss of crystal momentum conservation. If the phonon eigenstates are now designated by β, Eq. (1) yields for a fixed solid angle of collection $\Delta\Omega$ and fixed $\Delta\omega_S$, a Stokes first-order intensity that varies as

$$I^{(1)}_{\{\alpha\}}(\omega) \propto \omega_I^4 \left(1 - \frac{\omega}{\omega_I}\right)^3 \frac{A}{\cos\theta_I}$$
$$\times T_{\{\alpha\}}(\omega_I) \sum_\beta \left\{ \frac{n(\omega_\beta) + 1}{\omega_\beta} C^{\{\alpha\}}(\omega_\beta) \right\} \delta(\omega - \omega_\beta). \tag{3}$$

Here it is assumed that $T_{\{\alpha\}}(\omega_I, \omega_S) \simeq T_{\{\alpha\}}(\omega_I)$, given that $\omega_S \simeq \omega_I$ and that the optical constants are locally smooth in frequency for an amorphous solid. The matrix-element factor $C^{\{\alpha\}}(\omega_\beta) = |\langle \partial P_{ij}/\partial u_\beta^* \rangle|^2_{\{\alpha\}}$, which represents polarizability modulations by modes β, is referred to as the coupling parameter (Shuker and Gammon, 1970). If the phonon density of states $\rho(\omega)$ can be decomposed into distinct, weakly overlapping bands δ, then the relation

$$\rho_\delta(\omega) = \sum_\beta \delta(\omega - \omega_\beta) \tag{4}$$

may be employed to rewrite the sum in Eq. (3). In particular, if it is assumed that the $\{\ \}$ factor is smoothly varying over δ, then the summation in Eq. (3) may be simplified so that

$$I^{(1)}_{\{\alpha\}}(\omega) \propto \left(1 - \frac{\omega}{\omega_I}\right)^3 \sum_\delta \frac{n(\omega) + 1}{\omega} \tilde{C}^{\{\alpha\}}_\delta(\omega)\rho_\delta(\omega). \tag{5}$$

Implicit in Eq. (5) is that different modes β of common frequency ω have a similar coupling parameter $\tilde{C}_\delta(\omega)$. Equation (5) also assumes nonresonant scattering conditions. Resonance effects may be included in an approximate manner by multiplying Eqs. (3) and (5) by

$$\left|\frac{d\tilde{\varepsilon}}{d\omega_I}\right|^2 = \left|\frac{d\varepsilon_1}{d\omega_I}\right|^2 + \left|\frac{d\varepsilon_2}{d\omega_I}\right|^2,$$

where ε_1 and ε_2 are the real and imaginary parts of $\tilde{\varepsilon}$ (Bermejo *et al.*, 1977). A further phenomenological simplification of Eq. (5) is often employed in which the band index δ is omitted, corresponding to a smooth variation of the coupling parameters over the entire spectral range so that

$$I^{(1)}_{(\alpha)}(\omega) \propto \left(1 - \frac{\omega}{\omega_{\mathrm{I}}}\right)^3 \left[\frac{n(\omega) + 1}{\omega} \overline{C}^{(\alpha)}(\omega)\right] \rho(\omega), \tag{6}$$

where $\overline{C}^{(\alpha)}(\omega)$ represents an effective or average coupling parameter, which may be obtained if both $I^{(1)}_{(\alpha)}(\omega)$ and $\rho(\omega)$ are known either from experiment or theory. The Raman spectra are thus related to a coupling-parameter weighted density of phonon states.

2. COUPLING PARAMETERS

Earlier Raman studies in a-Si and a-Ge, as well as some recent studies in other amorphous solids, have compared the reduced Raman spectra $\rho^{(\alpha)}_r(\omega) \equiv \overline{C}^{(\alpha)}(\omega)\rho(\omega)$ with the phonon density of states. This comparison, in which

$$\rho^{(\alpha)}_r(\omega) = I^{(1)}_{(\alpha)}(\omega)\omega/[(n + 1)(1 - \omega/\omega_{\mathrm{I}})^3] \simeq I^{(1)}_{(\alpha)}(\omega)\omega/(n + 1),$$

is valid only if $\overline{C}^{(\alpha)}(\omega) = $ const. Such constancy is, however, inconsistent with the requirement that at $\omega = 0$ the polarizability fluctuations vanish. Experimental and theoretical studies in a-Si and a-Ge (Lannin, 1973; Martin and Brenig, 1974) indicate that at low frequencies $\overline{C}^{(\alpha)}(\omega) \propto \omega^2$. The range of frequencies for which this quadratic behavior occurs is not known in general. In a-Si and a-Ge it is found to occur over a range approximately corresponding to the elastic continuumlike or Debye-like regime in c-Si and c-Ge for which $\rho(\omega) \propto \omega^2$. In low-temperature rf sputtered films this corresponds to maximum frequencies of $\omega_{\max}(\mathrm{Si}) \simeq 64$ cm^{-1} and $\omega_{\max}(\mathrm{Ge}) \simeq 36$ cm^{-1} (Lannin, 1973, 1974).

With increasing frequency it is plausible to expect that a less rapid variation of $\overline{C}^{(\alpha)}(\omega)$ will occur. A comparison of the Raman scattering spectra at lower frequencies with the theoretical (Alben *et al.*, 1975) and experimental (Axe *et al.*, 1974) density of states for a-Ge suggests a less rapid variation of $\overline{C}^{(\alpha)}(\omega)$ in the vicinity of the lowest-frequency TA peak. In particular, these results, as well as specific heat measurements (King *et al.*, 1974), suggest a variation of order nearer to a linear than a quadratic dependence of $\overline{C}^{(\alpha)}(\omega)$ in this spectral range (Lannin, 1977a). A more detailed estimate of $\overline{C}^{(\alpha)}(\omega)$ has been obtained in bulk a-Se based on a comparison of neutron (Gompf, 1979) and Raman measurements (Lannin and Carroll, 1982). The results indicate an approximately linear variation in $\overline{C}^{(\alpha)}(\omega)$ extending from beyond the Debye regime to above the lowest peak in the density of states. Perhaps of most importance is that either an extrapolation of the low-frequency quadratic variation or the assumption of constant coupling parameters implicit in $\rho_r^{(\alpha)}(\omega)$ yields spectra that differ considerably from the theoretical phonon density of states (Alben *et al.,* 1975).

An improved Raman spectral representation for comparison with $\rho(\omega)$ may be obtained by noting that the terms $\overline{C}^{(\alpha)}(\omega)$ and $\omega^{-1}(n+1)$ tend to cancel in the Debye regime. In addition, if at intermediate frequencies $\overline{C}^{(\alpha)}(\omega)$ is approximately linear in frequency, then again local cancellation occurs as $n + 1 \approx 1$. This *partial* cancellation of the coupling parameter and phonon displacement factors implies that the Raman spectra at 300°K yield a reasonable, convenient spectral comparison with the theoretical phonon density of states. Here the $n + 1$ factor is neither too small nor too large, as occurs at very low or high temperatures. In addition, the relative area of the TA band, which is separated by a fairly deep minimum from the intermediate and higher frequency bands (see Fig. 3), provides constraints on the coupling parameters. In a disorder-broadened pseudocrystal model for the amorphous state, this implies that the area of the TA band in the density of states is one half that of the intermediate plus higher frequency spectrum. A more rigorous justification of this for a two-parameter Born model, in the limit of small noncentral interactions, has been given by Alben *et al.* (1975). If the minimum gap or pseudogap frequency is designated by $\omega_{\min}$ and $\omega_{\max}$ is the maximum phonon frequency, this implies that

$$\int_0^{\min} \rho(\omega)\, d\omega \simeq \frac{1}{2} \int_{\min}^{\max} \rho(\omega)\, d\omega. \tag{7}$$

In terms of the Raman spectra this implies constraints on $\overline{C}^{(\alpha)}(\omega)$, because for $\omega \ll \omega_1$, Eq. (6) yields

$$\int_0^{\omega_{\min}} I^{(\alpha)}(\omega) \left[\frac{\omega/(n+1)}{\overline{C}^{(\alpha)}(\omega)} \right] d\omega \simeq \frac{1}{2} \int_{\min}^{\max} I^{(\alpha)}(\omega) \left[\frac{\omega/(n+1)}{\overline{C}^{(\alpha)}(\omega)} \right] d\omega. \tag{8}$$

Equation (7) is found to be accurate to within 5–20% in the density of states of c-Si (Dowling and Cowley, 1966) and c-Ge (Nelin and Nilsson, 1972) and in the theoretical densities of states of a-Si and a-Ge (Alben *et al.,* 1975; Beeman and Alben, 1977), indicating the utility of this constraint on possible variations of $\overline{C}^{(\alpha)}(\omega)$. Similar qualitative trends of $\overline{C}(\omega)$ in a-Si$_{1-x}$H$_x$ alloys may also be expected for much of the lower-frequency Si host band, although in-band resonant effects may also occur.

3. Second-Order Raman Scattering

In crystalline solids second-order Raman scattering may involve sum or difference processes for which crystal momentum is conserved. Energy conservation implies that the two phonon wave vectors satisfy the relation $\mathbf{q} + \mathbf{q}' \simeq 0$. This results in two distinct scattering terms associated with overtone or combination processes. The Stokes second-order sum processes in crystals may be represented by a scattering intensity (Hayes and Loudon, 1978) that varies as

$$I_{\{\alpha\}}^{(2)}(\omega) \propto \sum_{\sigma,\sigma'} \sum_{\mathbf{q}} \left| \left\langle \frac{\partial^2 P_{ij}}{\partial u_{\sigma,\mathbf{q}} \, \partial u_{\sigma',-\mathbf{q}}^*} \right\rangle \right|_{\{\alpha\}}^2$$
$$\times \frac{(n_\sigma + 1)(n_{\sigma'} + 1)}{\omega_\sigma \omega_{\sigma'}} \delta(\omega_\sigma + \omega_{\sigma'} - \omega), \qquad (9)$$

where σ and σ' label the phonon branches for phonons $\mathbf{q}$ and $\mathbf{q}' \simeq -\mathbf{q}$. Equation (9) results in overtone and combination scattering processes. For overtone (OT) scattering, in which $\omega_\sigma = \omega_{\sigma'}$ and $\sigma = \sigma'$, Eq. (9) yields, in a manner similar to that of Eqs. (3)–(6), the phenomenological relation

$$I_{\{\alpha\}}^{(2)}(\omega) \propto \overline{C}_{\mathrm{OT}}^{(\alpha)}\left(\frac{\omega}{2}\right) \left[\frac{n(\omega/2) + 1}{\omega/2}\right]^2 \rho\left(\frac{\omega}{2}\right), \qquad (10)$$

where the term $\overline{C}_{\mathrm{OT}}^{(\alpha)}(\omega/2)$ represents the effective second-order coupling parameter. Experimental studies in crystalline elemental semiconductors (Temple and Hathaway, 1973; Weinstein and Cardona, 1973; Carroll and Lannin, 1983) and semimetals (Lannin et al., 1975) yield second-order Raman spectra that are similar in the form to that of the phonon density of states. This implies the predominance of overtone processes in these materials as well as the smooth variation of $\overline{C}_{\mathrm{OT}}(\omega/2)$ within a given phonon band.

In noncrystalline solids the absence of crystal momentum conservation requires that Eq. (9) be modified to the approximate form (Lannin and Carroll, 1982)

$$I_{\{\alpha\}}^{(2)}(\omega) \propto \int_0^{2\omega_{\max}} d\omega' \, \frac{n(\omega') + 1}{\omega'} \frac{n(\omega - \omega') + 1}{\omega - \omega'}$$
$$\times \rho(\omega')\rho(\omega - \omega')\overline{C}^{(\alpha)}(\omega,\omega'). \qquad (11)$$

This result indicates that the second-order scattering in noncrystalline solids or liquids is a coupling-parameter-weighted self-convolution of the phonon density of states. In contrast to crystalline solids, this implies that all phonons may combine in second-order Raman processes, whereas in crystals the constraint $\mathbf{q} + \mathbf{q}' \simeq 0$ exists. If overtone processes dominate, a result similar to Eq. (10) is obtained. Convolution results for second-order infrared absorption spectra that are similar to Eqs. (10) and (11) readily follow and emphasize the role of additional combination processes in amorphous solids that are caused by the absence of crystal momentum conservation.

In a-Si and a-Ge, distinct broad high-frequency features indicate second-order Raman scattering (Smith *et al.,* 1971a). It is often implicitly assumed that such scattering is overtonelike, so that designations such as 2LA and 2TO have been employed. As Eq. (11) indicates, however, overtone scattering will not in general dominate, so that different phonons within and between different vibrational bands may combine in the second-order process. The experimental results discussed in Section 6 indicate that such combination processes may be significant in a-Si and a-Ge and their alloys with H.

III. Polarizability and Phonon Density-of-States Theory

4. Raman Polarizability Theory

Present Raman scattering calculations do not directly employ the electronic states, but rather assume local models for the polarizability tensor. Both direct (Alben *et al.,* 1975) and equation-of-motion methods (Beeman and Alben, 1977) have been employed to determine the tensor components. Beeman and Alben (1977) have represented the Raman cross section in terms of atomic-displacement-induced modulations of the polarizability tensor in the form

$$P_{ij} = \sum_{\gamma,f} D^{ij}_{\gamma f} u_f(\gamma),$$ (12)

where $u_f(\gamma)$ represents the displacement of the γth atom, f the Cartesian coordinate, and $D^{ij}_{\gamma f} \equiv \partial P_{ij}/\partial u_f(\gamma)$.

Because the dynamical polarizability projected out by the incident and scattered fields of polarizations $\mathbf{e_I}$ and $\mathbf{e_S}$ is

$$P = \sum_{ij} e^i_S P_{ij} e^j_I,$$ (13)

this yields a first-order differential cross section that varies as

$$\frac{d^2 I^{(1)}(\omega)}{d\omega_S \, d\Omega} \propto \left(\sum_{ij} \sum_{\gamma f} e^i_S D^{ij}_{\gamma f} e^j_I \right)^2.$$ (14)

Alben *et al.* (1975) simplify the polarizability by assuming three local bond symmetry constraints so that $D^{ij}_{\gamma f}$ has only three components, which for simplicity we label as $_k D$. The relative weighting factors $_k B$ of these components are not known and may in principle be obtained from experiment if interference terms such as $_2 D \times {}_3 D$ are neglected. The $_k D$ values have been calculated for the model of Polk and Boudreux (1973) for a-Ge by Lottici and Rehr (1980), whose results are shown in Fig. 2 along with the phonon density of states. Mechanism $_2D$ is found to be qualitatively rather similar to $\rho(\omega)$, whereas $_3D$, which also exhibits major low- and high-frequency bands, indicates a shape that is somewhat modified at low frequencies from that of $\rho(\omega)$. In contrast, the $_1D$ mechanism yields primary high-frequency scattering. These results in a-Ge are also qualitatively applicable to a similar a-Si model because the force constants are similar to those of a-Ge. Although the total scattering will be a function of the weighted sum of the $_kD$ values, only the individual polarized terms have to be evaluated, since the forms of mechanisms $_1D$ and $_2D$ have been chosen to yield depolarization ratios of 3/4. In contrast, $_3D$ yields polarized scattering. In the limit of perfect tetrahedral symmetry, the forms of $_2D$ and $_3D$ were also chosen to vanish. The form of the $_kD$ terms is also such that two of these, $_1D$ and $_3D$, have only contributions from bond stretching motions, whereas $_2D$ has both bend and stretch motions contributing. This is consistent with the $_2D$ scattering component more nearly resembling, to first order, the form of the vibrational density of states. The relation between the phenomenological cou-

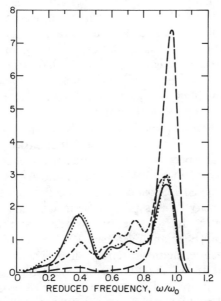

FIG. 2. Theoretical results of Lottici and Rehr (1980) for $_kD$ mechanisms and $\rho(\omega)$ for the Polk–Boudreaux model of a-Ge. ———, $_1D$; ———, $_2D$; ----------, $_3D$; · · · ·, $\rho(\omega)$. [Reprinted with permission from *Solid State Communications,* Vol. 35, by P. P. Lottici and J. J. Rehr, A connection between Raman intensities and EXAFS Debye–Waller factors in amorphous solids. Copyright 1980, Pergamon Press, Ltd.]

pling parameters $\overline{C}^{(\alpha)}(\omega)$ and the reduced $_kD$ mechanisms can be represented for the depolarized, (VH), and polarized (HH) spectra in the form

$$\overline{C}^{VH}(\omega) \sim \tfrac{3}{4}[(_1B_1\tilde{D})^2 + (_2B_2\tilde{D})^2]/\rho(\omega), \tag{15}$$

$$\overline{C}^{HH}(\omega) \sim [(_1B_1\tilde{D})^2 + (_2B_2\tilde{D})^2 + (_3B_3\tilde{D})^2]/\rho(\omega), \tag{16}$$

where the $(_k\tilde{D})^2$ represent the contributions of the $_kD$ in Eq. (14) and Fig. 2.

5. PHONON DENSITY OF STATES

The qualitative similarity between $_2D$ and $\rho(\omega)$ in Figs. 2 and 3 for the Polk–Boudreaux model makes it useful to consider theoretical variations in the phonon density of states of a-Si and a-Ge, particularly because these results allow trends in the Raman spectra to be clarified. In crystalline- (c-) Si and Ge the phonon spectra consist of four bands that partially overlap and are denoted by TA, LA, LO, and TO (Nelin and Nilsson, 1972). Of interest in a-Si and a-Ge is whether corresponding bands occur in $\rho(\omega)$ and the Raman spectra. A considerable number of calculations of the phonon density of states have been performed for a range of structural models and force constants, including Bethe-lattice, computer, and hand-built models (Weaire and Alben, 1972; Thorpe, 1973; Alben *et al.,* 1975; Beeman and Alben, 1977). Most calculations have employed two-parameter Hamiltonians that consist of central stretch-type interactions as well as noncentral interactions. Although models using Born, Keating, and valence-force interactions differ in detail (Weaire and Taylor, 1980), their results are qualitatively similar in yielding the main low-frequency TA band and the high-frequency TO band with $\omega_{TA}/\omega_{TO} \simeq 0.4$ for the peak frequencies. In contrast, a broadened crystalline spectrum yields $\omega_{TA}/\omega_{TO} \simeq 0.3$, indicating limitations on the two-parameter models at low frequencies.

Bethe-lattice calculations have suggested that closed rings of bonds are required to obtain a spectral feature at intermediate frequencies near the crystalline LA band (Thorpe, 1973; Beeman and Alben, 1977). The question of a crystal-related LO band has also been considered, although the position and intensity of this peak even in c-Si and c-Ge is such that significant broadening due to structural disorder may result in the absence of a distinct feature (Beeman and Alben, 1977; Sen and Yndurian, 1977). Detailed Raman scattering measurements in a-Si shown in Section 6 indicate a weak shoulder in the LO region. With increasing disorder this shoulder increases in intensity as a consequence of broadening, implying that its origin in these films is different from that of c-Si. In addition, more ordered a-Si and a-Si$_{1-x}$H$_x$ alloys do not indicate distinct structure in this spectral range, suggesting that its apparent absence is a separate feature in films prepared to date. A possible exception at high pressures is discussed in Section 7 for a-Si$_{1-x}$H$_x$.

More detailed lattice dynamical calculations by Meek (1977), on a number of models of a-Ge, employed a three-parameter modified bond-charge model (Weber, 1974). Meek's results for $\rho(\omega)$ are shown in Fig. 3 for a crystalline model and a number of amorphous Ge models with a range of bond-angle fluctuations $\Delta\theta$. The model of Steinhardt *et al.* (1974) has $\Delta\theta = 6.8°$ whereas the even-membered ring model by Connell and Temkin (1974) has $\Delta\theta = 9.1°$. Models B–F, in which $\Delta\theta = 10.2–13.3°$, were constructed by variations in topology and bond-angle statistics by Beeman and Bobbs (1975). As Fig. 3 indicates, the three-parameter model yields a reduction of ω_{TA} so that ω_{TA}/ω_{TO} is nearer to that of c-Si or c-Ge than are the two-parameter models. Figure 3 also indicates that the TA band, for fixed force constants, is relatively insensitive in intensity to variations in structural order, particularly bond-angle fluctuations. This is also observed for the relative intensity of the LA band. In contrast, the peak intensity and

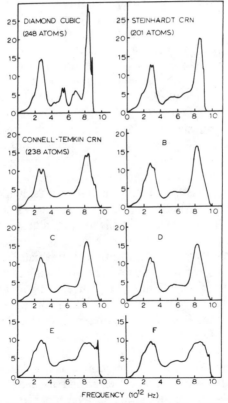

FIG. 3. Phonon densities of states of Meek (1977) for crystalline (248 atom) and amorphous Ge models of Beeman and Bobbs (1975).

width (full width at half maximum) of the TO band are found to be particularly sensitive to bond-angle variations. Similar trends are also observed in the results of Beeman and Alben (1977).

It is useful to note that the Steinhardt *et al.* (1974) model of highest short-range order indicates a shoulder corresponding to the position of the crystalline LO band. The value of $\Delta\theta$ in this model is, however, considerably smaller than the estimated minimum of $\sim 9°$ in anneal-stable a-Si or a-Ge (Lannin *et al.*, 1982). Although the models of Fig. 3 were initially constructed and studied to obtain a best single fit to the structural, optical, and vibrational spectra of a-Ge, they constitute a range of possible structures when the amorphous state is viewed as being structurally variable with preparation, as is discussed in Section 6. The similarity in Fig. 2 between $_2D$ and $\rho(\omega)$ implies that changes in the depolarized Raman spectra, as may arise with modifications of structural order, will primarily reflect variations of $\rho(\omega)$.

Recent inelastic neutron scattering measurements employing the time-of-flight method have been obtained for a-Si prepared by rf sputtering (Kamitakahara *et al.*, 1983). The spectra, which yield the approximate phonon density of states, indicate an enhanced TA peak intensity relative to that predicted by the theoretical results indicated in Fig. 3 or those of other studies (Alben *et al.*, 1975; Beeman and Alben, 1977). This may be a consequence of coherency effects associated with the range of neutron wave-vector transfers employed (Alben *et al.*, 1975) or with limitations of the theory. The neutron spectra also indicate an unexpectedly large shoulder at ~ 400 cm^{-1} not predicted theoretically. Further confirmational studies would be useful because this would also imply both a rapid variation for the VH Raman coupling parameters at higher frequency and a less distinct TO peak in $\rho(\omega)$. Alternatively, the HH Raman spectra (as will be seen in Fig. 7), which also exhibit an enhanced shoulder in this interval, may better reflect the form of the phonon spectra at higher frequencies. Similar inelastic neutron measurements performed on a-Si$_{0.88}$H$_{0.12}$ indicate H local and resonant modes as well as a-Si hostlike modes in which H primarily follows the Si motions at lower frequencies. The observation of this essentially H-projected density of states is a consequence of the fact that the neutron cross section for H is ~ 40 times greater than for Si. This enhanced H scattering implies that Raman scattering measurements may provide a more useful measure of the low frequency phonon density of states in the hydrogenated alloys.

Detailed theoretical studies of the phonon density of states of a-Si$_{1-x}$H$_x$ or a-Ge$_{1-x}$H$_x$ models have not been performed to date, although cluster Bethe-lattice calculations of the former have been performed by Pollard and

Lucovsky (1982). The theoretical results for the local densities of states indicate high-frequency local modes in a-Si$_{1-x}$H$_x$, as well as substantial resonant mode contributions at low frequencies in more concentrated alloys. In particular, polysilane configurations such as (SiH$_2$)$_3$ were found to yield substantial modifications of the low-frequency spectra derived from a-Si due to considerable H atom participation in these modes. As discussed in Section 7, the lower-frequency Raman spectra in concentrated alloys do not, however, indicate a substantial change in form upon alloying.

IV. Experiment

The opaque, thin film nature of a-Si, a-Ge, and their H alloys requires backscattering Raman measurements, which are schematically shown in Fig. 4 for incident electric fields E_I^V and $E_I^{H'}$. The prime on H corresponds to horizontal polarization in the scattering plane for nonnormal incidence, whereas V denotes vertical polarization perpendicular to the scattering plane, which is defined by $\mathbf{k}_I$ and the surface normal. The angles of incidence and refraction are shown as θ_I and θ_T, respectively, although for absorbing media the latter, as defined by the generalized Snell's law, is complex. The true backscattering geometry corresponds to $\theta_I = 0$. For high-refractive-index systems such as a-Si or a-Ge, however, the transmitted field is nearly parallel to the surface so that $E_T^{H'} \simeq E_T^H$, where H is specified by the analyzer to be along the x direction. For finite θ_I the transmission factors of Eq. (2) for

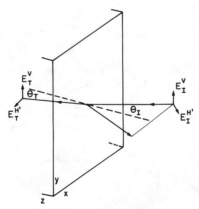

FIG. 4. Pseudobackscattering Raman geometry for incident fields E^V or $E^{H'}$, perpendicular or parallel to the scattering plane, respectively. The analyzer of the scattered field E_S is oriented parallel to x(H) or y(V). Here $\varepsilon_1 \gg \varepsilon_2$ is assumed.

incident H′ or V polarizations are (Klein, 1970)

$$|\tau_I^{H'}(\omega_I, \theta_I)|^2 = \frac{4|\tilde{\varepsilon}(\omega_I)|\cos^2 \theta_I}{[\tilde{\varepsilon}(\omega_I) \cos \theta_I + (\tilde{\varepsilon}(\omega_I) - \sin^2 \theta_I)^{1/2}]^2}, \tag{17}$$

$$|\tau_I^{V}(\omega_I, \theta_I)|^2 = \frac{4 \cos^2 \theta_I}{[\cos \theta_I + (\tilde{\varepsilon}(\omega_I) - \sin^2 \theta_I)^{1/2}]^2}, \tag{18}$$

whereas the scattered transmission factor $|\tau_S|^2$ for an isotropic solid at near normal incidence appropriate here is (Klein *et al.*, 1978)

$$|\tau_S|^2 = 4|\tilde{\varepsilon}(\omega_S)|/|1 + [\tilde{\varepsilon}(\omega_S)]^{1/2}|^2. \tag{19}$$

A schematic of the Raman experimental apparatus employed in the author's laboratory for either Ar or Kr excitation is shown in Fig. 5. Because the total photon counting intensity is the sum of inelastic and elastic processes plus photomultiplier dark counts, considerable rejection of the elastic or stray light contribution due to diffuse film reflectance is required. In a-Si and a-Ge this requires the use of a third monochromator or iodine filter or a combination of the two for an accurate estimate of low-frequency scattering, in which $\omega \lesssim 300$ cm^{-1}. The inelastic scattering, in addition to having first- and second-order Raman contributions, may contain a luminescence contribution, particularly in high H concentration a-Si$_{1-x}$H$_x$ alloys. Additional inelastic contributions from relaxation or tunneling processes, Brillouin scattering, or true Rayleigh scattering may also occur at very low frequencies.

Of experimental interest is the true backscattering depolarization ratio

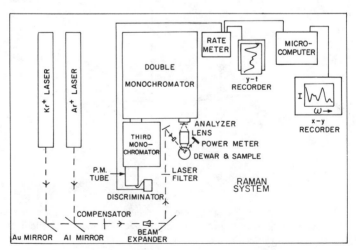

FIG. 5. Schematic of the Raman apparatus for Ar or Kr laser excitation.

$dp(\omega)$, which, for a fixed horizontal analyzer parallel to the x axis of Fig. 5, is defined as $dp(\omega) = I_{\text{VH}}(\omega)/I_{\text{HH}}(\omega)$. For the pseudobackscattering geometry in opaque solids this may be approximated by the relations

$$dp \simeq \frac{I_{\text{VH}}/|\tau_{\text{I}}^{\text{V}}|^2}{I_{\text{H'H}}/|\tau_{\text{I}}^{\text{H}}|^2}, \tag{20}$$

$$dp \simeq \left[\frac{1 - R_{\text{H'}}(\omega_{\text{I}}, \theta_{\text{I}})}{1 - R_{\text{V}}(\omega_{\text{I}}, \theta_{\text{I}})}\right]\frac{I_{\text{VH}}}{I_{\text{HH}}}, \tag{21}$$

where $R_{\text{H'}}(\omega_{\text{I}},\theta_{\text{I}})$ and $R_{\text{V}}(\omega_{\text{I}},\theta_{\text{I}})$, which represent the reflectivities for V and H' incident fields, may be determined experimentally. In sputtered a-Si the expression in square brackets in Eq. (21) has, e.g., a value of 1.1 for a typical $\theta_{\text{I}} \simeq 35-40°$ and an Ar laser excitation wavelength of 5145 Å.

An alternative means of studying the Raman spectra of very thin films of thickness $d \simeq 50-200$ Å employs the method of interference-enhanced Raman scattering (IERS) (Connell *et al.*, 1980). The IERS method employs a trilayer geometry to enhance the local Raman electric field so that enhancements by factors of about 5–20 in *total* counts relative to thick films have been observed. The condition for IERS corresponds to that of minimizing the reflectance of the film + dielectric spacer + metal reflector trilayer to obtain an antireflection coating condition at the incident laser frequency. This involves a computer program to determine the dielectric and film thicknesses for minimum reflectance. The enhanced signal in the IERS method has allowed small changes in the Raman spectra of a-Si$_{1-x}$H$_x$ with H evolution to be studied (Tsai and Nemanich, 1980), as well as the effect of annealing conditions on structural order in a-Ge (Yehoda and Lannin, 1983).

V. Results

6. RAMAN SCATTERING IN a-Si AND a-Ge

a. Raman Spectra

First-order Raman scattering measurements of a-Si and a-Ge were first reported by Smith *et al.* (1971a,b), who noted the correspondence of features in the reduced Raman spectra to the broadened crystalline densities of states. The similarities in the short-range order of the amorphous and crystalline phases and the predominance of short-range interactions on the lattice dynamics of c-Si and c-Ge make this result physically reasonable. High-frequency scattering as well as small changes with temperature of the reduced Raman spectra also indicated weak second-order scattering. Although small differences between peak positions in $\rho_{\text{r}}^{(\alpha)}(\omega)$ and $\rho(\omega)$ for the crystalline phases were initially attributed to changes in long-range interactions, subsequent analysis has suggested that these differences are a consequence of the factor $\bar{C}^{(\alpha)}(\omega)$ in Eq. (6). These initial measurements also suggested similar VH and HH components with a constant depolarization ratio of 0.8, implying that the $_3D$ coefficient in Eq. (14) might be neglected (Alben et al., 1975). Subsequent studies by Bermejo et al. (1977) indicated a ratio of 0.55 ± 0.05.

Attempts by Smith et al. (1971b) to determine changes in the Raman spectra with a variety of deposition conditions yielded similar spectra in both a-Ge and a-Si. Given the relatively small changes in short-range order that occur with deposition variations (Paul et al., 1973), this result might appear rather plausible. More recent detailed studies of the VH and HH Raman spectra of a-Si and a-Ge have, however, demonstrated that very substantial modifications of the spectra may occur as a consequence of changes in deposition conditions (Lannin et al., 1982; Maley et al., 1983a). The latter may arise from variations in substrate temperature, substrate bias, sputtering pressure, and more generally, the method of preparation. An example of this is exhibited in Fig. 6 in which the VH and HH Raman spectra at 300°K of a low-temperature sputtered film are compared to a high-temperature, heterogeneous CVD film (Kshirsagar and Lannin, 1981, 1982). The sputtered film was deposited near the substrate-support plate temperature† $T_p \simeq 120°C$, whereas the CVD film had $T_p \simeq 525°C$. Because CVD films with 0.2–0.7% H have similar spectra, they may be viewed as being highly dilute H-alloys that primarily reflect the a-Si host spectrum. Figure 6 indicates that the CVD film exhibits substantial changes in the high-frequency TO peak width Δ for both HH and VH components relative to the sputtered film, as well as large modifications in the relative intensity of the TA band. Figure 6 also illustrates significant differences in the VH and HH Raman components, emphasizing the importance of polarization analysis of the scattered light.

A systematic study of the influence of sputter deposition temperature on the Raman spectra of rf diode sputtered films deposited at low pressures indicates (Fig. 7) that a continuous variation of the TO widths Δ_{VH} and Δ_{HH} occurs (Lannin et al., 1982). For the depolarized component, variations in Δ_{VH} are attributed primarily to changes in the width of the high-frequency phonon density of states with deposition temperature. Such changes are ascribed to modifications of short-range order, particularly the bond-angle distribution. This is based on the similarity to changes in theoretical phonon densities of states of a-Ge given by Meek (1977) and shown in Fig. 3, wherein the TO band is observed to broaden substantially with an increase

† The actual film temperature is difficult to determine and may differ significantly from T_p.

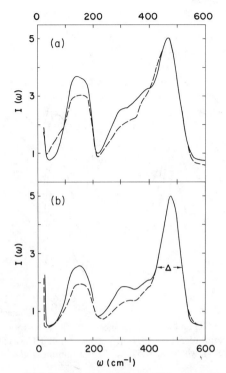

FIG. 6. Comparison of (a) low-T_p sputtered a-Si VH and HH Raman spectra and (b) heterogeneous CVD a-Si at 300°K. ———, HH; ————, VH. [From Kshirsagar and Lannin (1982).]

in the width of the bond-angle distribution. The qualitative similarity noted in Fig. 2 between $\rho(\omega)$ and $_2D$ and $_3D$ Raman mechanisms suggests that changes in Δ_{VH} are due primarily to variations in $\rho(\omega)$ rather than the Raman coupling parameters. Substantial changes in Δ_{HH} as well as in the relative intensity of the TA and LA peaks are also suggested to arise from modifications of the density of states as well as from possible variations in the Raman coupling parameters.

A more detailed analysis of the background in sputtered a-Si indicates that Δ_{VH} is directly proportional to the peak intensity ratio $(I_{TA}/I_{TO})_{VH}$ (Maley and Lannin, unpublished). This suggests that the relative changes in the low- and high-frequency intensities are primarily due to modifications of the TO band, which decreases in intensity on broadening. This implied relative constancy of the TA contribution is consistent with the theoretical density of states calculations of Fig. 3 which indicate rather small changes in the TA peak intensity up to relatively large disorder values of $\Delta\theta$. A comparison of the related changes of the VH Raman spectra of rf sputtered

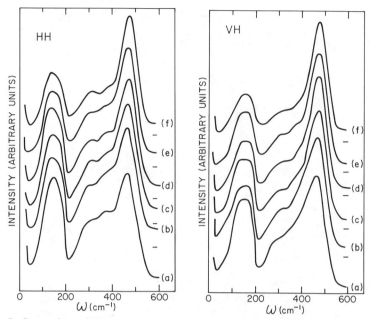

FIG. 7. Comparison of HH and VH Raman spectra at 300°K for sputtered a-Si as a function of substrate plate temperature T_p; (a) 30°C, (b) 200°C, (c) 400°C, (d) 500°C, (e) 560°C; (f) corresponds to CVD a-Si in which $T_p = 525$°C. [From Lannin *et al.* (1982).]

a-Ge studied by using the trilayer geometry (Yehoda and Lannin, 1983) indicates trends with annealing temperature very similar to those observed in a-Si. This is shown in Fig. 8 in which the a-Ge VH results indicate a relatively monotonic decrease in Δ_{VH} and I_{TA}/I_{TO} with the annealing temperature T_a up to the onset of crystallization at ~ 300°C. It is interesting to note that the Raman spectra in a-Ge also indicate that during crystallization the amorphous component continues to increase its short-range order. As in a-Si, VH Raman spectra of a-Ge indicate a proportionality between the Δ_{VH} and I_{TA}/I_{TO}. This primarily implies a decrease in the TA/TO ratio as Δ_{VH} narrows and may be understood as a consequence of an approximate conservation of the TO band area in the density of states with disorder. Figure 8 indicates that both the TO width and TA/TO intensity ratio decrease monotonically with T_a, implying a corresponding continuous change in the bond-angle distribution. Thus these results confirm experimentally that the VH Raman spectra reasonably indicate the form of the phonon density of states (Martin and Galeener, 1981). As such, the parameter Δ_{VH} provides a quantitative measure of the structural order of tetrahedrally coordinated amorphous solids. In addition, Δ_{VH} provides a means of comparing the local order in films prepared under different conditions and

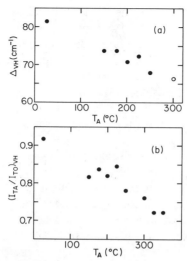

FIG. 8. Variation of (a) Δ_{VH} and (b) I_{TA}/I_{TO} with annealing temperature of an rf sputtered low-T_p a-Ge trilayer. [From Yehoda and Lannin (1983).]

by different methods if appropriate background subtraction and signal-to-noise ratio are obtained.

The depolarization ratio for the TO peaks in Fig. 7 is found from Eq. (21), and the measured intensities appear to be relatively constant with T_p, having a value $dp_{(TO)} \simeq 0.55 \pm 0.05$. This result, which is similar to the value obtained by Bermejo *et al.* (1977) on an rf sputtered film, implies relative constancy of this ratio with variations in structural order. This further suggests that the changes in the TO peaks for both VH and HH components for varying structural order primarily manifest changes in the phonon density of states. Alternatively, changes in the $_kD$ terms may be similar with disorder. The spectra in Fig. 7 further indicate that the depolarization ratio is not a constant with phonon frequency, as suggested by earlier studies (Smith *et al.*, 1971b; Bermejo *et al.*, 1977), but decreases at lower frequencies.

A comparison of the $_kD$ contributions of the Polk–Boudreaux model in Fig. 2 with the experimental results for the most ordered anneal-stable CVD a-Si spectra of Fig. 6 allows an estimate of the relative weighting factors $_kB$ discussed in Section 4 to be obtained. From Eqs. (15) and (16) the depolarization ratio may be obtained as $dp(\omega) = \overline{C}^{VH}(\omega)/\overline{C}^{HH}(\omega)$. If the depolarization ratios of the major TA and TO peaks are employed, one obtains the ratios $_1B^2 : _2B^2 : _3B^2 = 0.32 : 0.43 : 1.0$. However, for the LA band region these values yield a depolarization ratio about a factor of two smaller than the observed $dp_{(LA)} \simeq 0.39$. It is possible that this is a consequence of omission of interference terms $_iD \times _jD$ or the need to include an additional $_4D$ mechanism that relaxes the cylindrical symmetry constraint on $_kD$ (Alben *et al.*, 1975). Alternatively, the structural model, whose $\Delta\theta = 6.7°$ value is significantly below the estimated $\Delta\theta \simeq 9-9.5°$ of anneal-stable films, yields inappropriate $_kD(\omega)$ values. Figure 2 also indicates that the coupling parameters $\overline{C}^{(\alpha)}(\omega)$ obtained from Eqs. (15) and (16) are a relatively weak function of frequency except at lower ω, for which a more rapid decline is indicated. This is consistent with the present discussion and that of the coupling parameters in Section 2.

The Raman scattering measurements discussed in this and the following sections have primarily involved the use of 5145 Å Ar laser excitation. Bermejo *et al.* (1977) have noted that the integrated Raman intensity at this wavelength is greater by a factor of 20 than that of c-Si after corrections for optical transmittance differences. This is a consequence of resonant enhancement effects discussed in Section 1. Figure 9 illustrates that the resonance enhancement factor $|d\tilde{\varepsilon}/d\omega_I|^2$ for the integrated TO band intensity in a sputtered a-Si film agrees with the theoretical curve. Bermejo *et al.* also have noted a small variation in the TA/TO and LA/LO intensity ratios with excitation wavelength. Further background-reduced measurements at lower frequency are required, however, to determine the extent of this variation. Such a variation, if present, would formally require a generalization of Eq. (3) to include within the summation a resonance factor for each mode β.

b. Relation to Optical Gap and Structural Order

In Fig. 10 the inverse TO Raman widths Δ_{VH}^{-1} of a-Si and a-Ge films prepared by a number of methods are compared with their optical gaps E_0 (Maley *et al.*, 1983a). The latter were obtained from the Tauc extrapolation procedure in which the relation $(\alpha\hbar\omega_I)^{1/2} = C(\hbar\omega_I - E_0)$ was fit for small $(\alpha\hbar\omega_I)^{1/2}$ values between ~ 2 and 4 $(eV/\mu m)^{1/2}$. Because Δ_{VH} has been shown to be a measure of disorder, its inverse provides a convenient measure of phonon order to which the electronic order, via E_0, may be compared. Given the linear variation of Δ_{VH}^{-1} and E_0, Fig. 10 indicates a direct correspondence between changes in the vibrational spectra and the electronic, near-band-edge spectra for both a-Si and a-Ge. A similar variation of Δ_{HH}^{-1} versus E_0 in a-Si is also obtained after an appropriate subtraction of stray light background (Lannin *et al.*, 1982). The similar slopes in Fig. 10 for a-Si and a-Ge also imply an approximate scaling relation between changes in phonon and electron order parameters in these systems. These results also indicate, for the relatively high-mass-density sputtered films represented here, that changes in E_0 are also primarily structural in origin and not a consequence of changes in extrinsic, void-associated effects as has been

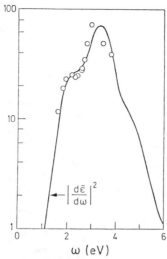

FIG. 9. Resonance enhancement of TO peak in rf sputtered a-Si compared to theory (solid line). [From Bermejo *et al.* (1977).]

suggested (Connell, 1979). The striking correspondence between the Raman and optical spectra, although requiring more detailed theoretical analysis on models of a-Si and a-Ge, is qualitatively consistent with the effect of narrowing the bond-angle distribution on increasing E_0 (Ching *et al.*, 1977) and narrowing the TO band. It has alternatively been suggested that E_0 may be a function of the dihedral angle distribution $P(\phi)$ (Cohen *et al.*, 1980). Because $P(\phi)$ is generally a function of $P(\theta)$, variations of the former may also be considered to occur with changes in structural order. Physically, the result noted in Fig. 10 may be viewed as a consequence of fluctuations in sp^3 orbital interactions with bond-angle distribution that result in modifications of the electronic and phonon densities of states. These fluctuations may also modify lower lying electronic gap states that are, however, expected to be more sensitive to changes in the tails of $P(\theta)$. Changes in the valence-band tails in selected ultraviolet photoemission measurements in a-Si (Fischer and Erbudak, 1971) are qualitatively consistent with variations of band tailing with structural order.

The correspondence between Δ_{VH}^{-1} and E_0 in Fig. 10 may also be employed to estimate the approximate variation in the width of the bond-angle distribution as a function of deposition conditions. Radial distribution function studies (Paul *et al.*, 1973) in low-pressure triode sputtered a-Ge films are found to yield thermally corrected $\Delta\theta$ values of $9.5° \pm 0.4°$ and $10.2° \pm 0.4°$ for films deposited at $T_p = 350$ and $150°C$, respectively. The former film had $E_0 = 0.9$ eV whereas the latter had $E_0 \simeq 0.7$ eV, so that a

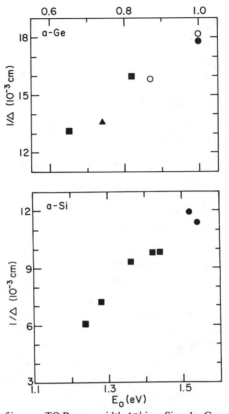

FIG. 10. Variation of inverse TO Raman width Δ_{VH}^{-1} in a-Si and a-Ge versus optical gap. The open symbols in a-Ge correspond to evaporated films. [From Maley *et al.* (1983a).]

gap variation here of ~ 0.2 eV corresponds to a change in $\Delta\theta$ of $\sim 0.7°$. The error range on the individual $\Delta\theta$ values is, however, rather large so that this estimate is clearly rather approximate. By using the Raman width and optical gap results of Fig. 10 and assuming a linear variation of Δ_{VH}^{-1} with $\Delta\theta$, a value $\Delta\theta \simeq 9.1°$ is implied for the most ordered anneal-stable evaporated a-Ge films (Thèye, 1974). Similarly, an extrapolation in highly disordered a-Ge films (Lannin *et al.*, 1983) in which $E_0 \simeq 0.5$ eV, indicates $\Delta\theta \simeq 10.9°$. Thus a total variation of $\sim 20\%$ in the bond-angle width is suggested for a linear variation of $\Delta\theta$ with Δ_{VH}^{-1}. The similar variation of Δ_{VH}^{-1} with E_0 in a-Si and a-Ge, as well as the experimental result that in evaporated a-Si $\Delta\theta \simeq 10°$ (Moss and Graczyk, 1970), suggests that a similar scale may be applicable as a first approximation in a-Si. More accurate radial distribution function measurements in the a-Ge and a-Si are required to obtain an improved estimate of the range of structural order in these systems. Raman studies on films of more accurately known $\Delta\theta$ values will then allow an improved determination of Δ_{VH}^{-1} with structural order.

7. ALLOYS

a. Nonhydrogenated Systems

Raman scattering in amorphous alloys of Si and Ge is primarily a function of short-range order and its modifications. The latter may involve changes in bond distances, bond angles, and coordination number as well as modifications of the electronic states that influence force constants and polarizability derivatives. In general, dilute alloys result in local or in-band resonant modes (Barker and Sievers, 1975), the former yielding relatively narrow features in the density of states. Resonant modes, which in contrast have a more delocalized character, particularly in crystalline solids, may also yield relatively narrow features in the scattering response. In more concentrated alloys the general tendency is to form distinct impurity-associated bands due to the broad range of local environments (Dean, 1972; Bell, 1972; Payton and Vischer, 1968). For example, for a random $A_{1-x}B_x$ semiconducting alloy, the opticlike bands may be viewed as arising approximately from A–A, A–B, and B–B bonds. An example of this is seen in rf sputtered low-T_p a-$Ge_{1-x}Si_x$ alloys whose reduced Raman spectra (Lannin, 1974) are shown in Fig. 11. The scattering configuration corresponds to an unanalyzed polarization that yields approximate averages of the VV and VH configurations for these spectra. In this alloy system the relatively similar force constants in Ge and Si are such that changes in the vibrational density of states are primarily a consequence of mass disorder. Figure 12 indicates the theoretical phonon density of states of a-$Ge_{1-x}Si_x$ calculated by Yndurian (1976) using a cluster Bethe-lattice method. Although the theoretical spectra are sharper than those of $\rho_r(\omega)$, due most likely to the absence of sufficient bond-angle disorder, the agreement between the three high-frequency peaks associated with Ge–Ge, Ge–Si, and Si–Si bonds and the trend with x is very good. At low frequencies the differences in the reduced Raman peaks and those of the density of states are attributed primarily to the substantial frequency dependence of the coupling parameter. This results in the lower TA band intensity as well as in a shift to higher frequency of the reduced spectra relative to $\rho(\omega)$ as discussed in Section 2. This is also noted in Fig. 11 by the small arrows that correspond to the crystalline TA peak positions obtained from inelastic tunneling studies (Logan *et al.*, 1964). The Raman spectra, in providing evidence of local bonding and cluster effects, also indicate that simplified single site effective-medium models for the alloy lattice dynamics are not satisfactory (Lannin, 1974). This conclusion is independent of $\overline{C}^{(\alpha)}(\omega)$, which primarily modifies relative

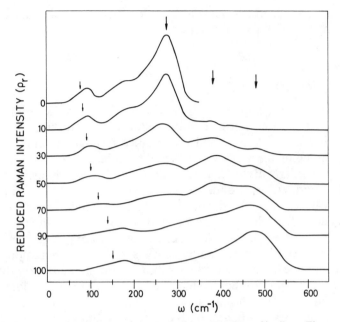

FIG. 11. Reduced Raman spectra of rf sputtered, low-T_p a-Ge$_{1-x}$Si$_x$ alloys. The numbers on the left indicate the atomic percent of Si. The large arrows indicate Ge–Ge, Ge–Si, and Si–Si optic bands. [From Lannin (1974).]

peak intensities at high frequency. In alloys, $\overline{C}^{(\alpha)}(\omega)$ at high frequency will also be inversely proportional to the reduced masses appropriate to the individual modes or bands (Leath and Goodman, 1969).

Similar Raman results for concentrated alloys indicating multiple bond formation have been obtained in a-Ge$_{0.5}$Sn$_{0.5}$ (Lannin, 1972) and a-Si$_{1-x}$As$_x$ (Nemanich and Knights, 1980) alloys, although in the latter the coordination of As is reduced to three. In these systems and in other alloys such as a-Si$_{1-x}$C$_x$ force-constant changes are more significant than in the a-Ge$_{1-x}$Si$_x$ system. In the Ge$_{1-x}$Si$_x$ system a correspondence between first-order amorphous and second-order crystalline Raman peaks is also observable (Lannin, 1974, 1977b) because a complete range of crystalline solid solubility exists. In contrast, the Ge–Sn system is essentially unique to the amorphous state. The Si–As system, as well as alloys of a-Si and a-Ge with elements not in group IV, may also be expected to have significant nonrandom local chemical ordering effects (Lucovsky, 1974; Brodsky, 1975).

Amorphous and microcrystalline alloys of Si with F and As have also been studied by Tsu et al. (1980). Theoretical analysis of a-Si–F and a-Si–F–H alloys (Agrawal, 1981) has indicated that the heavier mass of F results in local and resonant modes. The latter overlaps the Si host bands, complicat-

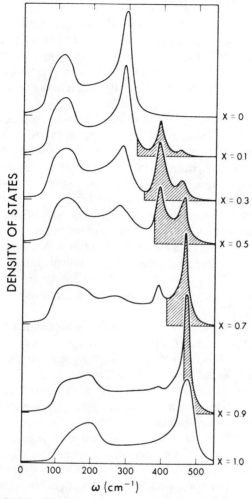

FIG. 12. Theoretical density of states of a-Ge$_{1-x}$Si$_x$. The shaded regions correspond to localized states in the theory. [From Yndurian (1976).]

ing an analysis of structural order. In addition, detailed low-frequency scattering has not been performed to date to ascertain the microcrystalline nature of these materials.

b. Hydrogen Alloys with a-Si and a-Ge

(1) *Low-Frequency Regime.* The low-frequency regime of a-Si$_{1-x}$H$_x$ and a-Ge$_{1-x}$H$_x$ alloys, which corresponds to the maximum host first-order phonon frequency and below, may be modified by the formation of in-band

resonant modes as well as by changes in short- or intermediate-range structural order about Si or Ge atoms. Because H atoms do not appear to form bridging bonds in these alloys (Wang *et al.*, 1982), as evidenced by the absence of A–H$\cdots$A infrared or Raman bending modes, its structural role is quite different from that of conventional alloys. This is a consequence, in chemical terms, of small electronegativity differences between Si or Ge and H as well as the monovalent nature of H.

Raman scattering evidence for in-band resonant modes has been obtained by Tsai and Nemanich (1980), who obtained a weak feature at ~ 210 cm^{-1}. This feature is shown in Fig. 13, in which their smoothed results are compared to Raman spectra of CVD a-Si containing less than 0.7% H (Kshirsagar and Lannin, 1982). A similar, though considerably more intense resonant mode is observed in infrared studies of a-Si$_{1-x}$H$_x$ and a-Ge$_{1-x}$H$_x$ alloys by Shen *et al.* (1980). The frequency of the resonant modes occurs at the minimum in the host density of states between the TA and LA bands as predicted theoretically by these authors. A number of additional resonant modes have been predicted by theoretical cluster models (Pollard and Lucovsky, 1982), although they have not been observed, possibly indicating small Raman coupling parameters or limitations on these models at low frequencies.

In addition to introducing local and resonant modes, it has been suggested that H atoms might increase structural order and reduce network strains. Direct evidence for this is provided by Raman scattering measurements in more disordered a-Si films as well as in H alloys. In particular, the high-frequency band width Δ_{VH} is found to be reduced at higher H concentration, as Fig. 13 indicates, suggesting a narrowing of the bond-angle fluctuations. A similar, though more substantial narrowing is also observed in more disordered sputtered a-Si as the H concentration is increased (Bermejo *et al.*, 1977; Ishidate *et al.*, 1982). A reduction of fluctuations in bond angle results in a decrease of local noncentral contributions to the strain free energy that may also explain the corresponding reduction in macroscopic strain in hydrogenated films. Because CVD films of low H concentration, as well as high-temperature annealed a-Si films, have Δ_{VH} values that are only $\sim 10\%$ below that of glow-discharge and CVD films of higher H concentrations, this implies that the role of H on structural order and strain relief may also be a function of the deposition process as well as the presence of H. In sputtered films, ionic and neutral atom bombardment result in increasing film disorder (Pilione *et al.*, 1983). In contrast, the CVD process, and to a lesser extent the glow-discharge method, results in reduced bombardment and associated network disorder.

The essential constancy of the frequency of the peaks of the TA and TO bands with H concentration indicates that the average force constants of neighboring Si atoms of a common tetrahedron joined to the H atom are

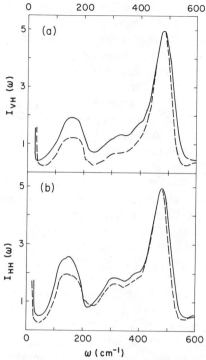

FIG. 13. Comparison of the (a) VH and (b) HH Raman spectra of GD a-Si$_{1-x}$H$_x$ ($x = 0.08$) (Tsai and Nemanich, 1980) and CVD a-Si (Kshirsagar and Lannin, 1982). ————, CVD; — — —, GD. [After Tsai and Nemanich (1980). Copyright North-Holland Publ. Co., Amsterdam, 1980.]

also not substantially modified. This is consistent with the predominance of local bending and stretching interactions that involve Si triads of atoms (Weber, 1974). The addition of H to a-Si also results, as Figs. 7 and 13 indicate, in a modification of the relative intensities of the TA and TO peaks for both VH and HH components in a manner similar to that of high-T_p sputtered or CVD a-Si films.

Raman scattering measurements (Maley *et al.,* 1983b) performed on a series of low H concentration a-Si$_{1-x}$H$_x$ alloys prepared by the homogeneous CVD process indicate that for $x \lesssim 0.04$ the VH width decreases with increasing H concentration. Because the H concentration increases in this CVD process as the substrate temperature decreases, the increased order is not a thermally associated effect but is attributed to the presence of H. At higher concentrations of H the Raman spectra for the TO band indicate, however, that the structural order in homogeneous CVD films is relatively constant. However, with increasing H concentration, changes do occur at lower frequency for the TA and LA bands. For example, Bermejo *et al.* (1977) have suggested that changes in the LA band may be associated with a decrease in the number of rings. The observation of changes in the TA band may imply a decrease in the number of Si triads necessary to sustain bond-bending modes. A decrease in the number of such triads would also naturally result in a decrease in the number of rings in the system. More systematic studies of the H concentration dependence are required to determine the possible origin of changes in the low-frequency spectra and address questions relating to changes in the LA band that may or may not be due to changes in intermediate-range order.

The influence of high pressure on the Raman spectra of a sputtered a-$Si_{0.76}H_{0.24}$ alloy has been studied by Minomura (1981) and Ishidate *et al.* (1982). The spectra were obtained in a diamond anvil device with the polarization of the scattered radiation unanalyzed. With increasing pressure, the low-frequency TA band was observed to shift to lower frequency and broaden. This shift was attributed to a softening of the bond-bending force constant with pressure. The high-frequency TO band, in contrast, was observed to shift to higher frequencies as well as to considerably narrow by $\sim 35\%$. While the authors attribute this narrowing to a reduction in bond-length fluctuations, it is more likely that a reduction in bond-angle fluctuations has occurred as in high T_p sputtered or CVD films, because bond-length fluctuations, $\Delta r_1/r_1 \simeq 1\%$, are relatively small in a-Si. In addition, the high-pressure film indicates a peak in the LO region near 410 cm^{-1}. It is thus possible that the influence of pressure in shifting the TO band, as well as the increasing short-range order, has allowed the observation of a distinct LO-like feature that has not been observed to date in a-Si, a-Ge, or their H alloys at normal pressures.

The introduction of H into a-Si is also found to produce structureless low-frequency Raman scattering below the TA peak (Lyon and Nemanich, 1983). It has been suggested that this structure does not arise from phonon processes but rather that it involves tunneling or two-level relaxation modes that have also been suggested for a-SiO_2 (Winterling, 1975). This additional scattering appears to be significant in higher H concentration films, suggesting modes that involve H atom displacements. The discussion here and in previous sections has assumed that voids do not appreciably modify the Raman spectra. In high H concentration films it is often found that a substantial low-density voidlike fraction is present unless appreciable particle bombardment occurs during film growth. Whether microscopic voids may play a role, particularly for very low-frequency excitations, is unclear. Thin films intentionally prepared under conditions so as to yield a large void fraction result in spectra that are similar to those of more dense films (Pilione *et al.*, 1983). This indicates that voids do not significantly modify the phonon spectra in these films.

A comparison of the first-order TO widths in the Raman spectra of Tsai and Nemanich (1980) with the corresponding second-order overtone frequency range indicates significant combination scattering in the H alloy. This is indicated by a peak width for the broad second-order band at 940 cm^{-1} that is $\sim 40-50\%$ greater than $2\Delta_{VH}$. Similar results are also clearly observed in more ordered a-Si films. This implies, as discussed in Section 3, that the high-frequency second-order scattering is not dominated by overtone processes, but contains significant combination contributions.

Structural effects of H in a-Ge$_{1-x}$H$_x$ that are similar to those discussed for Si alloys may also be expected. However, the increased stray light contribution in studies to date (Bermejo and Cardona, 1979) precludes an evaluation of changes in short-range order or of low-frequency spectral variations with H concentration.

(2) *Higher-Frequency Regime.* The addition of H to a-Si or a-Ge results in a series of high-frequency local modes with narrow bandwidths (Brodsky *et al.*, 1977; Tsai and Nemanich, 1980). A detailed discussion of this spectral range, which has been studied extensively by infrared absorption, is given in Chapter 4 by Zanzucchi and reviewed by Lucovsky (1981). In films of a-Si$_{1-x}$H$_x$ with lower H concentration, in which $x \lesssim 0.1$, Raman peaks are observed at ~ 620 and 2000 cm^{-1}, which are attributed, respectively, to local bending and stretching modes of Si–H bonds. With increasing H, additional features are observed that have been attributed to local vibrations of SiH$_2$ and (SiH$_2$)$_n$ or SiH$_3$ configurations. Of particular interest is the peak at 2100 cm^{-1} whose intensity relative to the peak at 2000 cm^{-1} increases with H concentration. An etching study (Nemanich *et al.*, 1980) of a glow-discharge prepared film of high H concentration indicates a decrease in the relative intensity of the 2100 cm^{-1} peak. This has been attributed to the reduction of (SiH$_2$)$_n$ configurations in intercolumnar regions of the microscopically inhomogeneous film. It has been noted (Paul, 1980) that the peak at 2100 cm^{-1} may also have contributions from other special Si–H configurations.

Resonant Raman scattering studies (Bermejo and Cardona, 1979) have also been performed in the high-frequency stretching mode regime, indicating normal resonant behavior at lower H concentrations and nonresonant behavior at higher concentrations. These resonance effects demonstrate that Raman scattering measurements are less satisfactory than infrared absorption measurements in yielding approximate film H concentrations. In contrast to the infrared studies, Raman scattering measurements also indicate a slower variation in the apparent cross sections of the 2000 and 2100 cm^{-1} peaks. In high H concentration films, the Raman spectra clearly exhibit both peaks, whereas the infrared spectra are dominated by the peak at 2100 cm^{-1} (Brodsky *et al.*, 1977).

8. Microcrystalline Materials

Microcrystalline Si and Ge provide interesting systems, along with crystalline alloys, for the study of disorder effects on Raman scattering. Raman measurements also provide a sensitive means of studying the crystallization of amorphous solids. In microcrystalline systems, for which the phonon eigenstates are represented by a wave vector $\mathbf{q}$, the Raman intensity in Eq. (3) is modified so that the first-order Stokes intensity is (Nemanich et al., 1981; Hayes and Loudon, 1978)

$$I_{(\alpha)}^{(1)}(\omega) \sim \sum_{\sigma,\mathbf{q}} C(\mathbf{q}, \omega_\sigma) \frac{n(\omega_\sigma + 1)}{\omega_\sigma} |F(\mathbf{q} - \mathbf{k})|^2 \delta(\omega - \omega_\sigma(\mathbf{q})). \quad (22)$$

Here the wave vector transfer $\mathbf{k}$ is not equal to $\mathbf{q}$. The factor $C(\mathbf{q},\omega_\sigma)$ represents the wave-vector- and frequency-dependent coupling parameter. The function $F(\mathbf{q} - \mathbf{k})$, which represents wave-vector relaxation, is the interference factor

$$F(\mathbf{q} - \mathbf{k}) = \int \exp[i(\mathbf{k} - \mathbf{q}) \cdot \mathbf{r}] \, d^3r. \quad (23)$$

For the diamond structure with cubic crystallites of size $\bar{L}$ this becomes (James, 1948)

$$|F(q - k)|^2 = |\sin^3 x/ \sin^3 (x/N^{1/3})|^2, \quad (24)$$

where $x = (k - q)\bar{L}/2$ and N equals the number of unit cells of size a. The interference factor has its minimum at $x = \pi$ or $q = k \pm 2\pi/\bar{L}$. It is thus reasonable to expect qualitatively, given the phonon dispersion in c-Si and c-Ge, that when $2\pi/\bar{L} \gtrsim (\frac{1}{2})\bar{q}_{BZ}$ a substantial contribution from all modes to the Raman scattering occurs. For an average Brillouin zone wave vector, $\bar{q}_{BZ} \approx 2$ Å^{-1}; this yields $\bar{L} \approx 6$ Å. In the limit of large $\bar{L}$, $F(q - k) = $ const, and crystal momentum conservation is recovered with $\mathbf{q} = \mathbf{k} - \mathbf{k}_S$.

The effect of wave vector relaxation in Eq. (20) is to admix Fourier components with $\mathbf{q} \neq 0$ into the Raman response. In c-Si and c-Ge the form of the dispersion curves of $\omega(\mathbf{q})$ for the optic branches is such that a shift and an asymmetry of the allowed Raman peak to lower frequencies occur. For values of $\bar{L}$ that are not too small, an approximate symmetrical Lorenztian is predicted (Nemanich et al., 1981; Richter et al., 1981). The asymmetry, however, will be significant for small values of $\bar{L}$ because the scattering response will have a high-frequency peak that approaches the position of the TO band density of states (Kanellis et al., 1980). This was observed in theoretical calculations by Alben et al. (1975) of a wurtzite microcrystallite of 64 atoms, corresponding to $\bar{L} \approx 11$ Å. Alben et al. also found that the width of the TO Raman peak was ~20% less than that of the narrow density-of-states peak, for which $\Delta \simeq 33$ cm^{-1}. Although this calculation involved a small number of atoms and did not include possible grain-boundary strain field effects, the results qualitatively indicate the trend in the limit of small crystallites. In particular, the results suggest that unless large grain-boundary strains that yield significant bond-angle distortions are present, the TO width of the depolarized Raman spectra for very small crystallites will be substantially less than that of true amorphous solids.

In the limit of small microcrystallites and large strain field distortions, the differences between the density of states of amorphous and microcrystalline material may become small. Structural and dynamical correlation effects may, however, provide for differences in the Raman and infrared response. At the present time, the smallest microcrystals studied of Si or Ge appear to have $\bar{L} \simeq 25-30$ Å (Iqbal et al., 1981), so that smaller values may be required to explore this boundary region. For example, it is possible that smaller $\bar{L}$ values are relatively unstable in presently prepared films because of the tendency to reduce noncentral local bond-angle distortions. An alternative thermodynamic criterion based on bond-length distortions has been discussed by Iqbal and Veprek (1982). The preceding discussion of line asymmetry and the variability of the amorphous state implies that attempts to fit the Raman line shape of microcrystalline films as a sum of crystalline and amorphouslike peaks are not well founded. Such a decomposition would be meaningful only if the material were a mixture of well-defined amorphous and microcrystalline systems.

For very small $\bar{L}$, disorder-induced Raman scattering from lower-frequency TA and LA phonons should occur. At the present time in Si and Ge no such features appear to have been reported in material that is solely microcrystalline, i.e., not a possible mixture with amorphous material. It is reasonable to expect, by analogy with the case of alloy-induced disorder discussed in Part I, that the relative scattering intensity of the TA band will be considerably less than that of the main allowed Raman band, except in the limit of small $\bar{L}$. This is a consequence of small anticipated $C(\mathbf{q}, \omega)$ values at low frequency since $_1D$ is weak here (Fig. 2) and $_2D$ and $_3D$ vanish as $\Delta\theta \rightarrow 0$. Alben et al. (1975) find for small $\bar{L} \simeq 11$ Å a theoretical reduced Raman intensity ratio for the TA band that is somewhat comparable to that calculated for a-Si and a-Ge.

It is useful to note that the width Δ_c of the Raman peak in microcrystalline Si in the studies of Iqbal et al. (1981) for $\bar{L}$ between 45 and 130 Å may be reasonably fitted by a simple relation of the form $\Delta_c \sim 1/\bar{L}$. A similar dependence of $\Delta_c \sim 1/\bar{L}$ is also observed in microcrystalline BN (Nemanich et al., 1981). An extrapolation of the c-Si results to lower values of $\bar{L}$ yields a value of $\Delta_c \simeq 33$ cm^{-1} for $\bar{L} = 11$ Å. This value is comparable to the theoretical Raman result of $\Delta_c \simeq 27$ cm^{-1} of Alben et $al.$ (1975) for the 64 atom microcrystalline model. In contrast, Fig. 6 indicates for CVD a-Si a minimum value of $\Delta_{VH} \simeq 83$ cm^{-1}. These results suggest that unless grain-boundary disorder effects are large, microcrystalline Si and Ge will yield Raman spectra that differ considerably from their amorphous counterparts.

VI. Conclusions

This review has emphasized Raman scattering as a sensitive probe of changes in the phonon states in a-Si, a-Ge, and their alloys, particularly those arising from small changes in short-range order. These changes, which are difficult to measure accurately in diffraction studies, have theoretically been shown to modify substantially the phonon density of states and thus the Raman spectra. As such, the Raman spectra provide a means of studying amorphous solids prepared under a wide range of deposition conditions in order to assess local structural order and its variation. The observation of a correspondence between changes in the depolarized VH Raman spectra in a-Si and a-Ge and the optical gap further emphasizes the importance of small changes in short-range order on near-band-edge electronic states. However, further theoretical studies of the relation between changes in phonon and electron states due to modifications of short-range order are required. In addition, the influence of variations in the Raman coupling parameters with short-range order requires further theoretical analysis. Experimental studies that determine more accurate changes in the radial distribution function, particularly in a-Si, as well as the phonon density of states, are of fundamental importance. Such studies will allow a more quantitative measure of structural disorder variations to be determined in future Raman scattering studies.

The role of H on structural order in alloys with a-Si and a-Ge has been discussed in somewhat less detail. Recent results imply that the particular thin-film deposition process determines the degree of local disorder, which the presence of H may directly or indirectly affect. Thus the role of H may differ in films deposited under conditions of enhanced high-energy particle bombardment relative to low-energy CVD prepared materials, for example. Further theoretical studies in such alloys that additionally address the influence of H clustering on the Raman spectra, as well as changes in intermediate-range configurations such as rings, are clearly of interest. Experimental studies on films of different micromorphologies may also be of value in addressing the effects of structural heterogeneities.

Microcrystalline Si and Ge films, as prepared to date, appear to differ qualitatively in the form of their Raman spectral response because of limitations on grain size that result in incomplete crystal-momentum relaxation. Further studies on small $\overline{L}$ systems are of basic interest in terms of understanding the role of disorder on the density of states and the Raman coupling parameters. There are, however, potential pitfalls here if mixtures of amorphous and microcrystalline systems are studied, because a separation of the scattering response of the two phases is not in general unique. It is possible that future preparation methods will allow the role of disorder in small microcrystallite systems to be explored. More detailed low-frequency measurements on small-$\overline{L}$ material currently available will also be of value.

REFERENCES

Agrawal, B. K. (1981). *Phys. Rev. Lett.* **46**, 774–778.
Alben, R., Weaire, D., Smith, J. E., Jr., and Brodsky, M. H. (1975). *Phys. Rev. B* **11**, 2271–2296.
Axe, J. D., Keating, D. T., Cargill, G. S., III, and Alben, R. (1974). *In* "Tetrahedrally Bonded Amorphous Semiconductors"(M. H. Brodsky, S. Kirkpatrick, and D. Weaire, eds.), pp. 279–283. Amer. Inst. Phys., New York.
Barker, A. S., Jr., and Sievers, A. J. (1975). *Rev. Mod. Phys.* **47**, *Suppl. 2*, 1–180.
Beeman, D., and Alben, R. (1977). *Adv. Phys.* **26**, 339–361.
Beeman, D., and Bobbs, B. L. (1975). *Phys. Rev. B* **12**, 1399–1403.
Bell, R. J. (1972). *Rep. Prog. Phys.* **35**, 1315–1409.
Bermejo, D., Cardona, M., and Brodsky, M. H. (1977). *In Proc. Int. Conf. Amorphous and Liquid Semicond. 7th* (W. Spear, ed.), pp. 343–348. Univ. of Edinburgh, CICL, Edinburgh.
Bermejo, D., and Cardona, M. (1979). *J. Non-Cryst. Solids* **32**, 405–419.
Brodsky, M. H. (1975). *In* "Light Scattering in Solids" (M. Cardona, ed.), Vol. I, pp. 208–253. Springer-Verlag, Berlin and New York.
Brodsky, M. H., Cardona, M., and Cuomo, J. J. (1977). *Phys. Rev. B* **16**, 3556–3571.
Carroll, P. J., and Lannin, J. S. (1983). *Phys. Rev. B* **27**, 1028–1036.
Ching, W. Y., Lin, C. C., and Guttman, L. (1977). *Phys. Rev. B* **16**, 5488–5498.
Cohen, M. H., Fritzsche, H., Singh, J., Yonezawa, F. (1980). *J. Phys. Soc. Jpn. Suppl.* **49**, *Suppl. A,* 1175–1178.
Connell, G. A. N. (1979). *In* "Amorphous Semiconductors" (M. H. Brodsky, ed.), pp. 73–111. Springer-Verlag, Berlin and New York.
Connell, G. A. N., and Temkin, R. J. (1974). *Phys. Rev. B* **9**, 5323–5326.
Connell, G. A. N., Nemanich, R. J., and Tsai, C. C. (1980). *Appl. Phys. Lett.* **36**, 31–33.
Cowley, R. A. (1971). *In* "The Raman Effect" (A. Anderson, ed.), Vol. 1, pp. 95–182. Dekker, New York.
Dean, P. (1972). *Rev. Mod. Phys.* **44**, 127–168.
Dowling, G., and Cowley, R. A. (1966). *Proc. Phys. Soc. London* **88**, 463–494.
Fischer, T. E., and Erbudak, M. (1971). *Phys. Rev. Lett.* **25**, 861–864.
Gompf, F. (1979). *In* "The Physics of Selenium and Tellurium" (E. Gerlach and P. Grosse, eds.), pp. 64–67. Springer-Verlag, Berlin and New York.
Hayes, W., and Loudon, R. (1978). "Scattering of Light by Crystals." Wiley, New York.
Iqbal, Z., and Vepřek, S. (1982). *J. Phys. C* **15**, 377–392.
Iqbal, Z., Vepřek, S., Webb, A. P., and Capezzeto, P. (1981). *Solid State Commun.* **37**, 993–997.
Ishidate, T., Inoue, K., Tsuji, K., and Minomura, S. (1982). *Solid State Comm.* **42**, 197–200.

James, R. W. (1948). "The Optical Principles of the Diffraction of X-Rays," p. 4. Bell, London.
Kamitakahara, W. A., Shanks, H. R., McClelland, J. F., Buchenau, V., Pintschovius, L., and Gompf, F. (1983). *Bull. Am. Phys. Soc.* **28**, 533.
Kanellis, G., Morhange, J. F., and Balkanski, M. (1980). *Phys. Rev. B* **21**, 1543–1548.
King, C. N., Phillips, W. A., and de Neufville, J. P. (1974). *Phys. Rev. Lett.* **32**, 538–541.
Klein, M. V. (1970). "Optics." Wiley, New York.
Klein, M. V., Holy, J. A., and Williams, W. S. (1978). *Phys. Rev. B* **17**, 1546–1556.
Kshirsagar, S. T., and Lannin, J. S. (1981). *J. Phys. Colloq. Orsay, Fr.* **42**, *Suppl. 12,* C6-54–56.
Kshirsagar, S. T., and Lannin, J. S. (1982). *Phys. Rev. B* **25**, 2916–2919.
Lannin, J. S. (1972). *Solid State Commun.* **11**, 1532–1527.
Lannin, J. S. (1973). *Solid State Commun.* **12**, 947–950.
Lannin, J. S. (1974). *In Proc. Int. Conf. Amorphous Liquid Semicond., 5th* (J. Stuke and W. Brenig, eds.), pp. 1245–1250. Taylor and Francis, London.
Lannin, J. S., Calleja, M., and Cardona, M. (1975). *Phys. Rev. B* **12**, 585–593.
Lannin, J. S. (1977a). *Phys. Rev. B* **15**, 3863–3871.
Lannin, J. S. (1977b). *Phys. Rev. B* **16**, 1510–1518.
Lannin, J. S., and Carroll, P. J. (1982). *Philos. Mag.* **45**, 155–165.
Lannin, J. S., Pilione, L. J., Kshirsagar, S. T., Messier, R., and Ross, R. C. (1982). *Phys. Rev. B* **26**, 3506–3509.
Lannin, J. S., Pilione, L. J., Maley, N., and Kshirsagar, S. T. (1983). *Bull. Am. Phys. Soc.* **28**, 531.
Leath, P. L., and Goodman, B. (1969). *Phys. Rev.* **181**, 1062–1069.
Logan, R. A., Rowell, J. M., and Trumbore, F. A. (1964). *Phys. Rev.* **136A**, 1751–1755.
Lottici, P. P., and Rehr, J. J. (1980). *Solid State Commun.* **35**, 565–567.
Lucovsky, G. (1974). *In Proc. Int. Conf. Amorphous Liquid Semicond., 5th* (J. Stuke and W. Brenig, eds.), pp. 1099–1120. Taylor and Francis, London.
Lucovsky, G. (1981). *In* "Fundamental Physics of Amorphous Semiconductors" (F. Yonezawa, ed.), pp. 87–103. Springer-Verlag, Berlin and New York.
Lyon, S. A., and Nemanich, R. J. (1983). *Physica Amsterdam* **117B/118B**, 871.
Maley, N., Pilione, L. J., Kshirsagar, S. T., and Lannin, J. S. (1983a). *Physica Amsterdam* **117B/118B**, 880.
Maley, N., Lannin, J. S., and Scott, B. A. (1983b). *J. Non-Cryst. Solids* **59/60.**
Martin, A. J., and Brenig, W. (1974). *Phys. Status Solidi B* **64**, 163–172.
Martin, R. M., and Galeener, F. L. (1981). *Phys. Rev. B* **23**, 3071–3081.
Meek, P. E. (1977). *In* "The Physics of Non-Crystalline Solids" (G. H. Frischat, ed.), pp. 586–590. Trans. Tech. Publications, Aedermansdorf, Switzerland.
Mills, D. L., Maradudin, A. A., and Burstein, E. (1970). *Ann. Phys.* (N.Y.) **56**, 504–555.
Minomura, S. (1981). *J. Phys. Colloq. Orsay, Fr.* **42**, *Suppl. 10,* C4-181–188.
Moss, S. C., and Graczyk, J. F. (1970). *In Proc. Int. Conf. Phys. Semicond., 10th* (S. P. Keller, J. C. Hensel, and F. Stern, eds.), pp. 658–662. U.S. Dept. of Energy, Washington, D.C.
Nelin, G., and Nilsson, G. (1972). *Phys. Rev. B* **5**, 3151–3160.
Nemanich, R. J., and Knights, J. C. (1980). *J. Non-Cryst. Solids* **35/36**, 243–248.
Nemanich, R. J., Biegelsen, D. K., and Rosenblum, M. P. (1980). *J. Phys. Soc. Jpn.* **49**, 1189–1192.
Nemanich, R. J., Solin, S. A., and Martin, R. M. (1981). *Phys. Rev. B* **23**, 635–636.
Paul, W. (1980). *Solid State Commun.* **34**, 283–285.
Paul, W., Connell, G. A. N., and Temkin, R. J. (1973). *Adv. Phys.* **22**, 529–633.
Payton, D. N., III, and Vischer, W. M. (1968). *Phys. Rev.* **175**, 1201–1207.
Pilione, L. J., Maley, N., Lustig, N., and Lannin, J. S. (1983). *J. Vac. Sci. Technol.* **A1**, 388–391.

Polk, D. E., and Boudreaux, D. S. (1973). *Phys. Rev. Lett.* **31**, 92–95.
Pollard, W. B., and Lucovsky, G. (1982). *Phys. Rev. B* **26**, 3172–3180.
Richter, H., Wang, Z. P., and Ley, L. (1981). *Solid State Commun.* **39**, 625–629.
Sen, P. N., and Yndurian, F. (1977). *Phys. Rev. B* **15**, 5076–5077.
Shanabrook, B. V., and Lannin, J. S. (1981). *Phys. Rev. B* **24**, 4771–4780.
Shen, S. C., Fang, C. J., Cardona, M., and Genzel, L. (1980). *Phys. Rev. B* **22**, 2913–2919.
Shuker, R., and Gammon, R. W. (1970). *Phys. Rev. Lett.* **25**, 222–225.
Smith, J. E., Jr., Brodsky, M. H., Crowder, B. L., and Nathan, M. I. (1971a). *In* "Light Scattering in Solids" (M. Balkanski, ed.), pp. 330–334. Flammarion, Paris.
Smith, J. E., Jr., Brodsky, M. H., Crowder, B. L., Nathan, M. I., and Pinczuk, A. (1971b). *Phys. Rev. Lett.* **26**, 642–646.
Steinhardt, P., Albert, R., and Weaire, D. (1974). *J. Non-Cryst. Solids* **15**, 199–214.
Suzuki, K., and Maradudin, A. A. (1983). *Physica Amsterdam* **117B/118B**, 529–531.
Temple, P. A., and Hathaway, C. E. (1973). *Phys. Rev. B* **7**, 3685–3697.
Thèye, M. L. (1974). *In Proc. Int. Conf. Amorphous and Liquid Semicond., 5th* (J. Stuke and W. Brenig, eds.), pp. 479–498. Taylor and Francis, London.
Thorpe, M. F. (1973). *Phys. Rev. B* **8**, 5352–5356.
Tsai, C. C., and Nemanich, R. J. (1980). *J. Non-Cryst. Solids* **35/36**, 1203–1208.
Tsu, R., Izu, M., Ovshinsky, S. R., and Pollak, F. H. (1980). *Solid State Commun.* **36**, 817–822.
Wang, Z. P., Ley, L., and Cardona, M. (1982). *Phys. Rev. B* **26**, 3249–3258.
Weaire, D., and Alben, R. (1972). *Phys. Rev. Lett.* **29**, 1505–1508.
Weaire, D., and Taylor, P. C. (1980). *In* "Dynamical Properties of Solids" (G. K. Horton and A. A. Maradudin, eds.), Vol. 4, pp. 1–61. North-Holland Publ., Amsterdam.
Weinstein, B. A., and Cardona, M. (1973). *Phys. Rev. B* **7**, 2545–2551.
Weber, W. (1974). *Phys. Rev. Lett.* **33**, 371–374.
Winterling, G. (1975). *Phys. Rev. B* **12**, 2432–2440.
Xinh, N. X. (1968). *In* "Localized Excitations in Solids" (R. F. Wallis, ed.), pp. 167–176. Plenum, New York.
Yehoda, J. E., and Lannin, J. S. (1983). *J. Vac. Sci. Technol.* **A1**, 392–394.
Yndurian, F. (1976). *Phys. Rev. Lett.* **37**, 1062–1065.
Zeyher, R., Ting, C. S., and Berman, J. L. (1974). *Phys. Rev. B* **10**, 1725–1740.

CHAPTER 7

Luminescence in a-Si:H

R. A. Street

XEROX CORPORATION
PALO ALTO RESEARCH CENTER
PALO ALTO, CALIFORNIA

I. Introduction

Luminescence is a technique widely used for studying localized states within the band gap of a semiconductor. The technique is therefore particularly applicable to amorphous semiconductors, since much of the interest in these materials relates to the localized states introduced by the intrinsic disorder or by specific defects. Indeed, in the past 15 years, such luminescence studies have proved to be one of the most successful means of exploring the detailed electronic properties of localized states. In this chapter we shall discuss what is presently known about luminescence in

197

hydrogenated amorphous silicon (a-Si:H). Particular emphasis will be given to the physical phenomena that are characteristic of the noncrystallinity of a-Si:H and that may be generalized to other amorphous semiconductors. The topics that remain controversial and that need further work will also be highlighted.

In early studies of amorphous silicon, luminescence was undetectable, because the density of defects was so large that nonradiative recombination was dominant. Hydrogenation by glow discharge or by reactive sputtering reduced the defect density, and it was quickly discovered that this material had strong luminescence (Engemann and Fischer, 1973). Since then, many studies of the emission have been reported, and from these, a clear picture of the various low-temperature recombination processes has emerged, including the different nonradiative processes. The recombination kinetics have also been studied and found to have distinctive properties related to the disorder. Luminescence has also proved to be a useful characterization tool for studying the effects of deposition conditions and postdeposition treatment on a-Si:H.

The luminescence process can generally be considered as comprising three distinct events in sequence. First, an electron–hole pair is excited, usually by the absorption of a photon. Second, the electron and hole relax down in energy by emitting a series of phonons, and usually end up in band-edge localized states. Finally, there is recombination, either by luminescence or by some nonradiative mechanism. Various aspects of the luminescence experiment explore each of these different events. In particular, the relaxation process, which is important in many studies of a-Si:H, can be studied in some detail.

Electron spin resonance (ESR) is another powerful technique for studying localized states (Biegelsen, 1980), and therefore it is not surprising that the two experiments have been closely coupled. Much of the ESR data are obtained with illumination at low temperature under the same experimental conditions as for luminescence. The two experiments, in fact, merge into one in measurements of spin dependent properties as is described briefly in Part V and in more detail by Morigaki in Chapter 4 of Volume 21C. Three different types of localized states are observed by ESR. The defect resonance at $g = 2.0055$ has been identified as Si dangling bonds (Brodsky and Title, 1969; Biegelsen, 1980). There are also resonances attributed to band-tail electrons at $g = 2.004$ and holes at $g \simeq 2.012$ (Dersch et al., 1981). The band-tail ESR is observed in doped a-Si:H and also in illuminated undoped material (Street and Biegelsen, 1980). As will become apparent, both the band-tail states and the dangling bonds are involved in the recombination. Luminescence and ESR data have proved to be a powerful combination, and together they have built a detailed model of the localized states in a-Si:H and their influence on the trapping and recombination.

1. EXPERIMENTAL ASPECTS

Some of the experimental requirements for studying luminescence in a-Si : H are different from those typically needed for crystalline semiconductors. The luminescence band is broad and generally featureless, extending from 0.8 up to 1.7 eV. Consequently, a low-resolution, low-numerical-aperture monochromator is required for greatest sensitivity. Since the spectrum extends into the infrared, a PbS or Ge detector is the most appropriate. Because the spectrum covers a wide range in energy, normalization for the response of the detection system is essential both to locate the peak correctly and to avoid being misled by structure in the spectrum that is really in the detector response. Such structure has been reported on a few occasions (Morigaki *et al.,* 1978).

The details of the sample growth have been described in many papers and are reviewed in Volume 21A. Samples need to be about 1 μm thick to ensure complete absorption for excitation energies of 2 – 3 eV. The physical structure of the sample is very important because it can be the origin of extraneous peaks in the luminescence spectrum. For example, it was quickly found that the luminescence spectrum was modulated by interference fringes (Engemann and Fischer, 1973). This effect can easily be removed by depositing the sample on ground glass. A less obvious effect is that the substrate absorption can also influence the spectrum (Street, 1981a). This is particularly noticeable for a crystalline silicon substrate that has its optical absorption edge in the middle of the a-Si : H luminescence band and results in gross distortion of the spectrum. The origin of this effect is that much of the luminescence is emitted after multiple reflections, and this component is quenched if the substrate is absorbing. As a result of these phenomena, the apparent luminescence intensity is very sensitive to the type of substrate. For example, using ground glass increases the intensity by about a factor of 3, whereas an absorbing substrate can reduce it by a similar amount. To compare the intensity of two samples, it is therefore important that similar substrates be used.

2. TUNNELING PROCESSES

Before describing the luminescence itself, one very important general property of amorphous semiconductors will be discussed — electronic tunneling. As we shall see, tunneling dominates the luminescence of a-Si : H and is a direct consequence of the localization of all the carriers in band-tail or defect states, so that recombination occurs between spatially separated sites. For the same reason, tunneling is important in many other electronic processes, of which hopping conductivity is probably the best known (Mott and Davis, 1979). It is therefore useful to outline the tunneling process and to analyze the effect of the distribution of the pair separation distances.

Tunneling can be either a radiative or a nonradiative process. An example of the former is the recombination of an electron and a hole in band-tail states, and the latter is typified by the capture of an electron into a defect. In either case, the tunneling rate p is given by

$$p = \omega_0 \exp(-2R/R_0), \tag{1}$$

where R is the separation of the electron and hole (or defect), R_0 the effective Bohr radius, and ω_0 a prefactor. This formula assumes that the wave functions fall off isotropically as $\exp(-R/R_0)$. In the simplest case, the appropriate value of R_0 in Eq. (1) is the larger of the values for the two states. Although the formula is an oversimplification, it has been widely applied and is probably valid for an amorphous semiconductor except when $R \lesssim R_0$.

The principal distinction between radiative and nonradiative tunneling is in the value of ω_0 in Eq. (1). For the radiative process ω_0 is of order $\sim 10^8$ $\sec^{-1}$. This value corresponds to the typical radiative rate for fully overlapping wave functions and for an allowed transition. The actual rate depends on the details of the wave functions, about which very little is known, and ω_0 could be up to an order of magnitude different. An allowed transition has always been assumed for these estimates, and this seems to be justified by experiment. The momentum selection rules are presumed to be relaxed so that only spin selection rules remain. Magnetic properties (see Part V) show that although both singlet and triplet states occur, the actual radiative rate is very close to the allowed singlet rate.

For a nonradiative transition, the prefactor ω_0 in Eq. (1) is assumed to be $10^{12}-10^{13}$ $\sec^{-1}$, corresponding to a phonon frequency. Depending on the exact mechanism of the transition, this rate could vary by many orders of magnitude, as has been discussed by several authors (Mott *et al.,* 1975). There are very few direct measurements of this prefactor for a-Si:H, and none at low temperature. This remains an important area for further study, although, as we shall see, the use of an inverse phonon frequency does give consistent results.

It is easy to see from Eq. (1) that a substantial distribution in the separation R for tunneling processes leads to a very broad distribution in the recombination rates. There are two important distribution functions that will be considered here. The first is that of the nearest-neighbor distances for a randomly distributed sea of electrons and holes or defects. Physically, this corresponds to recombination between states that are uncorrelated. Recombination will always take place with the nearest neighbor because of the strong R-dependence of Eq. (1). The distribution function is (Reiss, 1956)

$$G(R) = 4\pi R^2 N \exp(-4\pi R^3 N/3), \tag{2}$$

where N is the density of each type of state. This distribution is shown in

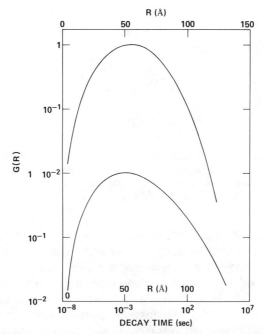

FIG. 1. The distribution function $G(R)$ corresponding to (upper curve) the nearest-neighbor distance for a random distribution of 10^{18} cm^{-3} states, and (lower curve) the geminate pair distances, assuming a Gaussian distribution with a mean of 50 Å. The decay times according to Eq. (1), assuming $\omega_0 = 10^8$ sec^{-1} and $R_0 = 10$ Å, are also shown.

Fig. 1 for $N = 10^{18}$ cm^{-3}. According to Eq. (1), R maps onto $\ln(\tau)$, where τ is the recombination time p^{-1}, so that Eq. (2) gives the distribution function for $\ln(\tau)$ for a particular value of R_0 and ω_0.

A particular application of this distribution function is for nonradiative tunneling of an electron to a defect (Tsang and Street, 1978). If this process occurs in competition with a radiative process of lifetime τ_R, then a *critical transfer radius* R_c can be defined such that

$$\tau_R = \omega_0^{-1} \exp(2R_c/R_0). \tag{3}$$

Therefore, R_c represents the distance at which the nonradiative and radiative rates are equal. Thus all those states for which $R < R_c$ will be predominately nonradiative, and all those with $R > R_c$ will be radiative. The luminescence efficiency is therefore given by

$$y_L = \int_{R_c}^{\infty} G(R) \, dR = \exp(-4\pi R_c^3 N/3). \tag{4}$$

The application of this formula will be discussed in Part IV.

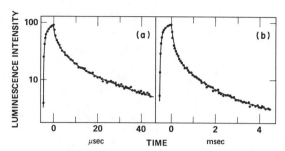

FIG. 2. Examples of the luminescence decay at 10°K, illustrating that a broad distribution of decay times results in a time dependence that apparently varies with the excitation pulse length. (a) Pulse length = 5 μsec. (b) Pulse length = 500 μsec. Note that the time scales in the two plots differ by a factor of 100. [From Tsang and Street, (1979).]

There are several other distribution functions that could be considered. For example, a Coulomb interaction between otherwise randomly distributed states will modify the nearest-neighbor distribution function at small R (Reiss, 1956). At the furthest extreme from the random distribution is a situation in which the electrons and holes are strongly correlated. The example we consider is that of a single electron and hole that are excited and then diffuse apart a small distance R and form a geminate pair. The distribution function is expected to be Gaussian (Street and Biegelsen, 1982):

$$G'(R) = (4R^2/\sqrt{\pi} R_s^3) \exp[-(R/R_s)^2], \qquad (5)$$

where R_s is the separation at the maximum of the distribution. This function is also shown in Fig. 1 for a value of R_s of 50 Å. It is interesting to note that the two functions discussed here are rather similar, although their physical origins are very different.

In view of these distribution functions and the tunneling mechanism, it is no surprise that the recombination times in a-Si:H span a very wide range. The measurement of the luminescence decay under these conditions can be very misleading and requires care. For example, Fig. 2 shows two examples of the decay after pulsed excitation. The results look very similar even though the time scales differ by a factor 100. It is a general property of a fairly flat distribution of recombination times that the apparent decay time scales with the excitation pulse length. It is therefore not sufficient to take the initial slope as an indication of the average decay time. The initial slope is closely related to the excitation pulse length and can be almost unrelated to the actual average decay time. Some early measurements reported decay times of $\sim 10^{-8}$ sec, whereas the peak of the distribution is in fact $\sim 10^{-3}$ sec, five orders of magnitude larger.

II. Radiative Recombination

3. LUMINESCENCE SPECTRA

Figure 3 shows examples of luminescence spectra of a-Si:H that are typical of the material. Figure 3a is a comparison of data at 10°K and high temperature for glow-discharge material. At low temperature, there is a single intense emission band near 1.4 eV with a full width at half height of 0.28 eV. When the temperature is increased, the luminescence intensity decreases, the peak moves to lower energy, and eventually a second peak near 0.9 eV is observed. Figure 3b shows some data for sputtered a-Si:H prepared at different bias voltages. The luminescence is at lower energy than that in Fig. 3a, the peak tends to be broader, and both energy and width of peak vary with the bias. It is generally the case that the dominant peak in both sputtered and glow-discharge material can lie between 1.2 and 1.4 eV, depending on the deposition conditions. This variation does not necessarily imply that the luminescence has many different origins but rather that the band gap and the disorder are varying due to changes in the microstructure, hydrogen content, etc. Several studies have shown how the structure and composition depend on the deposition process, but at present there is no detailed model to explain the changes. An extreme example of this variation is that of homogeneous CVD (homoCVD) films deposited at low temperature (Fig. 3c). The luminescence is now observed near 2.0 eV and has much weaker thermal quenching than for films deposited at a higher substrate temperature (Wolford *et al.*, 1983). Figure 3d shows the spectra of doped glow-discharge films. The 1.4 eV peak is much reduced in intensity and there is now a second peak at about 0.9 eV. The origin of the 0.9 eV luminescence is discussed further in Section 13.

Other luminescence bands have been reported, but will not be described in any detail here. For example, a weak peak at 1.1 eV, which is related to oxygen impurities, has been seen (Street and Knights, 1980), and presumably this could give information about deep oxygen impurity levels. However, at present there are insufficient data to extract this information. Luminescence above the gap has been reported (Shah *et al.*, 1980a), but subsequently this was shown to be due largely to surface hydrocarbon contamination (Wilson, 1981). More recently, a much weaker above-gap emission has been identified, which is attributed to recombination of hot carriers during thermalization (Nemanich and Lyon, 1984).

The discussion that follows will be concerned primarily with the 1.4 eV peak, since this dominates the spectrum and is the subject of most investigations. There is general (though not universal) agreement that this luminescence arises from transitions between band-tail states (Street, 1981a) and so is a consequence of the intrinsic disorder of the material. The energy of

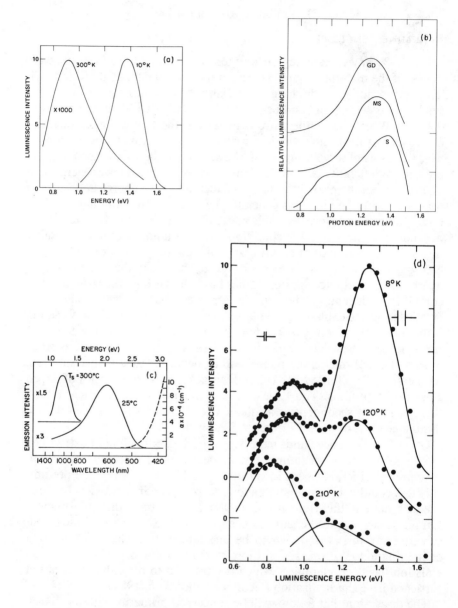

FIG. 3. Some examples of luminescence spectra of a-Si:H: (a) undoped glow-discharge material at 10°K and 300°K; (b) diode sputtered (S) and magnetron sputtered (MS) material compared with Dundee glow-discharge (GD) material (at 4°K) [from M. J. Thompson (1983), private communication]; (c) homoCVD (SiH$_4$) films prepared at room temperature [after Wolford et al. (1983)]; (d) doped glow-discharge films (3 × 10^{-3} [P]) [from Street (1980a)].

the emission relative to the band gap and the high density of recombination centers indicate that the states are close to the band edges. In addition, the peak is most intense in samples in which the defect density is lowest, which implies that intrinsic states (e.g., band tails) are involved. It is also clear from the long decay times that localized rather than extended states are involved, as is discussed in the next section. Aside from finding the origin of the radiative recombination centers, it is obviously important to understand the luminescence mechanism in terms of both the radiative event and the thermalization that precedes it. This information characterizes the electronic properties of the states and is relevant to many other measurements, such as electrical transport. Most of this chapter is in fact concerned with the recombination mechanisms. It will become apparent that many of the results support the interpretation of the 1.4 eV luminescence in terms of band-tail states.

4. DECAY RATES

The radiative decay rate is often the best means of determining the recombination mechanism because different processes have characteristically different rates. Figure 4 shows decay data for a-Si:H at low temperature, derived from pulsed measurements (Tsang and Street, 1979), and there is good agreement between this and many other sets of data (Austin et al., 1979; Kurita et al., 1979; Shah and DiGiovanni, 1981). The decay contains components that extend from 10^{-8} up to about 10^{-2} sec, and in fact the response can be observed up to 1 sec. Recent measurements have shown that the fastest decay component has a lifetime of ~ 8 nsec (Wilson et al., 1983). These data are from a low-defect-density sample in which the luminescence efficiency is high, and so the results represent the radiative decay times. The decay data can be inverted to give the distribution function for the lifetime, which is also shown in Fig. 4. The distribution has its peak near $10^{-3}-10^{-4}$ sec, with a width of 2–3 decades. Similar results have been obtained using a different technique in which the phase lag with respect to chopped excitation is measured as a function of frequency (Dunstan et al., 1982).

Considering that an exciton typically has a decay time of $10^{-8}-10^{-9}$ sec, the measured decays for a-Si:H are very long. The reason is that recombination occurs through well separated electrons and holes by a tunneling process, as given by Eq. (1). The recombination rate is decreased by the electronic overlap factor, which can be very small. Donor–acceptor pair recombination is the best known example of this type of decay (Thomas et al., 1965), and in fact the data show a striking resemblance to that of a-Si:H. Such a process is expected to occur when there is a large density of localized states at both band edges and so could be anticipated for a-Si:H. Within this

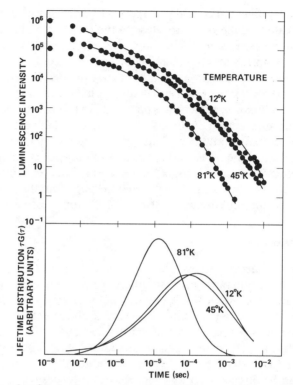

FIG. 4. Examples of the luminescence decay in a-Si: H (at an emission energy of 1.3 eV) after a short, weak illumination pulse, and the derived distribution of decay times. [From Tsang and Street (1979).]

model, the wide range of decay times is simply a consequence of the distribution of tunneling distances. It is easy to see a good qualitative similarity between the distribution in Fig. 4 and the examples in Fig. 1. In fact, the values of the parameters chosen in Fig. 1 are intended to match those of the radiative process.

There are two other explanations of the long decay times in a-Si : H, both of which can be ruled out fairly convincingly. The first is that of an exciton whose recombination is forbidden by spin selection rules. The best evidence against this comes from optically detected magnetic resonance (ODMR), which is described briefly in Part V and more fully by Morigaki in Chapter 4 of Volume 21C. The spin dependent behavior shows that although both singlet and triplet states are undoubtedly present, their exchange splitting is very small and the decay is overwhelmingly dominated by the allowed transition. The second model for the decay is that recombination occurs by thermal excitation from a shallow trap, so that the decay rate is characteristic of an activated excitation process rather than of the actual recombination, which can be much faster (Kurita *et al.*, 1979). The evidence against this model is that below about 30°K the decay rate is independent of temperature, showing that any thermal excitation mechanism has been frozen out (Street and Biegelsen, 1982). Above this temperature, the decay time decreases slightly, indicating that thermal excitation does have a weak effect.

The radiative tunneling model for recombination therefore appears to be firmly based. The observed distribution of decay times allows the electron–hole separations to be estimated from Eq. (1). Using a Bohr radius of 10 Å, which can be obtained from other luminescence data (Tsang and Street, 1979), results in an average electron–hole separation at the peak of the distribution of Fig. 4 of about 50 Å.

One conclusion of these results is that exciton recombination is insignificant in a-Si:H. The reason, presumably, is that the energy lowering that is achieved by the electron and hole moving apart to find deep tail states more than offsets the Coulomb attraction. The exciton binding energy can therefore be no more than about 0.1 eV. It is interesting that weakly disordered crystalline semiconductors generally exhibit the opposite situation—the recombination is by excitons, although the band is broadened by disorder and exhibits localization (Cohen and Sturge, 1982). Presumably, as the disorder is increased, a transition from excitonic to electron–hole pair behavior would be observed.

5. THERMALIZATION

The process of thermalization of the carriers down the band tails is a characteristic feature of amorphous semiconductors that is of importance in various experiments. For example, dispersive transport in a-Si:H is explained by progressively deeper trapping of carriers (Tiedje and Rose, 1981). Similarly, the luminescence properties will depend in part on how far and how fast the carriers move into the band tails. Several features of the luminescence exhibit the effects of thermalization. Figure 5 shows the variation of the peak energy with excitation energy E_x (Chen *et al.*, 1981; Shah *et al.*, 1982; Bhat *et al.*, 1983). Provided that E_x is greater than 1.8 eV, the luminescence peak energy is constant. However, when E_x is less than 1.8 eV, the peak energy decreases and the shift is roughly half of 1.8 eV $- E_x$. The general interpretation of these results is fairly obvious and can be described in terms of a "thermalization gap" of 1.8 eV. At high E_x, the electrons and holes are excited sufficiently far into the bands that thermalization always occurs, so that the luminescence peak shows no memory of the initial state. For low energy excitation, one or both carriers are excited into a deep localized state from which thermalization cannot occur, so that

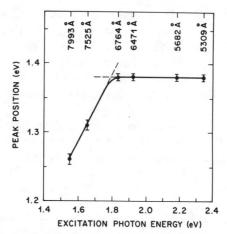

FIG. 5. Energy of the luminescence peak as a function of excitation energy E_x for measurements at 2°K in a-Si:H, showing the decrease of the peak when E_x is below 1.8 eV. [Reprinted with permission from *Solid State Communications,* Vol. 42, J. Shah, A. Pinczuk, F. B. Alexander, and B. G. Bagley, Excitation wavelength dependence of luminescence spectra of a-Si:H. Copyright 1982, Pergamon Press, Ltd.]

now the recombination retains a memory of the excitation energy. A detailed analysis of the process is complicated by the fact that the excitation involves the combined density of states of both band edges.

It is tempting, though probably incorrect, to associate the thermalization gap with the mobility edges (Shah *et al.,* 1982). Thermalization between localized states can easily occur within the radiative lifetime, provided that the density of states is at least 10^{18} cm^{-3} eV^{-1}. On the other hand, the density of states at the mobility edges is usually assumed to be $\sim 10^{20}$ cm^{-3} eV^{-1}. These values are all very uncertain, but they lead to the estimate that the mobility gap is roughly 0.2 eV above the thermalization gap.

Thermalization effects can also be seen in time-resolved luminescence spectra, examples of which are shown in Fig. 6. The luminescence moves to low energy with increasing time delay from 10^{-8} sec to 10^{-5} sec, and this has been interpreted in terms of thermalization of carriers (Street, 1980c; Wilson and Kerwin, 1982). Since thermalization to the mobility edge is expected to take only $\sim 10^{-12}$ sec, these longer times are associated with tunneling within band-tail states, as discussed previously.

After the thermalization occurs, the peak shift ceases and in some materials reverses direction. An example of this for sputtered a-Si:H is shown in Fig. 6 (Collins and Paul, 1982a), and a similar result is obtained in a-Si:O:H alloys (Street, 1980c). The increase has been attributed to the weak Coulomb interaction between the separated electron and hole, so that the luminescence energy is given by

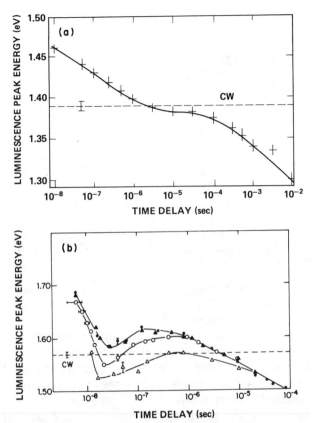

FIG. 6. Examples of the luminescence peak position observed in time-resolved measurements as a function of the delay time: (a) glow-discharge a-Si : H (12°K, E_x = 2.409 eV) (Tsang and Street, 1979); (b) sputtered a-Si : H, showing the effect of changing the excitation energy: E_x = 2.37 (●), 2.14 (▲), 1.97 (○), 1.88 (△) eV. [From Collins and Paul (1982a).]

$$E_p = E_0 + e^2/4\pi\varepsilon\varepsilon_0 R, \tag{6}$$

where E_0 is the energy in the absence of a Coulomb interaction. The reason that the Coulomb energy can be seen in the time-resolved spectra is that the decay time also depends on the pair separation R, according to Eq. (1). Note that for donor–acceptor pairs, the Coulomb interaction has the sign opposite to that of electron–hole pairs, because the donors and acceptors are charged in their lowest energy state. The peak energy decreases at much longer delay times, as seen in Fig. 6, because the deeper tail states have a smaller Bohr radius and therefore longer decay times (Street, 1980c).

The Coulomb interaction is not explicitly seen in the time-resolved spectra of glow-discharge a-Si : H (Fig. 6a), and its absence appears to be explained by the two other mechanisms that push the peak down. However,

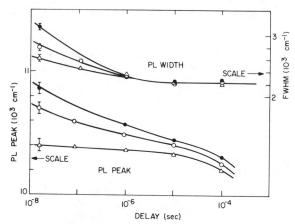

Fig. 7. Further measurements (at 12°K) of the time-resolved luminescence for glow-discharge a-Si:H at different excitation photon wave numbers: $\tilde{\nu}_x = 18{,}790$ (●), 13,560 (○), 13,170 (△) cm^{-1}. [From Wilson and Kerwin (1982).]

the results of some detailed experiments have failed to demonstrate the shift (Wilson and Kerwin, 1982). These measurements used low-energy excitation to remove the thermalization component, and the results in Fig. 7 show no sign of the Coulomb term. This result implies that either the effect is weaker than expected, because of a larger pair separation, or that one carrier (but not both) occupies a charged center. The latter possibility is in conflict with the intrinsic tail-state model and would require a large density of oppositely charged centers elsewhere in the gap for which there is little evidence. The conflicting results regarding this aspect of the time-resolved luminescence clearly require further clarification.

6. Geminate and Nongeminate Recombination

The concept of a geminate pair is particularly pertinent for an amorphous semiconductor and has been widely applied, particularly in regard to the photogeneration of photoconductivity (Knights and Davis, 1974). A geminate pair is an electron and a hole created by the same photon — recombination of this particular pair is termed geminate recombination. The alternative — nongeminate recombination — is a transition between carriers excited by different absorption events. The distinction sounds subtle but is highly significant for the detailed understanding of recombination, and in fact the two types of recombination can, in principle, be clearly distinguished. To understand how requires further discussion of the thermalization process.

During thermalization, the excited electron and hole will diffuse apart to a distance R_s — with distribution function $G(R_s)$ — that is characteristic of the thermalization process. The actual distribution expected for the separation is given by Eq. (5). If geminate recombination occurs, the decay times are given by Eq. (1) for this particular $G(R)$, so that information about the thermalization is obtained. The recombination is independent of all other absorption processes, so that it is independent of excitation intensity [see Eq. (5)], and the decay is therefore monomolecular.

On the other hand, if the excitation density is sufficiently large that different geminate pairs overlap, then the decay will not be geminate. An electron will recombine with the nearest hole at a distance D that depends on the density of electrons and holes. The appropriate distribution is that of nearest neighbors, and Eq. (2) gives the function for randomly distributed carriers. The decay now contains no information about the thermalization process [see Eq. (2)]. Moreover, because the separation D will vary with excitation intensity, so will the decay times, and the recombination will be bimolecular.

The distinction between geminate and nongeminate recombination is quite profound, and in fact the topic has been much debated. The first reported data looking at these effects are shown in Fig. 8, in which the luminescence decay at 1 msec after excitation is plotted versus excitation intensity I_0 for several excitation pulse widths (Tsang and Street, 1979). At low intensity, the decay is independent of I_0 but decreases with increasing I_0. These results are characteristic of the transition from geminate to non-

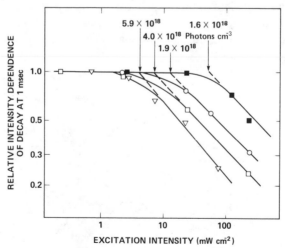

FIG. 8. Measurements of the luminescence intensity for a-Si:H at 12°K for a 1 msec decay time versus the excitation intensity. The change of slope indicates a transition from geminate (monomolecular) to nongeminate (bimolecular) recombination. The measurements are made for different excitation pulse lengths: 100 μsec (■), 500 μsec (○), 2 msec (□), 5 msec (▽). [From Tsang and Street, (1979).]

geminate recombination, as expected from the above discussion. The cross-over between the two types of recombination occurs at a pair density of about 10^{18} cm^{-3} for which the mean distance between pairs is about 50 Å. This measurement therefore provides a measure of the thermalization distance, because the change in mechanism occurs when $R_s = D$.

From the data in Fig. 8, it is seen that the geminate process occurs only at quite low intensities, ~ 1 mW cm^{-2}, mainly because the long lifetimes make it easy to generate a fairly large density of excited carriers. The use of short excitation pulses makes it easier to stay within the geminate regime, so that the data in Fig. 4 should apply to geminate pairs, although Dunstan has argued that this may not be so and has given a different interpretation to the data (Dunstan, 1982).

Two experiments have challenged the existence of the geminate regime. One is a different measurement of the decay time distribution, which claims to find no region in which the decay is independent of I_0 (Dunstan *et al.*, 1982). However, the shift of τ with I_0 gets very weak at low intensity, which would seem to indicate at least that geminate recombination is dominant. The other experiment is the measurement of light-induced ESR (LESR) (Boulitrop and Dunstan, 1982). As described in the introduction, low-temperature illumination reveals paramagnetic states characteristic of the band tails. These have been associated with the luminescence transition through such measurements as their temperature dependence (Street and Biegelsen, 1980). However, the data in Fig. 9 show that the LESR spin density depends very weakly on intensity, as is characteristic of a bimolecular process extending to very low intensities. No sign of geminate recombination is observed.

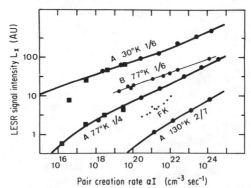

FIG. 9. The intensity dependence of the LESR showing behavior characteristic of bimolecular recombination. The LESR comprises the band-tail electron and hole resonances. [Reprinted with permission from *Solid State Communications,* Vol. 44, F. Boulitrop and D. J. Dunstan, Nongeminate radiative recombination of amorphous silicon. Copyright 1982, Pergamon Press, Ltd.]

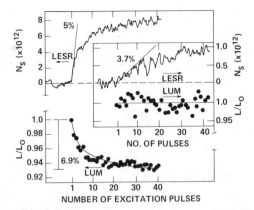

FIG. 10. The LESR and luminescence response to a sequence of excitation pulses separated by 12.5 sec. The data are shown for two values of the excitation intensity, with the right-hand data corresponding to the lower intensity. The quantum efficiency for generating LESR spins is indicated. [Reprinted with permission from *Solid State Communications,* Vol. 44, R. A. Street and D. K. Biegelsen, Distribution of recombination lifetimes in amorphous silicon. Copyright 1982, Pergamon Press, Ltd.]

Subsequently, further experiments were performed that seem to clarify the issue (Street and Biegelsen, 1982). Figure 10 shows the LESR and luminescence response to a series of weak excitation pulses, each separated by about 10 sec. The observation that the luminescence remains constant or decreases for succeeding pulses confirms that the process is indeed geminate. The slow increase in LESR exhibits properties of a nongeminate process, except that the generation efficiency for spins is only about 4% instead of 100%, as is expected for a purely nongeminate process.

A consistent explanation of the data can be made by using the distribution function for the geminate pair separation and decay time shown in Fig. 1. The important feature of this distribution is the very wide range of decay times. Even if the peak of the distribution corresponds to a decay time of 10^{-3}, about 7% of the pairs would have a decay time exceeding 10 sec, and these pairs are essentially metastable from the point of view of geminate recombination. With continued excitation, these metastable pairs build up in concentration until nongeminate recombination occurs. Under the excitation conditions shown in Fig. 10, the nongeminate component is about 5%. As the intensity increases, this fraction will also increase, but relatively slowly, so that the transition between the two regimes is very gradual. At most experimentally accessible intensities, components of both geminate and nongeminate recombination are present. The reason that LESR observes only the nongeminate component is because the time-averaged spin density is weighed by the decay time, and so the experiment is most sensitive to the slow bimolecular component. In this way the two experiments can measure the same type of states and yet find very different results.

The unusual results of these studies are a consequence of a particular feature of the decay distribution function, namely, that the average lifetime

$$\bar{\tau} = \int G(t)t \, dt$$

is very different from the median lifetime τ_m. For the distribution in Fig. 1a, $\tau_m \sim 10^{-3}$ sec, whereas $\bar{\tau}$ is many orders of magnitude larger. The two experiments, luminescence and LESR, in essence measure these two averages. It is interesting that the unusual properties of dispersive transport are also related to a similar effect in the distribution of thermally activated release times from an exponential band tail. This phenomenon seems to be unique to disordered systems.

7. THE DISTRIBUTION OF BAND-TAIL STATES

For reasons given in the next section, it is hard to obtain much information about the energy distribution of recombination centers directly from the luminescence spectrum. However, some conclusions can be drawn from the thermal quenching that occurs at temperatures above about 50°K. At these temperatures, carriers (probably electrons) are excited to the mobility edge and so can diffuse away to a nonradiative site. The thermal quenching is therefore a measure of the binding energy of the states. Figure 11 shows examples of the thermal quenching plotted according to the relation

$$[(1/y_L) - 1]^{-1} = y_0 \exp(-T/T_0), \tag{7}$$

where y_0 is the luminescence intensity in the low temperature limit and T_0 is a constant (Collins *et al.*, 1980b). This is an unusual temperature dependence in that the exponential term is not that of a Boltzmann activated process. In fact, this form of quenching is widespread in amorphous materials and was first observed in chalcogenide glasses (Street *et al.*, 1974).

The interpretation of Eq. (7) in a-Si : H is based on the activated release of carriers from an exponential band tail $\exp(-E/E_0)$, where E is the energy below the mobility edge and E_0 is a constant. For a particular value τ_R of the radiative decay time, there is an energy E_D given by

$$\tau_R^{-1} = \omega_0 \exp(-E_D/kT) \tag{8}$$

at which the thermal excitation rate equals the radiative rate. All states deeper than E_D will have a high probability of radiative recombination, whereas shallower states will be excited to the mobility edge and will diffuse away to a nonradiative center. Based on this simple model, it is easy to show that the luminescence efficiency is given by

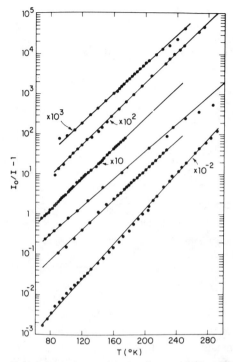

FIG. 11. The temperature dependence of the luminescence intensity plotted to show the fit to Eq. (7). Various samples of a-Si:H are shown. [Reprinted with permission from *Solid State Communications,* Vol. 34, R. W. Collins, M. A. Paesler, and W. Paul, The temperature depedence of photoluminescence in a-Si:H alloys. Copyright 1980, Pergamon Press, Ltd.]

$$y_L = \exp(-E_D/E_0). \tag{9}$$

Substituting for E_D from Eq. (8) yields an expression of the form of Eq. (7) (for $y_L \ll 1$) with

$$T_0 = E_0/k \ln(\omega_0\tau_R). \tag{10}$$

Assuming that $\tau_R = 10^{-3}$ and taking the observed value of $T_0 = 23°K$ from Fig. 11 give $E_0 \sim 0.04$ eV, which is of the expected magnitude for the band tails.

Further information concerning the distribution of states is obtained from the temperature shift of the luminescence peak, as shown in Fig. 12 (Tsang and Street, 1979). The peak shifts to lower energy much faster than the band gap, and again an explanation can be given in terms of the band tailing (Street, 1981b). Following the preceding discussion, the luminescence peak should have the same temperature dependence as E_D and

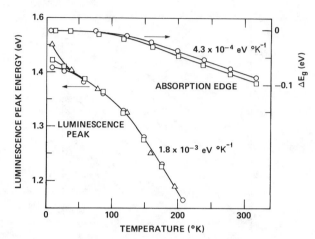

FIG. 12. The temperature dependence of the luminescence peak energy for a-Si:H compared to the much weaker dependence of the optical band gap. Absorption edge: $\alpha = 10^5$ cm^{-1} (O), 10^3 cm^{-1} (□). Luminescence data is for different samples and excitation energies. [From Tsang and Street (1979).]

therefore a temperature coefficient of $k \ln(\omega_0 \tau_R) \simeq 20°$K. This value of 1.7×10^{-3} compares well with the observed results. Although the model to explain these results is oversimplified to the extent of neglecting the distribution of decay times and ignoring the details of the diffusion of carriers (Hong *et al.*, 1981), the data seem to be good evidence that the carriers are within the exponential band tails.

8. PHONON INTERACTIONS

The reason that the luminescence energy cannot be used immediately to deduce the energy distribution of tail states is because of the complicating effects of phonon interactions. Luminescence generally occurs as the combined emission of a photon and some phonons. Since only the photon energy is measured in the spectrum, the luminescence does not necessarily correspond to the total energy difference between the ground and excited states. The same effect moves the optical absorption of the transition to high energy, and the difference between the luminescence and absorption energies is called the Stokes shift. The origin of the phonon component is the electron–phonon interaction. The effect has been known for many years and is well understood (Fowler, 1968). The simple semiclassical analysis in terms of a configurational coordinate is sufficient to understand most of the effects.

The magnitude of the phonon coupling in the luminescence can have a wide range. For example, shallow excitons in crystalline GaAs show almost no effect; most of the luminescence is in the zero-phonon transition, and there is essentially no Stokes shift. Under such conditions of weak coupling, the phonon side bands, if present, are usually seen as discrete lines corresponding to particular zone center phonons, of which the LO phonon usually dominates. At the other end of the scale are materials such as alkali halides in which the Stokes shift can be 1 eV or more. This shift corresponds to about 20 phonons, and typically the luminescence is a broad band in which the individual phonon lines are unresolved.

The difference between the two extremes is not so much in the strength of the electron–phonon interaction as in the degree of localization of the states. The magnitude of the Stokes shift depends on a trade-off between the electron–phonon interaction, which reduces the energy of the excited state by distorting the lattice, and the lattice energy, which increases quadratically with the distortion. Increasing the volume over which the distortion occurs results in a much larger lattice energy, which in turn reduces the equilibrium distortion. Hence the reason that excitons in GaAs are weakly coupled is that the radius of their wave function is ~ 100 Å compared with a few angstroms for an exciton in an alkali halide. It is generally the case that an increased electronic binding energy goes along with a larger Stokes shift. For this reason, one would expect a significant Stokes shift in amorphous silicon for which the binding energies of electrons and holes are in the range 0.1–0.5 eV.

Measuring the magnitude of the Stokes shift in a-Si:H has proved to be difficult and controversial. In general, there are two types of measurements that could be used. One method is to observe the phonon side bands directly as structure in the spectrum, and this has been performed in a-C:H (Lin and Feldman, 1982). Although the spectrum is featureless when excited at high energy, structure appears when the energy is reduced, and an example is shown in Fig. 13. The appearance of such structure is a well-known effect in systems with a large inhomogeneous broadening by disorder, and the use of resonant excitation in this way is often termed site selection spectroscopy. When excitation is below the thermalization gap (see Section 5), a narrow band of excited states is excited resonantly and recombines with little or no thermalization. In this way, the inhomogeneous broadening is reduced and structure in the homogeneous line (e.g., phonon side bands) can be seen. In a-C:H after resonant excitation, the structure is still very diffuse but seems to be associated with C–H vibrations and so is indicative of a substantial Stokes shift. To date, no such structure has been reported in a-Si:H.

The second method of evaluating the phonon interactions is to measure the Stokes shift — the energy difference between absorption and emission. At first sight, the existence of a Stokes shift seems obvious. The luminescence peak is at 1.4 eV, whereas at 10°K the absorption does not reach

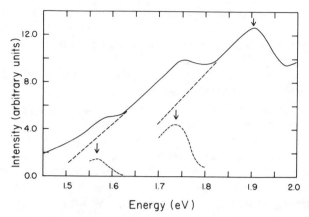

FIG. 13. The luminescence spectrum of a-C:H excited at 2.10 eV, showing side bands with separation 0.175 eV from the peak. The structure is not observed for excitation energies above 2.4 eV. The structure is interpreted as phonon side bands of C–H wagging modes. [From Lin and Feldman (1982).]

10^4 cm^{-1}, a value often associated with the mobility edge, until 2.0 eV. However, this comparison is not really valid. The luminescence is known to originate from electrons and holes below the mobility edge, but their exact energy is not known. The temperature dependence suggests that the electron binding energy is ~0.1 eV, but there is not much definitive information about holes. In addition, the absorption extends down in energy to well below the luminescence energy, although at very low values. The difficulty is therefore in determining how the comparison between absorption and luminescence should be made.

Several attempts have been made to explain the luminescence without a Stokes shift (Engemann and Fischer, 1974; Pankove et al., 1980; Collins et al., 1980a; Dunstan and Boulitrop, 1981; Dunstan, 1982). For example, the luminescence has been associated with a shoulder in the absorption of ~1.3 eV, although this was subsequently shown to be due to deep defects (Jackson and Amer, 1982) and alternatively to other features in the density of states, which were also subsequently questioned. Calculations of the absorption within a rigid band model (no phonon coupling) find reasonable agreement with the data (Dunstan and Boulitrop, 1981; Dunstan, 1982), but the uncertainties in the energies used to fit the absorption and luminescence are large enough to accommodate a substantial Stokes shift, so that the results are inconclusive.

The problem of a quantitative comparison of luminescence and absorption can be overcome in principle by applying detailed balance arguments. It has been shown that the absorption coefficient α is related to measurable luminescence quantities by the expression (Street, 1981a)

$$\int \alpha \, dv \geq y_L G(g_2/g_1)(\lambda^2/8\pi n^2), \tag{11}$$

where y_L is the luminescence efficiency, G is the generation rate, g_1 and g_2 are degeneracy factors, λ is the wavelength, and n is the refractive index. The integration is over the frequency (v) spectrum of the luminescence band. This expression is a consequence of the relation among the absorption strength, the radiative decay time, and the density of states, and is a statement of the fact that one cannot excite more electron–hole pairs than the density of states will allow. Here $y_L G$ is the absolute luminescence intensity, so the maximum experimental value of this quantity gives a lower limit on α.

Figure 14 shows the results of such a measurement (Street, 1978). Saturation of the luminescence was observed, and using Eq. (11), the equivalent lower limit on the absorption was calculated; this is shown for the known shape of the luminescence band in Fig. 14. The resulting peak absorption is

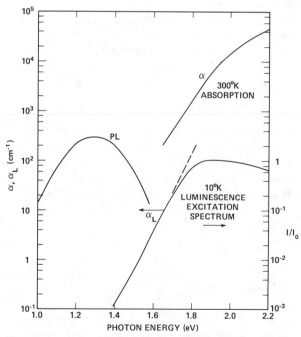

FIG. 14. Luminescence and excitation spectra compared with the room temperature absorption spectrum. The intensity of the luminescence band is obtained from detailed balance as discussed in the text. [From Street (1978).]

~ 300 cm^{-1}. Measurement of the actual absorption was made by using the luminescence excitation spectrum. This method was particularly appropriate since only absorption that can excite luminescence is required for Eq. (11). It was found by experiment that at 1.4 eV the absorption was only 0.1 cm^{-1}, which is a factor of 3000 below the lower limit given by the luminescence. The discrepancy can be resolved by postulating a Stokes shift. For example, if the shift was 0.4–0.5 eV, the absorption band corresponding to the luminescence would be shifted up to ~ 1.8 eV, and Fig. 14 shows that consistent results would be obtained. The conclusion was made that a Stokes shift of this magnitude indeed occurred.

The conclusion has been challenged, although no convincing alternative explanation of the above experiment has been given. For example, the results would be invalidated if there were carrier diffusion out of the generation region. However, a diffusion length greater than 10^{-2} cm would be needed, which is far above any measurements in a-Si:H. Nevertheless, the result is sufficiently important for further investigation to be needed. One implication of the Stokes shift is that electrons and/or holes must undergo substantial lattice distortion. Of the two, holes are the more likely candidates because of their greater localization. A band-tail hole corresponds to an electron taken out of a bonding orbital. A weakening of the bond and relaxation of the network are therefore reasonable expectations. The Stokes shift will also influence the absorption edge and so should be considered in any model to account for the Urbach tail.

III. Nonradiative Recombination

There are four types of nonradiative mechanisms that account for most of the nonradiative recombination in semiconductors. These are multiphonon transitions through deep states, Auger recombination, surface recombination, and thermal ionization. In a-Si:H, each of these mechanisms is observed, and together they seem to explain all the observations. As we shall see, the relative importance of each process is different from that for a typical crystalline semiconductor.

9. TUNNELING TO DEFECTS

In undoped a-Si:H, the dangling bond density, measured by the $g = 2.0055$ ESR line, can be made to vary over a wide range by a suitable choice of deposition conditions (Street *et al.*, 1978b). The typical defect density varies from 10^{15} up to 10^{19} cm^{-3}, with substrate temperature, gas composition, and rf power being the most important controlling variables for plasma deposited samples. Figure 15 shows the relation between the dangling-bond density N_S and the luminescence intensity I_L for measure-

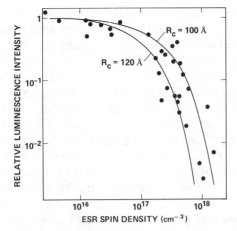

FIG. 15. The dependence of the luminescence intensity on the dangling-bond density as measured by ESR in undoped a-Si : H. Each data point corresponds to samples of different deposition conditions. [From Street *et al.* (1978b).]

ments at $10°$K. Provided that N_S is below 10^{-17} cm^{-3}, I_L is large and more or less independent of N_S. However, when N_S exceeds 10^{-17} cm^{-3}, I_L decreases rapidly, becoming undetectable for $N_S > 10^{18}$ cm^{-3}. Since unhydrogenated a-Si typically has a dangling-bond density of $\sim 10^{19}$ cm^{-3}, it is no surprise that this material does not exhibit luminescence.

The quenching of luminescence by a large defect density establishes dangling bonds as an important nonradiative center. The mechanism of recombination is deduced to be direct capture from band-tail states, because the luminescence intensity is independent of temperature up to about $50°$K. Since radiative recombination between band-tail states is by tunneling at low temperature, it is natural to expect a similar mechanism to apply to the nonradiative process. In Section 2, nonradiative tunneling for a random (uncorrelated) distribution of centers was analyzed, and this case applies directly to the present problem. The expression for the luminescence efficiency given by Eq. (4) is plotted in Fig. 15 and can be seen to give an excellent fit to the data. In particular, the tunneling model explains why there is an apparent threshold of 10^{17} defects before nonradiative capture becomes significant. The fit to the tunneling formula results in a critical transfer distance R_c of ~ 100 Å. This means that defects closer than 100 Å to an electron–hole pair will cause nonradiative recombination, whereas those farther away will be ineffective. Using Eq. (3), one obtains an effective Bohr radius of 10 Å for the tunneling process, and this value agrees well with the estimates of the radius in Section 4. The nonradiative tunneling is effective over a greater distance than is the radiative tunneling because the prefactor

ω_0 in the tunneling rate is 4–5 orders of magnitude larger. On the other hand, the luminescence of a-Si : H is less sensitive to defects than is generally the case in crystalline semiconductors because the capture rate for free, mobile carriers is much larger than for carriers in band-tail localized states.

It is not clear from the luminescence data whether the tunneling rate applies to the capture of electrons or holes. Since electrons are expected to have the larger Bohr radius, their capture rate should be greater than for holes, and so it was assumed that electron capture is the dominant process (Street *et al.*, 1978b). The similarity of the Bohr radius for both the radiative and nonradiative processes is also consistent with both rates being determined by the band-tail electrons. However, the complete recombination of carriers through the defect states requires the subsequent capture of a hole, and this process is not well documented. A weak luminescence band at 0.9 eV has been associated with recombination through defects, as discussed further in Part IV, but the full details of the process are not well established and require further investigation.

The relation between N_S and I_L provides a convenient means of characterizing samples through luminescence measurements (Engemann and Fischer, 1974; Paesler and Paul, 1980; Street *et al.*, 1979; Voget-Grote *et al.*, 1980; Biegelsen *et al.*, 1979). For example, Fig. 16 shows that annealing samples to high temperature quenches the luminescence (Pankove, 1978). The reason is that hydrogen is evolved and the defect density increases rapidly. Exposure of the sample to atomic hydrogen restores the luminescence, showing that hydrogen is taken up by the sample and that the dangling bonds are removed. Changes in the luminescence intensity by ion bombardment and prolonged illumination have also been associated with changes in the dangling-bond density. (See Chapter 11 by Schade.)

10. AUGER RECOMBINATION

The temperature dependence of luminescence between 10 and 100°K is shown in Fig. 17 for different excitation intensities (Street, 1980b). An unusual feature of the data is that at moderate excitation intensities, the luminescence increases with temperature. Evidently, at low temperatures there is a nonradiative process whose effect increases with the excitation intensity. The same nonradiative process is also observed from the luminescence intensity dependence under pulsed excitation (Shah *et al.*, 1980b). These results have been interpreted as Auger recombination (Street, 1980b). The Auger effect occurs when the recombination energy is given up to a third particle that is then excited deep into one of the bands, subsequently losing its energy by thermalization back down to the band edge. The requirement of having a third particle explains why the Auger effect increases with excitation intensity, since it could not occur for an isolated

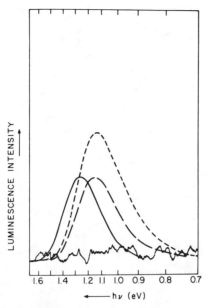

FIG. 16. The change in the luminescence spectrum following an anneal at 530°C and the subsequent rehydrogenation by atomic hydrogen. The variations of the intensity are a measure of the changes in the dangling-bond density. ———, initial, [H] ≈ 27%; —— ——, T_A = 450°C, [H] ≈ 13%; ᴧᴧᴧᴧᴧᴧ, T_A = 530°C, [H] ≈ 4%, gain 40×; — — —, rehydrogenated, [H] ≈ 13%. [From Pankove (1978).]

geminate pair. The intensity when the effect is observed corresponds roughly to the onset of significant nongeminate recombination. The origin for the temperature dependence in Fig. 17 is believed to be that the decay rate increases with temperature so that the density of carriers and their overlap decreases.

Further information about the Auger rate is shown in Figs. 10 and 18. As described in Section 6, Fig. 10 shows the results of a sequence of well spaced excitation pulses applied after the sample is cooled in the dark. The excitation causes a slow increase in the density of metastable electrons and holes, and this is observed by ESR. As the density increases, the luminescence intensity decreases, and the change is shown in Fig. 18 for various excitation conditions (Street and Biegelsen, 1982). The decrease in luminescence can again be attributed to the Auger interaction, with the metastable sea of electrons and holes providing the third particle. A spin density of only 10^{16} cm^{-3} is sufficient to cause a 10% reduction in the luminescence intensity. (The sample used for the data of Fig. 18 had a thickness of $\sim 10\ \mu$m.) For a random distribution of this density, there is a 10% probability of finding a carrier closer than 130 Å to a given electron–hole pair, and this distance

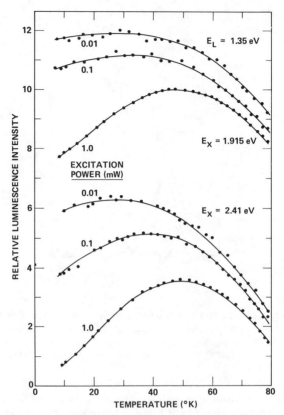

FIG. 17. The temperature dependence of the luminescence intensity for a-Si : H in the range 0–80°K for various excitation powers and wavelengths as shown. The data are normalized such that the maximum intensity is 10 units of the vertical scale, and the different curves have offset zeros. [From Street (1980b).]

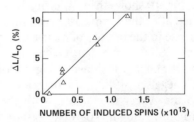

FIG. 18. Relative reduction in luminescence intensity at 30°K in a-Si : H as a function of the metastable density of excited electrons and holes as measured by LESR. [Reprinted with permission from *Solid State Communications,* Vol. 44, R. A. Street and D. K. Biegelsen, Distribution of recombination lifetimes in amorphous silicon. Copyright 1982, Pergamon Press, Ltd.]

gives an estimate of the range of the Auger interaction. The range is surprisingly large and is presumably a consequence of an efficient Auger mechanism coupled with a very slow radiative rate. One possible reason why the Auger rate should be relatively large is that the momentum selection rules for the excitation of the third particle are relaxed because of the disorder. However, no theory for the Auger rate in a disordered semiconductor has yet been developed.

11. SURFACE RECOMBINATION

Surface recombination can be detected in the luminescence of a-Si : H, although it is a weak effect. Some data demonstrating the process is shown in Fig. 19 (Dunstan, 1981). The experiment makes use of the observation that surface recombination is usually enhanced by band bending at the surface of a semiconductor and that band bending can be removed by strong illumination. Figure 19 shows that the luminescence efficiency decreases slightly for blue illumination compared with red illumination at low excitation intensity. The interpretation of the data is that the quenching is caused by surface recombination, which is enhanced by band bending. Red light excitation does not show the effect because the absorption depth is too large. Surface recombination has also been identified by the dependence of the initial luminescence decay on sample thickness (Rehm *et al.*, 1980).

It is easy to understand that surface recombination is weak because the

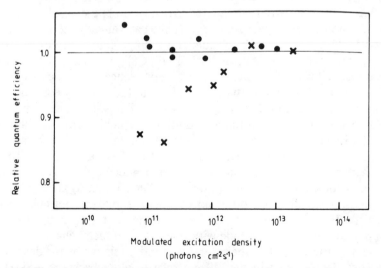

FIG. 19. The relative quantum efficiency at different excitation densities for a sample of a-Si : H annealed at 150°C. The circles are for red excitation and the crosses for blue, on the substrate side of the sample. [From Dunstan (1981).]

presence of band-tail states ensures that the diffusion length is very small. Estimates of the thermalization process lead us to conclude that the carriers will diffuse $50-100$ Å in a time $\sim 10^{-12}$ sec before being trapped in tail states (Davis, 1970). Within this time, the drift distance of free carriers under the electric field of the surface band bending is also ~ 100 Å. Thus only the carriers excited very near the surface are likely to be affected. At present, the mechanism of surface recombination is unknown. However, there is plenty of data showing that the surface of a-Si : H has a substantial defect density of $\sim 10^{12}$ cm^{-2} (Jackson and Biegelsen, 1983; Street and Knights, 1981). These defects are apparently dangling bonds and so could easily act as nonradiative centers.

Although surface recombination is very weak at low temperatures, it is apparently more important at higher temperatures and has been identified as the main cause of reduced blue response of a-Si : H solar cells. The increase of the effect is easily explained by thermal excitation of carriers out of band-tail states, which increases the diffusion length up to 1000 Å or more.

12. THERMAL QUENCHING

Thermal quenching is really not a separate nonradiative process. Since the carriers do not lose any energy, the actual nonradiative process occurs later and may be any of those described above. However, thermal quenching is a sufficiently characteristic effect to be treated separately. Usually quenching occurs because carriers are excited to a more mobile state, allowing them to find a nonradiative path more easily. We have already seen in Section 7 that a-Si : H has strong thermal quenching (Engemann and Fischer, 1973; Tsang and Street, 1979; Austin *et al.,* 1979; Collins *et al.,* 1980b) and that the unusual form of the temperature dependence can be explained by excitation from an exponential band tail up to the mobility edge, in agreement with expectations for the quenching mechanism.

Some more detailed information about the quenching process is shown in Fig. 20 (Street, 1980b). These measurements were made at 170°K, at which temperature the luminescence is quenched by $1-2$ orders of magnitude. The data show the excitation intensity dependence of the luminescence efficiency for samples of different defect density. When N_S is large enough, no intensity dependence is observed, except for a small heating effect at high intensity. However, in low-defect-density material, the efficiency increases by up to an order of magnitude with intensity. No such behavior is seen at low temperature—in fact the reverse is found, as shown by the data of Fig. 17. The effects shown in Fig. 20 are seen in glow-discharge a-Si : H but apparently are much weaker in sputtered material (Collins and Paul, 1982b).

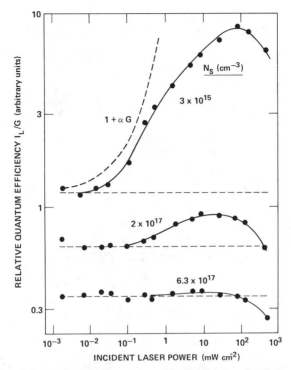

FIG. 20. Plot of the relative luminescence quantum efficiency at 170°K versus incident excitation power for a-Si:H samples of different spin density N_S. The dashed curve illustrates the relation predicted by Eq. (12). [From Street (1980b).]

The interpretation given to these data is that the thermally excited, mobile carriers can recombine in one of two ways. Either they can be captured by dangling bonds and recombine nonradiatively, or they can diffuse to a band-tail state of the opposite type and recombine radiatively. The radiative process is bimolecular and so will be enhanced at high excitation intensity. The defect recombination is monomolecular and will be greatest in samples of the highest defect-density material. These processes qualitatively account for the data in Fig. 20, and in fact a simple calculation based on this model is in reasonable quantitative agreement. Analysis of the rate equations of the recombination leads to an intensity dependence of the luminescence given by

$$y_L(G) = Y_0(1 + \gamma G/\beta^2 N_S^2),\qquad(12)$$

where γ and β are capture rate coefficients for the radiative and nonradiative processes. Figure 21 shows that the predicted dependence on N_S^2 is observed

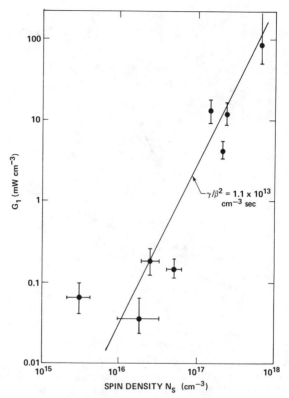

FIG. 21. A plot of G_1 versus spin density N_S for various undoped a-Si:H samples. G_1 is the value of G for which $\gamma G/\beta^2 N_S^2 = 1$ and is obtained by fitting intensity dependence data as in Fig. 20 to Eq. (12). The solid line is the dependence on N_S^2 predicted by Eq. (12) for the value of γ/β^2 as shown. [From Street (1980b).]

and also gives a value for γ/β^2 (shown in Fig. 21) that seems to be a reasonable value based on estimates of the two capture rates.

It is clear from the preceding discussion that the recombination of a-Si:H is complex, with the details depending on the precise experimental conditions. We have shown that the radiative recombination may be either geminate or nongeminate, with the geminate process strongest at low excitation intensity. Dangling-bond defects act as nonradiative centers either by direct tunneling from the band tail or by diffusion and capture, and probably also in their role as surface states. Auger recombination is important at high excitation intensities and surface recombination at high excitation energies. It appears that in the limit of low temperature, low defect density, low intensity, and weakly absorbed light, each of the nonradiative processes can be reduced to an insignificant level. If there are no other nonradiative processes, then under these circumstances the internal efficiency should be unity. Estimates of the efficiency have obtained values of 30% or more, but the accuracy is probably no better than a factor 2–3. An accurate value of the efficiency would be valuable to establish whether any unidentified recombination processes are present.

IV. Doped and Compensated a-Si : H

All of the preceding data apply to undoped a-Si : H. Doping has a large influence on the luminescence properties, but we shall see that the changes are largely dominated by defects, just as for the undoped material. Figure 22 shows the luminescence peak position, intensity, and linewidth for phosphorus and boron doping up to levels of 10^{-3}; gas phase concentrations of PH_3; and B_2H_6 relative to SiH_4 (Fisher *et al.*, 1980). Present evidence indicates that the relative concentrations of P and B in the films are higher than in the gas by a factor of 3–5 (Zesch *et al.*, 1980; Street *et al.*, 1981). At doping levels above 10^{-4}, there is a strong quenching of the 1.4 eV luminescence, and a new peak at 0.8–0.9 eV is observed. This peak is seen in Fig. 22 by the abrupt change in peak position and the increase in linewidth at the

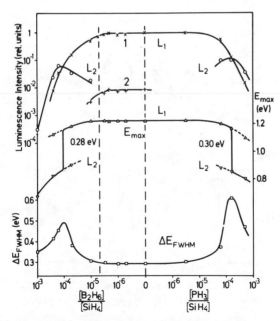

FIG. 22. The luminescence intensity, peak position, and linewidth as a function of doping. L_1 and L_2 refer to the 1.4 eV and 0.9 eV transitions. [From Fischer *et al.* (1980). Copyright North-Holland Publ. Co., Amsterdam, 1980.]

doping level when both peaks have similar intensity. At higher doping levels, both peaks are completely quenched.

To interpret the luminescence in doped a-Si : H, it is necessary to understand the defect properties, and these have been greatly clarified by light-induced ESR (LESR) measurements (Street and Biegelsen, 1980; Biegelsen *et al.*, 1981; Street *et al.*, 1981). The dark ESR of undoped samples is a measure of the dangling-bond density, because these defects are neutral and paramagnetic. When the Fermi energy is moved to the band edges by doping, the dangling bonds become doubly occupied in *n*-type and unoccupied in *p*-type material and therefore are not paramagnetic (Dersch *et al.*, 1981). However, illumination at low temperature causes excitation of carriers into the defects, revealing them in LESR. Examples of the LESR resonances, including that for undoped a-Si : H, are shown in Fig. 23. In undoped material, only the band-tail resonances are observed, as was discussed in Part I. However, the dangling-bond line is clearly present in the doped samples. In *p*-type material, the broad band-tail hole line is also found in the LESR, and it is presumed that the electron band tail is present in the *n*-type resonance, although this cannot be resolved explicitly.

The LESR data demonstrate that the dangling-bond density increases

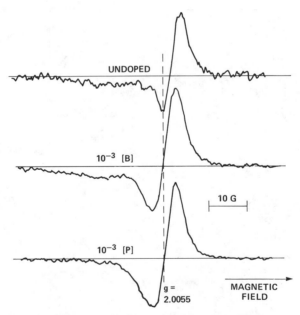

FIG. 23. LESR spectra for doped and undoped a-Si : H at 30°K. [Reprinted with permission from *Solid State Communications,* Vol. 33, R. A. Street and D. K. Biegelsen, Luminescence and ESR studies of defects in hydrogenated amorphous silicon. Copyright 1980, Pergamon Press, Ltd.]

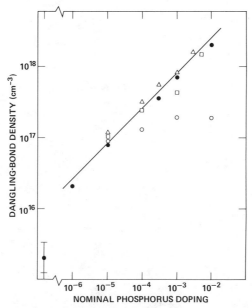

FIG. 24. The defect density versus phosphorus doping level for a-Si : H, estimated from luminescence, LESR, photothermal deflection spectroscopy, and photoconductivity. Boron doping data are similar [data of Street *et al.* (1981), Jackson and Amer (1982), and Wronski *et al.* (1982)]. △, luminescence; ○, ESR; ●, PDS; □, photoconductivity. [After Street *et al.* (1981).]

rapidly with doping, and Fig. 24 shows the variation with doping level. Light-induced LESR always provides a lower limit on the total density because it is a nonequilibrium process, and not all the dangling bonds are necessarily excited. The dangling-bond density N_S can also be determined from the extrinsic absorption shoulder below the gap (Jackson and Amer, 1982). This absorption has been measured using photothermal deflection spectroscopy, and the results are shown in Fig. 24. Recent measurements of the absorption made by using the photoconductivity excitation spectrum obtain essentially identical results (Wronski *et al.*, 1982). Finally, N_S can also be estimated from the luminescence intensity (Street *et al.*, 1981). In Section 9, Fig. 15 showed the relation between defect density and luminescence intensity for undoped samples. By assuming that the same dependence on N_S occurs for doped samples, the fourth set of data in Fig. 24 is obtained. The good agreement between the different techniques is convincing evidence that doping increases the dangling-bond density and that these defects are responsible for quenching the luminescence. It is seen that N_S reaches $10^{17} - 10^{18}$ cm^{-3} for heavy doping and in fact increases as the square root of the doping concentration.

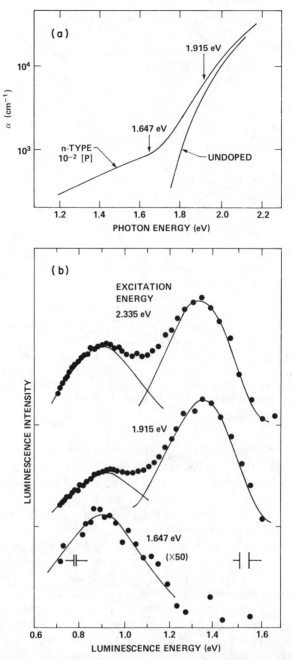

FIG. 25. (a) The absorption edge of doped and undoped a-Si:H. (b) The luminescence spectrum of phosphorus-doped a-Si:H at [P] = 3 × 10⁻³ and $T = 10°K$ for different excitation energies, showing that the 0.9 eV peak is excited by subgap illumination, unlike the 1.4 eV peak. [From Street (1980a).]

An explanation of the increase in dangling-bond density has been given in terms of an autocompensation mechanism (Street *et al.,* 1981; Biegelsen *et al.,* 1981). This model suggests that the defects are a consequence of the shift in the Fermi energy, which lowers the defect energy, rather than the result of any specific structural or chemical change in the deposition process. More recently it has also been argued that substitutional doping can occur only under conditions in which the dopant state is charged (Street, 1982a). The energy level of the *n*-type dopant must be above the Fermi energy and below it for *p*-type material. The Fermi energy is held down by dangling-bond defects whose density, to a first approximation, will therefore equal that of the dopant; the remainder of the impurities will be in inactive threefold coordinated sites. The bonding of phosphorus, for example, can therefore be described by the reaction

$$P_3^0 \rightleftarrows P_4^+ + D^-. \tag{13}$$

If it is further assumed that the deposition is sufficiently well approximated by an equilibrium process, then the law of mass action can be applied. This gives directly the result that the defect density increases as the square root of the total impurity content, in agreement with the observations. In the same way, the doping efficiency (the fraction of phosphorus in fourfold coordination) decreases as the square root of the concentration and is only about 10^{-2} at a doping level of 10^{-3}. Thus, according to this interpretation of the data, most of the phosphorus is in threefold coordination with a relatively small number of active substitutional donors.

The quenching of the luminescence by doping can therefore be understood in terms of the increased defect density. The shift of the 1.4 eV peak to lower energy is probably a consequence of wider band tails due to the increased disorder from the dopants or from the charged defects. The other major effect of doping, the 0.9 eV luminescence band, also seems to be a result of the defect states, as is discussed in Section 13.

13. THE 0.9 eV LUMINESCENCE

An example of the 0.9 eV luminescence is shown in Fig. 25b for an *n*-type sample with excitation at different energies (Street, 1980a). A luminescence band at the same energy is observed in undoped samples, particularly those with a high defect density and at high temperature. Figure 26 shows the temperature dependence of the spectrum of an undoped sample, which illustrates the emergence of the lower energy peak (Engemann and Fischer, 1977). The reason that this luminescence is seen best at high temperature is that its thermal quenching is much weaker than that of the 1.4 eV line (Nashashibi *et al.,* 1977; Street, 1980a). The 0.9 eV luminescence is also observed at low temperature in undoped samples, particularly after electron

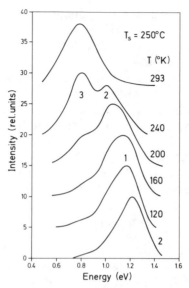

FIG. 26. The temperature dependence of the luminescence spectrum of glow-discharge a-Si: H, showing the emergence of the 0.9 eV peak at high temperatures. [From Engemann and Fischer (1977).]

bombardment, which is known to induce dangling bonds. Just as for the 1.4 eV luminescence, the actual energy of the peak varies among samples (compare Figs. 25 and 26). Nevertheless, it is generally presumed that the variation is due to differences in the composition and disorder.

Because of its association with samples containing dangling bonds, it has been proposed that the 0.9 eV luminescence is a transition through the defect state. Further evidence for the association with defects is that the thermal quenching of the dangling bond LESR (Fig. 23) is identical to that of the luminescence (Street and Biegelsen, 1980). Furthermore, the luminescence can be excited by the subgap extrinsic absorption, as shown by Fig. 25a, and this absorption is known to originate from dangling-bond transitions (Jackson and Amer, 1982).

A model for the 0.9 eV luminescence is shown in Fig. 27. From the discussion of undoped a-Si: H in Part III, it is believed that electrons are captured by dangling bonds by a phonon assisted capture process. As indicated in Fig. 27, the subsequent capture of a hole by the dangling bond may be the radiative process. Since the energy levels of the dangling bonds are still not well established, it is hard to know whether the luminescence energy is consistent with the model. The probable existence of a Stokes shift also complicates a comparison of the energy levels.

In n-type a-Si: H, essentially the same model as that in Fig. 27 can apply.

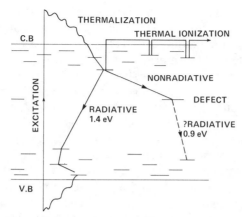

FIG. 27. A model for the luminescence transitions showing the 1.4 eV band-tail transition, nonradiative tunneling to defects, and the subsequent recombination that is proposed to explain the 0.9 eV luminescence.

The dangling bonds are negative in equilibrium, so that it is not necessary to capture an electron first. The luminescence can occur simply by direct capture of a hole. This process also agrees with the LESR results, since after the transition the dangling bond is paramagnetic. The interpretation of *p*-type samples is more complicated, because in equilibrium the dangling bonds are positively charged. One possibility is that the luminescence occurs by the capture of a band-tail electron into the charged defect. There is, in fact, a small difference in the luminescence energy between *n*-type and *p*-type samples that is consistent with this interpretation. However, it is hard to account for the very similar thermal quenching in the two cases because, presumably, the temperature dependence is governed by hole emission in one case and electron emission in the other, and these are unlikely to have the same rate. An alternative suggestion is that the transition is the same in *p*-type material as in undoped and *n*-type. This requires that the excitation quickly saturate the dangling bonds, yielding neutral states, so that the same mechanism can occur regardless of the doping. This is a plausible mechanism at low temperature because LESR data show that saturation of a large fraction of the defects does indeed occur at very low excitation intensity. It is unclear whether the same mechanism can account for the results at higher temperatures. Evidently a lot more work is needed to resolve the details of the 0.9 eV luminescence.

14. COMPENSATION

Compensated a-Si:H containing both boron and phosphorus is found to contain fewer defects than singly doped material (Street *et al.*, 1981). This result is, in fact, the principal evidence for the autocompensation model for defects. The reduced defect density is observed by ESR and is also reflected by an increase in the luminescence intensity. However, other more complicated effects of compensation are also observed. For example, Fig. 28 shows the luminescence peak position versus doping level in some compensated samples. The peak energy decreases from 1.4 down to 0.9 eV as the doping is increased up to 1%. There is also an equivalent but much smaller shift in the optical gap. Although the luminescence reaches 0.9 eV, its origin is not the same as the defect luminescence at 0.9 eV discussed in the preceding section. This is best seen by the thermal quenching behavior, which is characteristic of the band-tail transition rather than the defect band. In addition, the luminescence clearly evolves continuously from the 1.4 eV line.

The LESR of compensated samples is also unlike that of singly doped material (Street et al., 1981). In particular, there is a broad holelike resonance with a g-value of 2.017, compared with $g = 2.012$ for the usual band-tail holes. Based on this data, it is suggested that compensation introduces new states above the valence band that have the effect of broad-

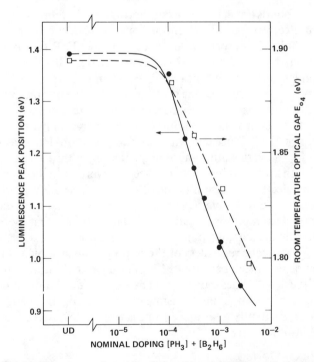

FIG. 28. The luminescence peak position (left scale) and optical gap E_{04} (right scale) versus doping level in nominally compensated a-Si:H samples at $10°K$ and $E_x = 2.41$ eV. [From Street et al. (1981).]

ening the band tail. As this happens, holes can thermalize into the new states that will shift the luminescence to lower energy. It is speculated that the new states are caused by boron–phosphorus pairs. Some evidence for this model is the observation of a chemical reaction between B_2H_6 and PH_3 during deposition that makes the concentration of each dopant sensitive to both gases in the plasma.

V. Electric and Magnetic Field Dependence

The luminescence of a-Si : H is modified by an electric or a magnetic field. For example, an electric field of about $2-3 \times 10^5$ V cm^{-1} will quench the luminescence intensity almost completely (Nashashibi et al., 1977). The quenching is observed to depend on the excitation wavelength and on the type of contacts in a complicated way. It is generally assumed that the mechanism is the field ionization of electron–hole pairs, but there have been few experiments performed to test this hypothesis.

Electric field quenching is usually observed in $p-i-n$ cells in reverse bias. When the cell is in forward bias, electroluminescence can be observed (Pankove and Carlson, 1976; Nashashibi et al., 1982a). Figure 29 gives an example of the spectrum, which is found to be shifted down in energy compared to the photoluminescence. However, in this case the shift is interpreted as an interference effect within the thin film rather than a real change in the spectrum. The electroluminescence quantum efficiency is

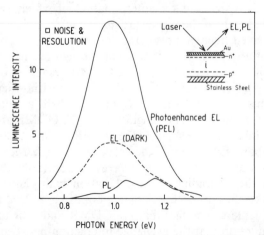

FIG. 29. Photoluminescence (PL), electroluminescence (EL), and photoenhanced EL spectra for a p^+-i-n^+ sample measured at 140°K. The same vertical scale is used in each measurement. The PEL data have the same forward current as for EL, and the same illumination intensity as the PL data. [From Nashashibi et al. (1982b).]

very low—presumably at the high fields needed to observe electroluminescence ($\sim 10^5$ V cm^{-1}), the transit time is less than the radiative lifetime, so that the probability of emission is small. Figure 29 also shows the interesting phenomenon in which the electroluminescence is enhanced by illumination (Nashashibi *et al.*, 1982b). The effect is not simply an added photoluminescence but is instead a real increase in the electroluminescence efficiency. The combined effect of illumination and an electric current gives greater luminescence than either individually. At present, the explanation of the effect is unclear.

A magnetic field of only 20 G is observed to increase the low-temperature luminescence by about 1%, although higher fields reduce the intensity (Street *et al.*, 1978a). Similar low-field changes are observed in the conductivity of a-Si:H (Mell, 1974). Both effects seem to be related to spin-dependent recombination or hopping processes. Recombination is spin dependent because the recombination rate will depend on the relative spin orientation of the particles in such a way that triplet transitions will be forbidden and singlet transitions allowed. It will be obvious from the preceding discussion that the luminescence and ESR properties of a-Si:H are closely associated, because each of the states involved in the recombination has identifiable ESR resonances. The technique of optically detected magnetic resonance (ODMR) measures the spin dependence of the recombination and provides a direct link between luminescence and ESR (Cavenett, 1981). The technique ODMR is discussed in detail in another chapter; here, only a brief discussion will be given of the main results. In fact, ODMR is a complicated experiment the interpretation of which is difficult. At present, several aspects of the results are quite controversial.

Optically detected magnetic resonance measures the modulation of the luminescence of a sample while it is swept through resonance in an ESR cavity. Figure 30 shows examples of the ODMR response for samples of different defect density (Street *et al.*, 1982). These measurements are made at 85°K, and in each case the ODMR corresponds to a reduction of the luminescence at resonance. The magnitude of the ODMR increases with defect density, and in fact the shape of the ODMR is almost identical to that of the dangling-bond ESR when the defect density is high. Measurements of the ODMR response to pulsed microwaves demonstrates that the effect occurs through the nonradiative recombination path. Thus the ODMR provides a nice confirmation that dangling bonds are indeed nonradiative recombination centers. The fact that the ODMR and ESR are so similar leads to the deduction that the exchange interaction between the states is very weak (Depinna *et al.*, 1982a; Biegelsen *et al.*, 1978). This result follows from the localization of the carriers and from their relatively large separa-

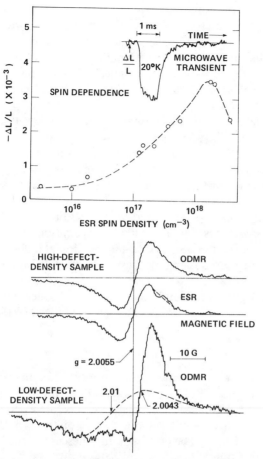

FIG. 30. ODMR data for various a-Si : H samples measured at 85°K. Upper curves show the increase in ODMR intensity with increasing dangling-bond density and an example of the microwave pulse response. Lower curves show ODMR spectra compared to ESR measurements. [From Street *et al.* (1982).]

tion, and is therefore consistent with the tunneling model for recombination.

The ODMR modulation is typically 0.1 – 1% of the luminescence intensity. On one hand, this can be viewed as a small effect. The size of the change is a measure of how much the triplet states influence the luminescence to increase the lifetime and reduce the efficiency. The small ODMR supports the hypothesis that in the general consideration of decay times, the triplet states can be neglected. In another sense, however, the ODMR is a very large effect. If the spins are assumed to be in thermal equilibrium, then the expected magnitude at 100°K is ~ 10^{-5}. The reason for the larger effect is that the spins are not thermalized—some of the luminescence occurs at a shorter time than the spin–lattice relaxation time T_1 of the spins, and it is this component that dominates the ODMR. In fact, measurements of the time-resolved ODMR allow the direct evaluation of T_1 and its temperature dependence (Street *et al.*, 1982).

At low temperatures and in samples of low defect density, the ODMR is more complicated (Biegelsen *et al.*, 1978; Morigaki *et al.*, 1978; Street, 1982b; Depinna *et al.*, 1982a,b). An enhancing line (corresponding to an increase of luminescence intensity at resonance) is observed in addition to one or more quenching lines. The enhancing line can be associated with the radiative process by its response to microwave pulsing. However, the ODMR has a *g*-value near 2.008 and cannot be directly identified with any of the known ESR lines. The most likely explanation seems to be that the ODMR is a combination of the band-tail electron and hole resonances but with a weak exchange coupling that results in the observation of an average *g*-value. If this is the correct explanation, then the results support the previous model of a transition between band-tail states. However, further experiments are needed to confirm the details of the ODMR process.

VI. Summary

Luminescence offers a detailed probe of localized states and recombination processes in a-Si:H. The studies described in this chapter show that several different recombination mechanisms operate and that the relative importance of each depends on the details of the sample and on the experimental conditions. The dominant luminescence near 1.4 eV is identified as a radiative tunneling transition between band-tail states. The other luminescence at 0.9 eV seems to originate from recombination at dangling bonds. Four different nonradiative processes can be identified. Dangling bonds quench the 1.4 eV luminescence by direct tunneling at low temperature and by thermal ionization to the mobility edge followed by trapping at high temperature. The two other nonradiative mechanisms are surface recombination and Auger transitions. A schematic diagram of these different processes is shown in Fig. 31.

Although the recombination mechanisms are the same as those commonly observed in crystalline semiconductors, the modifying effects of the amorphous network are plain. Localization of both band edges leads to tunneling recombination, both radiative and nonradiative, and to a very broad distribution of decay times. There is no indication that excitons play an important role in the recombination. Other consequences of the locali-

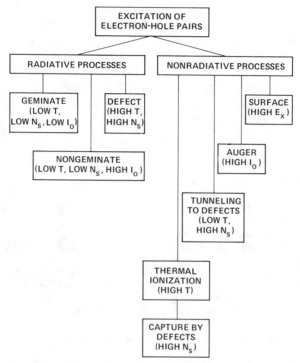

FIG. 31. Schematic diagram of the various recombination mechanisms in a-Si : H.

zation are that a relatively large defect density ($\sim 10^{17}$ cm^{-3}) can be introduced before the luminescence is quenched, and that surface recombination is weak at low temperature. The Auger process is also certainly influenced by the disorder that relaxes the momentum conservation requirement, but the theory of this process is still undeveloped.

Two further aspects of the recombination are emphasized by the disorder. One is the thermalization of carriers down the band-tail states, which can be observed to continue for up to 1 μsec. The second is geminate recombination, which is a consequence of the short mean free path in extended states and of the rapid localization. The role of geminate recombination and the transition to a nongeminate process can be observed and studied in a-Si : H.

Even though the main elements of the recombination are fairly well established, many aspects of the problem warrant further investigation. One controversial question is the magnitude of the Stokes shift. This is an important quantity because it relates to the nature of the band-tail states and to the description of optical absorption. The detailed origin of the 0.9 eV luminescence is another unresolved problem. A third example is the description of recombination at elevated temperatures. The luminescence data indicate that carrier diffusion and nonradiative capture by dangling bonds is an important process, and other studies show that surface recombination is significant. Clearly, direct radiative transitions are very weak. However, as yet there is no detailed understanding of the relative importance of the different mechanisms and whether Auger or other processes are significant. Finally, the possibility that time-resolved luminescence can be used as a direct probe of the mobility edge should be explored further.

REFERENCES

Austin, I. G., Nashashibi, T. S., Searle, T. M., LeComber, P. G., and Spear, W. E. (1979). *J. Non-Cryst. Solids* **32**, 373.

Bhat, P. K., Searle, T. M., Austin, I. G., Gibson, R. A., and Allison, J. (1983). *Solid State Commun.* **45**, 481.

Biegelsen, D. K. (1980). *Solar Cells* **2**, 421.

Biegelsen, D. K., Knights, J. C., Street, R. A., Tsang, C., and White, R. M. (1978). *Philos. Mag. B* **37**, 477.

Biegelsen, D. K., Street, R. A., Tsai, C. C., and Knights, J. C. (1979). *Phys. Rev. B* **20**, 4839.

Biegelsen, D. K., Street, R. A., and Knights, J. C. (1981). *AIP Conf. Proc.* **73**, 166.

Boulitrop, F., and Dunstan, D. J. (1982). *Solid State Commun.* **44**, 841.

Brodsky, M. H., and Title, R. S. (1969). *Phys. Rev. Lett.* **23**, 581.

Cavenett, B. C. (1981). *Adv. Phys.* **30**, 475.

Chen, W. C., Feldman, B. J., Bajaj, J., Tong, F. M., and Wong, G. K. (1981). *Solid State Commun.* **38**, 357.

Cohen, E., and Sturge, M. D. (1982). *Phys. Rev. B* **25**, 3828.

Collins, R. W., and Paul, W. (1982a). *Phys. Rev. B* **25**, 2611.

Collins, R. W., and Paul, W. (1982b). *Phys. Rev. B* **25**, 5263.

Collins, R. W., Paesler, M. A., Moddel, G., and Paul, W. (1980a). *J. Non-Cryst. Solids* **35/36**, 681.

Collins, R. W., Paesler, M. A., and Paul, W. (1980b). *Solid State Commun.* **34**, 833.

Davis, E. A. (1970). *J. Non-Cryst. Solids* **4**, 107.

Depinna, S., Cavenett, B. C., Austin, I. G., Searle, T. M., Thompson, M. J., Allison, J., and LeComber, P. G. (1982a). *Philos. Mag. B* **46**, 473.

Depinna, S., Cavenett, B. C., Searle, T. M., and Austin, I. G. (1982b). *Philos. Mag. B* **46**, 501.

Dersch, H., Stuke, J., and Beichler, J. (1981). *Phys. Status Solidi B* **105**, 265.

Dunstan, D. J. (1981). *J. Phys. C* **14**, 1363.

Dunstan, D. J. (1982). *Philos. Mag. B* **46**, 579.

Dunstan, D. J., and Boulitrop, F. (1981). *J. Phys. Colloq. Orsay Fr.* **42**, C4-331.

Dunstan, D. J., (1982). *Solid State Commun.* **43**, 341.

Dunstan, D. J., Depinna, S. P., and Cavenett, B. C. (1982). *J. Phys. C* **15**, L425.

Engemann, D., and Fischer, R. (1974). *Proc. Int. Conf. Amorphous and Liquid Semicond. 5th* (J. Stuke and W. Brenig, eds.), p. 947. Taylor and Francis, London.

Engemann, D., and Fischer, R. R. (1974). *Proc. Int. Conf. Phys. Semicond., 12th* (M. H. Pilkuhn, ed.), p. 1042.

Engemann, D. and Fischer, R. R. (1977). *Phys. Status Solidi B* **79**, 195.

Fischer, R., Rehm, W., Stuke, J., and Voget-Grote, U. (1980). *J. Non-Cryst. Solids* **35/36**, 687.

Fowler, W. B. (1968). "Physics of Color Centers." Academic Press, New York.

Hong, K. M., Noolandi, J., and Street, R. A. (1981). *Phys. Rev. B* **23**, 2967.
Jackson, W. B., and Amer, N. M. (1982). *Phys. Rev. B* **25**, 5559.
Jackson, W. B., and Biegelsen, D. K. (1983). *Appl. Phys. Lett.* **42**, 105.
Knights, J. C., and Davis, E. A. (1974). *J. Phys. Chem. Solids* **35**, 543.
Kurita, S., Czaja, W., and Kinmond, S. (1979). *Solid State Commun.* **32**, 879.
Lin, S., and Feldman, B. S. (1982). *Phys. Rev. Lett.* **48**, 829.
Mell, H. (1974). *Proc. Int. Conf. Amorphous and Liquid Semicond., 5th* (J. Stuke and W. Brenig, eds.), p. 203. Taylor and Francis, London.
Morigaki, K., Dunstan, D. J., Cavenett, B. C., Dawson, P., and Nicholls, J. E. (1978). *Solid State Commun.* **26**, 981.
Mott, N. F., and Davis, E. A. (1979). "Electronic Processes in Non-Crystalline Solids." Oxford Univ. Press, London and New York.
Mott, N. F., Davis, E. A., and Street, R. A. (1975). *Philos. Mag.* **32**, 961.
Nashashibi, T. S., Austin, I. G., and Searle, T. M. (1977). *Philos. Mag.* **35**, 831.
Nashashibi, T. S., Austin, I. G., Searle, T. M., Gibson, R. A., Spear, W. E., and LeComber, P. G. (1982a). *Philos. Mag. B* **45**, 553.
Nashashibi, T. S., Searle, T. M., Austin, I. G., Rhodes, A. J., Gibson, R. A., and LeComber, P. G. (1982b). *Philos. Mag. B* **45**, 573.
Nemanich, R. J., and Lyon, S. A. (1984). To be published.
Paesler, M. A., and Paul, W. (1980). *Philos. Mag. B* **41**, 393.
Pankove, J. I. (1978). *Appl. Phys. Lett.* **32**, 812.
Pankove, J. I., and Carlson, D. E. (1976). *Appl. Phys. Lett.* **29**, 620.
Pankove, J. I., Pollak, F. H., and Schnabolk, C. (1980). *J. Non-Cryst. Solids* **35/36**, 459.
Rehm, W., Fischer, R., and Beichler, J. (1980). *Appl. Phys. Lett.* **37**, 445.
Reiss, H. (1956). *J. Chem. Phys.* **25**, 400.
Shah, J., and DiGiovanni, A. E. (1981). *Solid State Commun.* **37**, 153.
Shah, J., Alexander, F. B., and Bagley, B. G. (1980a). *Solid State Commun.* **36**, 195.
Shah, J., Bagley, B. G, and Alexander, F. B. (1980b). *Solid State Commun.* **36**, 199.
Shah, J., Pinczuk, A., Alexander, F. B., and Bagley, B. G. (1982). *Solid State Commun.* **42**, 717.
Street, R. A. (1978). *Philos. Mag.* **37**, 35.
Street, R. A. (1980a). *Phys. Rev. B* **21**, 5775.
Street, R. A. (1980b). *Phys. Rev. B* **23**, 861.
Street, R. A. (1980c). *Solid State Commun.* **34**, 157.
Street, R. A. (1981a). *Adv. Phys.* **30**, 593.
Street, R. A. (1981b). *J. Phys. Colloq. Orsay Fr.* **42**, *Suppl. 10,* C4-575.
Street, R. A. (1982a). *Phys. Rev. Lett.* **49**, 1187.
Street, R. A. (1982b). *Phys. Rev. B* **26**, 3588.
Street, R. A., and Biegelsen, D. K. (1980). *Solid State Commun.* **33**, 1159.
Street, R. A., and Biegelsen, D. K. (1982). *Solid State Commun.* **44**, 501.
Street, R. A., and Knights, J. C. (1980). *Philos. Mag. B* **42**, 551.
Street, R. A., and Knights, J. C. (1981). *Philos. Mag. B* **43**, 1091.
Street, R. A., Searle, T. M., and Austin, I. G. (1974). *Proc. Int. Conf. Amorphous and Liquid Semicond., 5th* (J. Stuke and W. Brenig, eds.), p. 953. Taylor and Francis, London.
Street, R. A., Biegelsen, D. K., Knights, J. C., Tsang, C., and White, R. (1978a). *Solid State Electron.* **21**, 1461.
Street, R. A., Knights, J. C., and Biegelsen, D. K. (1978b). *Phys. Rev. B* **18**, 1880.
Street, R. A., Biegelsen, D. K., and Stuke, J. (1979). *Philos. Mag. B* **40**, 451.
Street, R. A., Biegelsen, D. K., and Knights, J. C. (1981). *Phys. Rev. B* **24**, 969.
Street, R. A., Biegelsen, D. K., and Zesch, J. (1982). *Phys. Rev. B* **25**, 4334.
Thomas, D. G., Hopfield, J. J., and Augustyniak, W. M. (1965). *Phys. Rev.* **140**, 202.

Tiedje, T., and Rose, A. (1981). *Solid State Commun.* **37,** 49.
Tsang, C., and Street, R. A. (1978). *Philos. Mag. B* **37,** 601.
Tsang, C., and Street, R. A. (1979). *Phys. Rev. B* **19,** 3027.
Voget-Grote, U., Kummelle, W., Fischer, R., and Stuke, J. (1980). *Philos. Mag. B* **41,** 127.
Wilson, B. A. (1981). *AIP Conf. Proc.* **73,** 273.
Wilson, B. A., and Kerwin, T. P. (1982). *Phys. Rev. B* **25,** 5276.
Wilson, B. A., Hu, P., Harbison, J. P., and Jedju, T. M. (1983). *Phys. Rev. Lett.* **50,** 1490.
Wolford, D. J., Reimer, J. A., and Scott, B. A. (1983). *Appl. Phys. Lett.* **42,** 369.
Wronski, C. R., Abeles, B., Tiedje, T., and Cody, G. D. (1982). *Solid State Commun.* **44,** 1423.
Zesch, J. C., Lujan, R. A. and Deline, V. R. (1980). *J. Non-Cryst. Solids* **35/36,** 273.

SEMICONDUCTORS AND SEMIMETALS, VOL. 21, PART B

CHAPTER 8

Photoconductivity

Richard S. Crandall

RCA/DAVID SARNOFF RESEARCH CENTER
PRINCETON, NEW JERSEY

I. Introduction

In this review we shall place most emphasis on those concepts particular to amorphous silicon. The physical concepts will be stressed, using experimental results for support. Photoconductivity is divided into three parts: (1) generation by an external source of electron–hole pairs, (2) transport of mobile carriers in either extended or localized states, and (3) recombination of the excited electron and hole. Each part has been the subject of much research and controversy in a-Si:H and only now is there perhaps some consensus about the important features of each.

In the simplest terms, generation consists of exciting electrons and holes across the forbidden gap from the valence band to the conduction band (in more general terms for an amorphous solid, from a high density of valence-

245

band tail states to a high density of conduction-band tail states). It is implied in the generation process that the electron–hole pairs are actually separated and are then free to contribute to the conduction process. If their Coulomb attraction is sufficiently strong, they will recombine geminately and not contribute to transport. Whether geminate recombination is important in a-Si:H is the subject of considerable debate.

The transport process in a-Si:H is also being debated at present. There is not a consensus as to whether conduction takes place predominantly in extended states by delocalized electrons or by hopping motion in band-tail states. Even though we favor the latter approach, we shall assume in this chapter that the predominant transport takes place at a well-defined mobility edge. Departures from this picture will occur if time-dependent photoconductivity is considered (see, for example, Chapters 9 by Tauc, and those by Scher and Tiedje in Volume 21C).

Geminate recombination limits the fraction of electron–hole pairs that actually contribute to the photoconduction process to a number less than unity. Nevertheless, an electron and hole that escape the geminate recombination process will eventually recombine, usually through defects in the forbidden gap. The details of this process are complicated and not well understood in a-Si:H.

Perhaps one of the most important aspects to be aware of is the role of the electrical *contacts* in determining the type of photoconductivity that is measured. Ohmic contacts are normally used. By ohmic contacts we mean that the contact can always supply the current required by the photoconductor. This mode of photoconduction is termed *secondary photoconduction* and is tacitly assumed in the majority of texts (Bube, 1960; Rose, 1963; Ryvkin, 1964).

However, not all contacts are ohmic. In fact the contacts to a photovoltaic solar cell are by definition "blocking" for the photogenerated carriers. By blocking we mean that the contact does not supply sufficient current for the photoconduction process. Increasing the electric field near the contact, or increasing the conductivity of the photoconductor does not cause more current to flow from the contact. Examples of this contact to a-Si:H are a high work-function metal–Schottky-barrier contact and the $p-i$ and $n-i$ interfaces in a $p-i-n$ junction solar cell. The energy-band diagram for the $p-i-n$ junction is illustrated in Fig. 1. Because the Fermi level in the p layer is well below midgap, an electron must surmount a considerable barrier, at least 1 eV, to pass from the p-type a-Si:H into the conduction band of the undoped insulating i layer. The photocurrent that flows when the contacts are blocking is referred to as a *primary* photocurrent. It is perhaps the earliest example of photoconduction and certainly comprised the dominant mode of conduction for the best studied examples of early photoconductors,

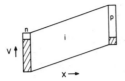

FIG. 1. Schematic representation of the potential energy diagram of a $p-i-n$ solar cell structure illustrating blocking contacts to the insulating i layer. The p and n layers contain the space charge necessary to support the uniform field in the i layer.

namely, the alkali halides (Gudden and Pohl, 1921). The photoconductive properties are qualitatively simple. If only one sign of photocarrier is excited in the bulk, then a steady-state photocurrent cannot be drawn. This is because the space charge arising from the immobile carrier will screen the applied field. If both carriers are free to move, then the maximum steady-state current is proportional to the rate of photogeneration.

We shall place considerable emphasis on the distinction between secondary and primary photocurrents in this review. The exploitation of these two modes of photoconduction can be quite useful in elucidating photoconductive properties. Even though the primary photoconduction process has the longest history, the theoretical framework has been devoted mainly to the time-dependent single-carrier process. Only recently have there been attempts to describe the two-carrier steady-state transport process. Because this is the dominant mode of transport in a solar cell, its complete understanding is a necessary adjunct to the advancement of solar cell design.

II. General Concepts

Many of the concepts of photoconductivity in insulators that we shall use in this chapter follow closely those laid down by Rose (1951, 1963). His ideas, which were presented to describe insulating photoconductors that contained large concentrations of poorly defined states in the gap, are particularly well suited to describe amorphous silicon. Even though there has been a great deal of investigation of a-Si: H, knowledge about the states in the gap is incomplete, and for that matter the role that they play as recombination centers is only beginning to be understood.

1. DISTRIBUTION OF STATES

The connection between Rose's phenomenological approach and more formal ideas can be seen by the sketch of the density of states of a-Si: H shown in Fig. 2. This is a composite, pieced together from various experiments (Crandall, 1980a; Lang et al., 1982; Jackson, 1982). The finer details of the density of states are not important for our discussion. The main point

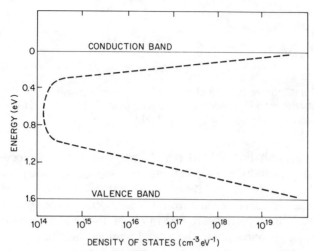

FIG. 2. Simplified density of states diagram for a-Si:H. The actual magnitudes are only qualitatively correct.

is that the density of states, i.e., number of states per unit volume per unit energy, $N(E)$ in the band gap connects smoothly with the density of states in the conduction band. Within the gap, away from the band edges, in the so-called band-tail region, the density of states decreases exponentially to low values some tenths of electron volts away from the band edges. The mathematical form of the band tail extending from the conduction band is

$$N(E) = N_c \exp(-|E_c - E|/E_0^c), \qquad (1)$$

where N_c is the density of states at the energy E_c of the conduction band edge, and E_0^c the characteristic decay energy of the band tail. There is a similar expression for the density of states near the valence-band edge with a characteristic energy E_0^v. The magnitude of E_0^c is about 0.025 eV and that of E_0^v is about 0.045 eV (Tiedje et al., 1981). Even though both these numbers show that the density of states decreases rapidly into the band gap, the valence-band tails are about twice as wide as the conduction-band tails. If there were no states in the gap other than these band tails, then the minimum density of states, using the above values for the characteristic energy and choosing a band gap of 1.7 eV, which is typical for glow-discharge a-Si:H, would be about $2 \times 10^{-11} N_c$. Choosing $N_c \sim 10^{22}$ cm^{-3} eV^{-1} gives a minimum $N(E)$ of 2×10^{11} cm^{-3} eV^{-1}. This is an unrealistically small number, so we must conclude that foreign impurities or fundamental defects such as dangling bonds (Brodsky and Title, 1969; Biegelsen, 1980) contribute to the density of states in this region. These exponential band tails have been well verified by transport (Tiedje et al., 1980) and optical absorption (Cody *et al.,* 1981) measurements, as discussed in other chapters of this book. These exponential band tails presumably arise from the absence of long-range order in the amorphous network. Thus one expects them to be a nearly universal property of amorphous silicon, irrespective of the residual impurity density. Their properties, such as characteristic energy, should vary only insofar as the density and/or disorder of a-Si:H varies with preparation conditions.

This picture of the states in the gap is very much like Rose's (1951) model of an insulator having a distribution of traps throughout the band gap. He considered low densities of states, deep in the forbidden gap, that function chiefly as recombination centers. They are so far from the band edges that thermal emission of a trapped charge is unlikely before one of the opposite sign is trapped at the same site and gives rise to recombination. Near the band edges there is a much higher density of traps having a population in thermal equilibrium with the conducting states. These "shallow traps" will not be recombination centers because an electron is thermally emitted before a hole is captured.

To explain the often observed noninteger power law dependence of photocurrent on light intensity, Rose (1963) postulated that the traps had an exponential distribution in energy of the form given by Eq. (1). This exponential distribution of states is just what the band-tail states in a-Si:H are thought to look like. In fact Tiedje and Rose (1981) used an exponential distribution of shallow band-tail states to give a simple description of dispersive transport in disordered materials. This model is discussed in detail by Tiedje in Chapter 7 of Volume 21C. An exponential distribution of band-tail states is also used to explain the spectral dependence of the optical absorption. For excitation below band gap the absorption coefficient decreases exponentially with decreasing photon energy (Cody *et al.,* 1981). This absorption coefficient is believed to reflect the valence band-tail density of states.

2. TRANSPORT

There is still considerable debate (Datta and Silver, 1981; Silver *et al.,* 1981; Tiedje and Rose, 1981) as to whether it is best to describe the electron and hole transport in amorphous silicon by a band model with completely delocalized states or a "hopping model" in which all carriers are localized within states in the forbidden gap. The actual situation is somewhere between these two extremes, making a calculation of the transport properties difficult. However, it is usually helpful to think about the transport, using one or the other of these limiting cases.

The textbook viewpoint of photoconductivity is that even in heavily disordered materials electron and hole transport takes place at the band edges. States in the gap function only as traps or recombination centers. To make a quantitative distinction between a trap and a recombination center, one often introduces the idea of a "demarcation level." This is an energy level for electrons such that an electron in a state above this level will have a high probability of being thermally excited into the conduction band. An electron in a state lying below this level has a much higher probability of capturing a hole than of being thermally excited to the conduction band.

The shallow states, in this model, serve only as traps. The idea of Tiedje and Rose (1981) and also Orenstein and Kastner (1981) is that electrons are trapped in these shallow band-tail states and can only move among the states by being thermally excited to the band edge where they can be transported through the lattice. This is referred to as the multiple trapping model of transport and is the general idea used by Rose (1963); it is perhaps the most common view used to describe the transport.

An alternative view is to consider that thermally assisted hopping (Mott, 1968; Döhler, 1979) between these shallow states is a more important process for electron redistribution than is thermal excitation to the band edge. Both these models can be true. Which one is more important depends on the circumstances. If the Fermi level is well below the band edge and the density of band-tail states is high enough, then hopping transitions are important because there is a large amount of wave-function overlap between adjacent sites. This mode then outweighs thermal excitation to the conduction band and transport takes place at the mobility edge. In the other extreme, the density of tail states is low so that there is little wave-function overlap and the hopping probability is low. Then thermal excitation to the band edge is the dominant mode of transport.

Whether we should consider transport in the band-tail states depends critically on the position of the Fermi level and the temperature, because the principal energy for transport will be determined by the maximum in the product of the Fermi function $f(E)$ and the differential conductivity $\sigma(E)$. The differential conductivity is the product of the electron charge, the energy-dependent mobility $\mu(E)$, and the density of states $N(E)$. The quantities $f(E)$ and $\mu(E)$ are sketched in Fig. 3 to illustrate the fact that they have opposite energy dependences. The mobility $\mu(E)$ decreases rapidly for energies below E_c, whereas $f(E)$ increases for states in the gap. In a-Si:H the Fermi energy most likely lies below the mobility edge. The differential current, the product of $\sigma(E)$ and $f(E)$, is represented by the bell-shaped curve. Döhler (1979) has interpreted measurements of the conductivity and thermopower on heavily doped a-Si:H by using this model and has concluded that at temperatures below 600°K the dominant transport is by hopping below the mobility edge. Since the quasi-Fermi level under photoexcitation in undoped a-Si:H would most likely lie at or below the values

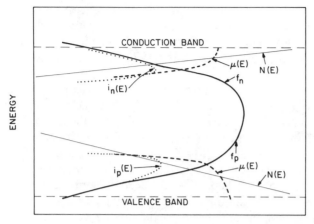

DENSITY OF STATES

FIG. 3. Schematic density of states showing occupancy functions f_n and f_p for electrons and holes, respectively. [These quantities correspond to $f(E)$ in the text.] The mobilities $\mu(E)$ and differential currents $i(E)$ are also shown.

for heavily doped n-type samples, it seems likely that photoexcited electrons would also be transported by hopping. In fact Döhler (1981) interprets photoconductivity in a-Si:H, using this model. An important consequence of these ideas is that recombination and trapping proceed through transitions between localized states instead of from a delocalized to a localized state. As we shall discuss in Section 10, there is now convincing evidence (Dersch et al., 1983) that recombination proceeds via the band-tail states. Because of this, we shall henceforth use E_c and E_v to represent the mobility edges or energy levels that separate conducting from nonconducting states irrespective of whether they are delocalized or not.

3. EXCITATION AND RECOMBINATION

In many crystalline photoconductors an electron and a hole, once excited, remain free until they are trapped and eventually recombine. However, in some materials the excitation remains localized, with a high probability that the electron and hole recombine before leaving the site of the excitation. It is well established that this process of geminate recombination limits the room temperature photoresponse of some molecular crystals and amorphous chalcogenides (Enck and Pfister, 1976; Mort, 1981). Geminate recombination takes place because the Coulomb attraction between the hole and electron is so strong that the excitation is localized long enough to permit the electron and hole to recombine before they can separate by diffusion or by the force of an applied electric field. In materials such as amorphous selenium in which this recombination mechanism has been studied in great detail, one has been able to separate the photogeneration efficiency from transport and eventual recombination to demonstrate that there is a strong dependence of the generation efficiency on photon energy, temperature, and applied electric field. Enck and Pfister (1976) give a good discussion of this phenomenon.

Figure 4 represents an electron–hole pair immediately following the absorption of a photon of energy $h\nu$ greater than the band-gap energy E_g. The attraction between the electron and hole place them in a potential well that is Coulombic at large distance. In order for the electron to escape this Coulomb well it must diffuse to a distance r_e where the attractive energy is kT (at room temperature this is 0.025 eV) before it thermalizes. For example, in amorphous silicon $r_e \approx 50$ Å at room temperature. Whether the electron can diffuse to this distance before losing its kinetic energy depends on the electron mobility and the hot carrier thermalization time. By using an estimate of the thermalization time (Ackley *et al.,* 1979) for an electron 0.3 eV above the band edge and a mobility of 10 cm^2 V^{-1} sec^{-1}, we find that the electron gives up all its energy in 32 Å. This distance is usually referred to as the thermalization length r_0. Since this is on the same order as r_e, there is a possibility of geminate recombination in a-Si : H. Of course, whether it will be important depends on the value of the electron mobility. Various measurements obtain values in the range 1 – 10 cm^2 V^{-1} sec^{-1}. However, there are recent speculations (Datta and Silver, 1981) that the mobility may be over an order of magnitude higher, which would make geminate recombination unlikely.

The large body of low-temperature luminescence data (Street, 1981 and

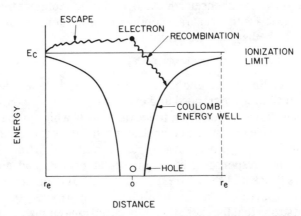

FIG. 4. Potential energy level diagram to represent an electron and hole immediately following excitation above the band edge. The hole is self-trapped and the electron remains free but bound in the Coulomb well. If the electron diffuses to a distance greater than r_e, it can escape. If not, it recombines with the hole.

references therein) is entirely consistent with the electron and hole recombining geminately. The near unit quantum efficiency for the luminescence shows that radiative recombination occurs before the electron and hole can diffuse apart and undergo nonradiative recombination. These views are discussed in detail in Chapter 7 by Street. It is not surprising that geminate recombination dominates at low temperature, because exciton effects are favored over pair separation. A quantitative comparison of luminescence data with a geminate recombination model, based on a solution of the time-dependent diffusion equation for the carrier distribution and assuming radiative recombination via tunneling, has been made by Noolandi *et al.* (1980), who find clear support for the model. At high temperature, however, thermally activated diffusion of the electron–hole pair competes with radiative recombination. At room temperature the efficiency of the radiative process is only about 10^{-4} (Street, 1981). This result therefore implies that geminate recombination via a radiative transition is unimportant at room temperature. If there is geminate recombination at room temperature it must occur by nonradiative multiphonon emission.

The experimental situation as to whether there is geminate recombination at room temperature remains unclear. Mort *et al.* (1981) obtained evidence for geminate recombination using delayed-field collection efficiency measurements. They predicted a zero-electric-field collection efficiency of 0.6. We made collection-efficiency measurements on *n*-type Schottky-barrier solar cells (Crandall *et al.*, 1979) and also concluded that there was an electric field dependence to the collection efficiency. An analysis of the data, based on field ionization of the electron, also led to a zero-field collection efficiency of about 0.5 (Crandall, 1980b).

These results are quite puzzling in view of the fact that many others have failed to observe a field-dependant collection efficiency (Madan *et al.*, 1980; Okamoto *et al.*, 1980; Spear, 1980; Carasco and Spear, 1983). This includes a large body of data on RCA *p–i–n* solar cells that show generation efficiencies of over 95% with an applied field of 10^4 V cm^{-1}. Based on the Onsager model of geminate recombination (Onsager, 1938), these results indicate that the zero-field generation efficiency would still be in excess of 90%.

A further complication with the geminate recombination model for a-Si:H at room temperature is that even when the field dependence of the generation efficiency indicates a geminate recombination mechanism the expected wavelength dependence is not observed. Both theory and experiment in materials other than amorphous silicon, in which the effect is unambiguously determined, show a rapid decrease of the photogeneration efficiency with decreasing energy of the absorbed photon. The reason for this is that the farther above the band edge that an electron is, the greater distance it diffuses before losing sufficient kinetic energy to thermalize. Experimentally, one observes that the generation efficiency does not track the absorption coefficient.

In recent measurements of the field and wavelength dependence of the generation efficiency of $p-i-n$ solar cell structures, Carasco and Spear (1983) failed to observe any reduction in the collection efficiency due to geminate recombination for photon energies above the energy of the optical gap. However, the data clearly showed a reduction in the generation efficiency for photon energies below the optical gap, corresponding to transitions between band-tail states. Their data are reproduced in Fig. 5, showing the electric field dependence of the photocurrent in a reverse-biased $p-i-n$ solar cell. The solid curves are the results of calculations based on the Onsager model for the values of the thermalization length, labeled r_0 in the figure. Since the Onsager theory does not include the effects of nongeminate recombination on the carrier lifetime, it does not predict the rapid decrease in the current at low fields that is shown by the data. This decrease is due to the eventual recombination of electron–hole pairs that have been freed following excitation. This effect will be discussed in detail later in Section 8 on primary photocurrent. All the experimental data for photon energies of 1.51 eV and above show a field-independent photocurrent above about 2×10^4 V cm^{-1}. This implies that the thermalization length is 400 Å or

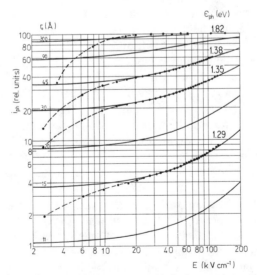

FIG. 5. Comparison of measured photocurrent (dashed curves) to calculated generation efficiency (solid curves) curves predicted by the Onsager theory. The curves correspond to different values of the thermalization length, labeled r_0 in the figure. [After Carasco and Spear (1983).]

greater, so that the generation efficiency would be in excess of 95% at zero electric field.

For photon energies below the optical gap, 1.65 eV, the data clearly show an electric field dependence due to geminate recombination. With decreasing photon energy, the optical transitions connect states deeper in the band tails. Carriers in these states, being more localized, have increasing difficulty in being freed by diffusion before they recombine geminately. This is reflected in the decrease in the thermalization length with decreasing photon energy.

However, this does not answer the question as to why some workers have observed a reduction in the number of collected electron–hole pairs at low field. Perhaps the only explanation, which is not completely satisfying, is the variation of samples with different growth techniques. Some evidence for this comes from Mort (1981), who found that the thermalization radius can vary with sample preparation.

III. Recombination

4. TRAPPING

Trapping or capture is the process by which an electron makes a transition from a conducting to a nonconducting state. This will be a recombination process only if the state is occupied by a hole. There are two general models that treat electron trapping in detail. One model applies to the regime in which the capture can be considered "ballistically" (Lax, 1960). This applies to the situation in which the electron mean free path for scattering is considerably longer than the capture radius of the trap. This is most often encountered in high-mobility solids such as silicon and germanium single crystals. Here, the bottleneck for electron capture is the ability of the electron to lose sufficient kinetic energy by phonon emission to form a bound state of the defect. Because these are high-mobility or long scattering mean-free-path materials, an electron loses kinetic energy slowly and will traverse many recombination centers before being captured.

For low-mobility solids the ability to lose energy is not the bottleneck to electron capture. Here, the bottleneck is the difficulty of diffusing to the trap or recombination center. Since most of the experimental evidence (Allan *et al.*, 1977; LeComber and Spear, 1970; Moore, 1977; Moore, 1980; Tiedje *et al.*, 1980) shows that electrons and holes have a low mobility in amorphous silicon, at best $1 - 10$ cm^2 V^{-1} sec^{-1}, this diffusive recombination should dominate all other types. Even if the high-energy tail of the electron distribution contains high-mobility carriers, most of the electrons will be in the band-tail states in which the mobility is low. As we shall discuss later, recent evidence (Dersch *et al.*, 1983) shows that recombination proceeds via band-tail states.

The earliest example of this type of capture is Langevin recombination, which treats the diffusive capture by a charged center (Langevin, 1903). It has been well established in low-mobility solids such as anthracene (Coppage and Kepler, 1967), sulfur (Dolezalek and Spear, 1975), and the alkali halides (Crandall, 1965). The following abbreviated argument outlines this diffusive capture of an electron by a positively charged trap. Consider a spherical surface of radius R_s surrounding the trap. The flux of electrons inward across this surface is the product of the drift velocity of the electrons, the surface area of the sphere, and the electron density. The drift velocity is the product of the mobility μ_n and the electric field E_+ due to the positive charge. This field at large distances is Coulombic, given by $E_+ = e/\varepsilon R_s^2$, where ε is the dielectric constant. Because any electron that drifts through the surface is assumed to be captured, the flux of electrons is equal to the product of the capture probability P_{cn} and the electron density. Therefore

$$P_{cn} = 4\pi R_s^2 \mu_n E_+ = 4\pi e\mu_n/\varepsilon. \tag{2}$$

This probability will be reduced by thermal emission from the trap. For deep traps this is not expected unless they have excited states well separated in energy so that multiphonon transitions weaken the transition between the ground and excited states (Crandall, 1965).

Mort *et al.* (1980a) have studied hole recombination in a-Si:H and concluded that Langevin or diffusive recombination is the most likely mechanism. A consequence of this type of capture is that, because the capture probability depends on the mobility, a slow carrier has a poorer chance of being captured than does a fast carrier.

5. RECOMBINATION

There are various pathways by which electrons and holes can recombine so as to reestablish the equilibrium that has been disturbed by photoexcitation. These can be divided into four main catagories: (1) direct bimolecular recombination of an electron in the conduction band with a hole in the valence band, (2) tail-state recombination, the process by which an electron in a region of high density of conduction-band tail states recombines with a hole in a high density of valence-band tail states, (3) monomolecular recombination through a trap or recombination center, and (4) recombination at a surface. The first of these, the direct mechanism, is not expected to be important as a recombination mechanism for a free electron and free hole. In fact, it is already accounted for in the generation process by the consideration of geminate recombination. Second, the density of free or delocalized electrons and holes is quite small since the high density of tail states contains most of the photoexcited charge. Therefore, we expect recombination from tail state to tail state to be the dominant bimolecular type of recombination. This should predominate at high excitation intensities. In the steady state, at lower light intensities the bulk of the carriers are in deep tail states or recombination centers so that the third process becomes important. It is thought that the important recombination center in a-Si : H is the dangling bond. However, other unspecified defects may become important for material with low densities of dangling bonds. Recombination at surfaces is quite likely if the surface contains a high density of surface states. This process can be important in solar cell operation.

Recombination through traps is the most familiar. It was outlined by Shockley and Read (1952) for semiconductors and is commonly referred to as Shockley–Read recombination. Their theory is quite general and applies over a wide range of temperatures and light intensities as long as only one trap dominates. If there is a distribution of traps, then the phenomenological approach of Rose (1951, 1963) is much more useful and physically appealing. Moreover, the general nature of the Shockley–Read model leads to complicated expressions that are not always physically transparent. Part of this complexity stems from the fact that thermal equilibrium is assumed to be the factor determining the occupation of the traps and recombination centers. This is not always the case in insulators. It is more likely, because the recombination centers are near midgap, that equilibrium is determined by electron and hole capture rather than by thermal emission. In this case, simple expressions can be obtained for the recombination rates of electrons and holes through deep traps.

We shall outline one situation in detail here since we wish to use the result later for detailed calculations of the primary photocurrent. Figure 6 represents a single trapping level located between the conduction- and valence-band mobility edges E_c and E_v, respectively. There are four possible transitions: (1) an electron is captured by the trap, (2) a trapped electron is

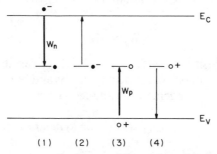

FIG. 6. Band diagram representing four possible transitions involved in a recombination event at a single level.

thermally excited to E_c, (3) an electron is thermally excited to the trap from the valence-band mobility edge, i.e., hole emission from the trap, and (4) the trapped electron falls into an empty state near the valence band, i.e., hole capture. Steps (1) and (4) taken together comprise a recombination event. If this trap is located near midgap, then it is nearly 0.8 eV from either conducting state. The thermal emission rate W_e to these states is roughly

$$W_e = \nu_0 \exp(-E_t\beta) \sim 10^{13} \times 10^{-14} \sim 10^{-1} \quad \text{sec}^{-1}. \qquad (3)$$

Here, ν_0 is the attempt-to-escape frequency, E_t the trap depth, $\beta = (kT)^{-1}$, where k is Boltzmann's constant, and T the absolute temperature. Since either process (1) or (4) is expected to have a rate greater than 10^3 sec^{-1}, the occupancy of the trap will be determined by capture and not by thermal emission. This is the difference between recombination in insulators and semiconductors. An example of this type of trap or recombination center in a-Si:H is the neutral or singly occupied dangling bond (Biegelsen, 1980) (Street, 1982a) that is located near midgap (Cohen et al., 1982; Jackson, 1982).

We now give a simple argument to calculate the recombination rate for the above process by considering that only steps (1) and (4) are important. Let W_n be the capture rate per electron when all the recombination centers of concentration N_r are empty and W_p be the similarly defined capture rate per hole. These rates depend on the concentration of recombination centers and the details of the capture process but not on the occupancy of the recombination centers. In terms of the capture probability expressed by Eq. (2), $W_n = N_r P_{cn}$. The rate of capture or trapping of conducting electrons will be proportional to the density of conducting electrons n, to W_n, and to $1 - f_n$, where f_n is the fraction of the N_r traps occupied by electrons. Thus the electron capture rate R_n is of the form

$$R_n = nW_n(1 - f_n). \qquad (4)$$

Similar arguments for the capture of holes lead to a capture rate for holes R_p of the form

$$R_p = pW_p f_n, \qquad (5)$$

where p is the density of conducting holes. If the excitation is constant so that a steady state can be established, electrons and holes must be trapped at the same rate since their generation and recombination occur in pairs. Thus $R_p = R_n$.

Equating these expressions permits one to determine f_n from Eqs. (4) and (5). We find that

$$f_n = nW_n/(nW_n + pW_p). \qquad (6)$$

This argument shows that the distribution function f_n is determined by the recombination kinetics rather than by thermal emission and capture as is the case for Shockley–Read recombination. Substituting Eq. (6) into Eq. (4) gives for the pair recombination rate $R = R_p = R_n$ the value

$$R = npW_pW_n/(nW_n + pW_p). \tag{7}$$

This expression for the recombination rate [Eq. (7)] is, in fact, identical to that derived by the thermal equilibrium arguments in the Shockley–Read theory, provided that the low temperature limit is taken in their expression (Shockley and Read, 1952) for the recombination rate. This is understandable because thermal emission is unimportant, even in narrow gap material, at low temperature.

It is useful at this point to define trapping times, because they will have special significance for primary photoconductivity. The electron trapping time is just the inverse of the electron capture rate W_n. Thus

$$\tau_n = W_n^{-1} \tag{8}$$

and

$$\tau_p = W_p^{-1}. \tag{9}$$

These trapping times depend only on the trap density and details of the capture, whereas the *recombination* times, which are defined in terms of the recombination rate, depend on n and p. The electron recombination time τ_n^R is given by

$$\tau_n^R \equiv \frac{n}{R} = \tau_n + \frac{n}{p}\tau_p, \tag{10}$$

and the hole recombination time τ_p^R is

$$\tau_p^R \equiv \frac{p}{R} = \tau_p + \frac{p}{n}\tau_n, \tag{11}$$

where R is the equilibrium recombination rate. These recombination times are constant only for the minority carrier and cannot both be constant at the same time.

It is convenient to rewrite the equilibrium recombination rate R in terms of these trapping times to give

$$R = np/(n\tau_p + p\tau_n). \tag{12}$$

If n is small, the electron is the minority carrier and it controls the recombination as can be seen by neglecting $n\tau_p$ in the above expression to give

$$R \sim n/\tau_n. \tag{13}$$

Similarly, the hole will control recombination if

$$p\tau_n < n\tau_p. \tag{14}$$

6. DEMARCATION LEVELS

The above argument for the recombination rate applies when a single defect with a well-defined energy is responsible for electron–hole recombination. However, in impure or doped a-Si:H there can be a broad distribution of states in the gap that act as traps and recombination centers. Even a single impurity such as phosphorus does not produce a single state in the gap but rather produces a broad band of states, nearly 0.8 eV wide, in the center of the gap (Lang et al., 1982). For this reason it may be better to use a quasi-Fermi level approach to describe the photoconductivity rather than a rate equation formalism that attempts to treat a distribution of energy levels in detail. The quasi-Fermi level approach was used by Rose (1951) to describe complex photoconductive properties in insulators in simple physical terms.

A steady-state quasi-Fermi level for electrons E_{fn} can be defined as that Fermi level consistent with a density n of conducting electrons under illumination by means of the relation.

$$n = N_c^* \exp(-|E_{fn} - E_c|\beta). \tag{15}$$

Here, N_c^* is the effective density of states at the electron mobility edge. Similar reasoning leads to an analogous definition for the hole quasi-Fermi level, E_{fp}, defined by

$$p = N_v^* \exp(-|E_{fp} - E_v|\beta), \tag{16}$$

where N_v^* is the effective density of states at the valence-band mobility edge.

In an insulator, in the absence of illumination, these quasi-Fermi levels lie near midgap. Under illumination they move toward their respective mobility edges, resulting in "electronic doping." As a first approximation, states lying between a mobility edge and a respective Fermi level act mainly as traps rather than as recombination centers. States lying between the two Fermi levels will be recombination centers. Thus with increasing illumination the number of recombination centers increases so that a sublinear dependence of the photoconductivity on light intensity is expected and is often observed in a-Si:H. These Fermi levels act as fiducial marks to separate recombination from trapping effects.

A quantitative criterion to distinguish traps from recombination centers is made by the introduction of a demarcation level. An electron at this level will have the same probability of being thermally excited to the mobility edge as of capturing a hole. States above this electron demarcation level D_n

will act predominantly as traps, those below as recombination centers. This criterion can be defined by the equality (Rose, 1963)

$$v_n \exp(-|D_n - E_c|/\beta) = pP_{cp}, \qquad (17)$$

where P_{cp} is the probability that the state at the demarcation level will capture a hole. The left-hand side of this equality, the probability per unit time that an electron will be thermally emitted to the mobility edge, is given by the product of frequency v_n and a Boltzmann factor. The attempt-to-escape frequency v_n is usually expressed in terms of the electron capture probability P_{cn}, using detailed balance arguments (Rose, 1963) to give the relation

$$v_n = N_c^* P_{cn}. \qquad (18)$$

The right-hand side of Eq. (18) is just the capture rate for holes. Solving for the demarcation level gives

$$|D_n - E_c| = kT \ln[N_c^* P_{cn}/N_p^* P_{cp}]. \qquad (19)$$

It is not possible to determine the actual value of D_n without solving for the hole density. However, this expression can be recast in a more symmetrical form by using the definition of recombination times for holes and electrons [Eqs. (10) and (11)] and eliminating N_c^* by means of the definition of the electron quasi-Fermi level to arrive at

$$D_n = E_{fn} + kT \ln[\tau_n^R P_{cn}/\tau_p^R P_{cp}]. \qquad (20)$$

This definition states that to a first approximation, the electron demarcation level is located at the steady-state quasi-Fermi level. The demarcation level is displaced by the small correction term. If the probabilities for hole or electron capture are nearly equal and, furthermore, the hole and electron recombination times are similar, this correction term is negligible.

By the same reasoning, the demarcation level D_p for holes is found to be

$$D_p = E_{fp} - kT \ln[\tau_n^R P_{cn}/\tau_p^R P_{cp}]. \qquad (21)$$

As with the electron demarcation level, the hole demarcation level, in first order, is located at the hole quasi-Fermi level. The second-order corection is the same as the correction to the electron demarcation level; thus the shifts of the demarcation levels away from their respective quasi-Fermi levels are the same and in the same direction.

The temperature dependence of the demarcation levels can be estimated from Eqs. (19)–(21). In general, as temperature decreases, they move toward their respective band edges. This increases the number of recombination centers and causes the decrease in the photoconductivity with decreasing temperatures that is often observed (Rehm et al., 1977; Wronski and Daniel, 1981). Another result of the movement of the demarcation level with temperature is the shift in the characteristic luminescence peak with temperature (Street, 1981). The luminescence peak position and intensity increase with decreasing temperature as the demarcation levels move apart and more band-tail states are converted from shallow traps to radiative recombination centers.

These demarcation levels are represented in Fig. 7 on a standard band diagram. All the states between D_n and D_p are recombination centers, because the probability of thermal emission to a mobility edge is weak compared with electron or hole capture. Since the demarcation level is determined by the values of capture probabilities and recombination times, each different type of trap should have its own set of demarcation levels associated with it. Thus in general there would be a spectrum of demarcation levels. However, there is only sparse evidence at best to assign parameters to defects in a-Si:H. Thus it is reasonable to consider one set of demarcation levels. In fact interpretation of the deep level transient spectroscopy (DLTS) spectra for phosphorus doped a-Si:H (Lang et al., 1982) indicates a single capture probability for all the states in the gap.

In Fig. 8, the dotted–dashed curve represents the continuous density of gap states superimposed on the band diagram. Along with this is a dashed curve that represents the occupancy function for these states. Because the rapid falloff of the Fermi functions near E_c and E_v outweighs the increase in $N(E)$, the density of electrons and holes, shown by the solid curve, will peak somewhere between midgap and E_{fn} and E_{fp}, respectively. The rate of electron–hole recombination depends not only on the densities of n and p but on the overlap of their wave functions. Because the density of localized states decreases rapidly toward midgap, the wave-function overlap decreases dramatically for states far removed from the band edges. In fact, the states most likely to be involved in recombination will be those just below the

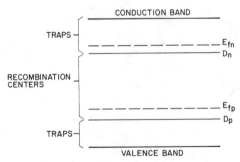

FIG. 7. Band diagram to represent the demarcation levels D_n and D_p for electrons and holes, respectively. The quasi-Fermi levels for electrons and holes, respectively, are E_{fn} and E_{fp}.

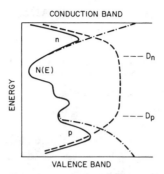

FIG. 8. Band diagram to represent the occupancy of the states in the gap under illumination. The dotted–dashed curve represents the continuous density of states and the dashed curve their occupancy function. The solid curve gives the actual electron and hole densities.

mobility edge. However, direct recombination between an electron and a hole may not be the most important recombination process. The reason is that a large amount of energy must be radiated by multiphonon transitions. Since we are considering the domain in which geminate recombination and hence luminescence transitions are weak, the electron and hole do not recombine radiatively. Note that the greater the number of phonons involved in a multiphonon transition, the weaker the transition. Therefore, recombination may proceed by a two-step process involving a recombination center near midgap. This reduces the number of phonons involved in each transition. The two-step process is the one that we considered above to calculate R.

Which of the two mechanisms dominates in any given situation depends on the matrix element for the transition and the density of final states. For the two-step process, the density of final states is the density of recombination centers, which is fixed by material properties. For direct recombination, the density of final states is proportional to the hole population, which can be varied with illumination level. Thus at low enough light intensity the two-step process will dominate and at high enough light intensity direct recombination will be most important. In fact, the work of Vardeny *et al.* (1982), using high excitation intensities to study time-resolved carrier relaxation, is consistent with a bimolecular recombination mechanism. A full discussion of this technique is given by Tauc in Chapter 9.

The use of demarcation levels is one of the features that distinguishes photoconductivity in insulators from that in semiconductors. In semiconductors, in which the number of thermally excited background carriers is much greater than the number of photogenerated carriers, all levels are in thermal equilibrium and occupation is determined by the equilibrium Fermi level.

IV. Photoconductivity Mechanisms

7. SECONDARY PHOTOCONDUCTIVITY

This is the most familiar type of photoconductivity and one that is described in most texts (Bube, 1960; Mort and Pai, 1976; Rose, 1963; Ryvkin, 1964). It occurs whenever the contacts do not in any way affect the photoconductivity. This means that in a homogeneous system that is uniformly illuminated the electron and hole densities must be uniform throughout the material. Transport of electrons and holes through the material in response to an applied electric field will not affect their distribution as it does in primary photoconductivity. Therefore, the steady-state electron density n is given directly in terms of the photogeneration rate G, which is the number of electron–hole pairs excited per unit volume per unit time, and the recombination time τ_n^R by the equation

$$n = G\tau_n^R. \tag{22}$$

The photocurrent is obtained by dividing this expression by the transit time t_n of a free electron between the electrodes. Thus the electron photocurrent density J_n is

$$J_n = eG\tau_n^R/t_n. \tag{23}$$

This expression shows the significant feature of secondary photoconductors, that is, they exhibit gain. The photoconductive gain g, given by

$$g = \tau_n^R/t_n, \tag{24}$$

is equal to the number of electrons passing through the photoconductor per absorbed photon. In this definition it is assumed that each absorbed photon produces a free electron–hole pair. In a long-lifetime high-mobility photoconductor the photoconductor gain can be quite large.

A similar expression for the free hole density p can be obtained in terms of the hole recombination time τ_p^R and the generation rate. It is

$$p = G\tau_p^R. \tag{25}$$

Since τ_p^R is not necessarily equal to τ_n^R, p and n are not equal. This feature distinguishes the insulator case, which we discuss here and throughout this chapter, from the semiconductor or trap-free insulator case in which the free-electron and free-hole densities are nearly equal. In these high purity photoconductors, with little trapping or few states in the gap, charge neutrality requires that $n = p$. In amorphous silicon there is such a large density of states in the gap that there are more electrons and holes in traps than in free states. This large reservoir of trapped electrons and holes can easily maintain charge neutrality, which permits the free-electron and free-hole densities to be quite different.

Combining the free-electron and free-hole contributions to the photocurrent gives for the total photocurrent density J the value

$$J = eG(\tau_n^R/\tau_n + \tau_p^R/\tau_p) = (eGF/L)(\mu_n\tau_n^R + \mu_p\tau_p^R), \qquad (26)$$

where the second expression follows from the definition of the transit time in terms of the mobilities μ_n and μ_p of electrons and holes, respectively. In Eq. (26) F is the electric field in the photoconductor and L the distance between the electrodes. From this expression the overall gain of the photoconductor can be represented as

$$g = (F/L)(\mu_n\tau_n^R + \mu_p\tau_p^R) = (l_n + l_p)/L, \qquad (27)$$

where l_n and l_p are the drift lengths of electrons and holes, respectively. These quantities represent the mean distance that a charge carrier drifts in the direction of the electric field before recombining. Because a charge carrier is replenished at the contacts, it can make more than one transit through the photoconductor during its lifetime. Since each time it makes a complete transit through the photoconductor it contributes one unit charge to the current, the gain is just the ratio of drift length to sample length.

There is always the problem of how to measure the recombination time. As the preceding expressions show, even when a single carrier dominates the current, one always measures a mobility–lifetime product. In a transient measurement, the time constant for the decay of the photocurrent is the *response* time τ_{RS}. It is always longer than the recombination time because the charge in shallow traps must be thermally excited to the mobility edge before it can recombine. Because of this,

$$\tau_{RS} = \tau_n^R/\theta_n, \qquad (28)$$

where θ_n is the ratio of the number of free to total electrons. The total number of electrons is the sum of free and trapped electrons. In a-Si:H, $\theta_n \ll 1$. This same factor θ_n connects the drift and microscopic mobilities by the relationship

$$\mu_n^D = \theta_n\mu_n, \qquad (29)$$

where μ_n^D is the drift mobility. By combining Eqs. (28) and (29), we see that the mobility–lifetime products are the same, i.e.,

$$\mu_n\tau_n^R = \mu_n^D\tau_{RS}. \qquad (30)$$

This equation permits one to use drift mobilities in Eq. (26), provided that the time is the response time rather than the recombination time. One should be careful about confusing these times, as is occasionally done in the literature. In a-Si:H, only the drift mobilities have been measured, so that the appropriate photoconductive time is the response time.

Before leaving this section we wish to touch on the subject of sensitization of a photoconductor (Rose, 1951, 1963). In narrow-gap pure semiconductors, in which the electron and hole densities must remain equal, the only mechanism for increasing the photoconductivity is to reduce the density of recombination centers. In wide-band insulators there is another method of increasing the photoconductivity. This is accomplished by increasing the gain of one sign of carrier at the expense of a decrease in the gain of the other. In an insulator there is no requirement that the gains must be the same for both carriers. Rose (1963) gives some examples of sensitization. The general idea is to add new recombination centers having very different cross sections for electron and hole capture. These new recombination centers, if in large enough concentration, will overwhelm the original recombination centers and hence determine the recombination traffic in the photoconductor. Since the recombination cross section for one carrier is small, it will have a long lifetime and thus a high gain. We mention this only as a caution to those who claim that their a-Si:H is quite pure because it has a high photoconductivity. The high photoconductivity may indicate just the opposite: that the material has a large number of defects that act as sensitization centers. In secondary photoconduction the large number of sensitizing centers is not a problem. However, when we consider primary photoconduction, the mode of operation of a solar cell, the space charge due to the trapping of one sign of carrier at these sensitizing centers will reduce the efficiency of the cell from its optimum.

8. PRIMARY PHOTOCONDUCTIVITY

The idea of primary photocurrents was introduced with the early systematic study of the alkali halides by the Göttingen photoconductivity school (Gudden and Pohl, 1921). Only a primary photocurrent was possible in these crystals, because their wide band gap precluded making a good ohmic contact to them. Until there was an industrial interest in photoconductors and the ensuing development of ohmic contacts to semiconductors and some insulators, this was the main area of study. Furthermore, the alkali halides are such wide band-gap insulators that the principal source of photoelectrons is from optical absorption by defects. Thus one need consider only single-carrier photocurrents.

However, single-carrier primary photocurrents can only be transient currents, because the space charge associated with the immobile charge will, at long times, completely screen the applied field. This photogenerated space charge is not distributed throughout the photoconductor but localized next to a contact within a distance referred to as the *schubweg* or drift length. This distance is the mean distance that a photocarrier drifts in the direction of the electric field before being trapped [see Mott and Gurney (1946) for a discussion of primary photocurrent].

Figure 9 contains a schematic of an insulating uniformly illuminated photoconductor with blocking contacts. These contacts are blocking because the thermal emission electron current that can flow from the metal contact into the insulator is much smaller than the photogenerated electron current moving away from the contact in the insulator. In this example, the holes are immobile. Thus there will be an exhaustion of carriers near the contact leaving a positive space charge. The width of this exhaustion region will be the mean distance that a free electron moves before being trapped. This is the drift length l_n defined by Eq. (27). At sufficiently large electric fields, the drift length approaches the sample thickness and all the photogenerated electrons are extracted. Under these conditions the time dependence of the current following a light flash can be studied to determine the drift mobility as described by Tiedje in Chapter 9 of Volume 21C. In the limit that the space charge within the distance l_n does not perturb the applied field, the field dependence of the number of extracted photoelectrons n_e can be used as a measure of l_n and hence the $\mu\tau$ product using the well known Hecht relationship (Mott and Gurney, 1946)

$$n_e = (N_0 l_n / L)[1 - \exp(-L/l_n)], \tag{31}$$

where N_0 is the number of absorbed photons. This expression applies only for a single carrier and only to transient currents. The Hecht relation, or variations of it (Bell, 1980), cannot be applied correctly to the steady state since it violates the fundamental continuity equations. However, the ideas behind the Hecht relation can be used reliably to describe charge motion following light flashes. This is described in the nuclear detection literature (Meyer, 1968).

Street (1982a) applied the Hecht relation to measurements of transient photocurrents in a-Si:H to determine trapping parameters of holes and

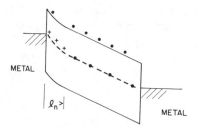

INSULATOR

FIG. 9. Potential energy diagram of a uniformly illuminated insulating layer between two blocking metal contacts. The exhaustion region, represented by positive charge, extends into the insulator a distance l_n. Electrons are represented in traps and conducting states by solid circles. Holes trapped throughout the insulator are not shown.

electrons. By using strongly absorbed light and drifting either electrons or holes, depending on the polarity of the applied field, he could measure the drift length of either carrier. Since he was able to obtain a close correlation between the drift length and the density of dangling bonds, he concluded that the dangling bond was an important trap. We shall discuss the nature of this trap latter. However, we wish to point out here that the Hecht relation is quite useful when applied correctly.

If both carriers are mobile, then steady primary photocurrents are possible. However, the blocking nature of the contacts changes the current flow significantly from the situation encountered for secondary photocurrents. The main difference is the nonuniformity of the electron and hole currents. Near the contact that is blocking for photogenerated electrons the current is carried mainly by holes. At the opposite contact, which is blocking for holes, the current is mostly an electron current. Of course, current continuity requires that the sum of electron and hole currents be the same throughout. The variation of electron and hole concentration across the insulator means that the recombination rate is space dependent. This introduces a great deal of complexity into the solution of the transport equations.

The space-charge regions that appear near the contacts distort the electric field and produce an additional complication to the solution of the transport equations by coupling them to the Poisson equation. Thus one must solve a set of coupled second-order nonlinear differential equations. In fact, for the general case, there is no closed form solution to these equations and one must resort to complicated numerical methods (Chen and Lee, 1982; Swartz, 1982).

Some progress (Crandall, 1981a,b, 1982, 1983a) has been made in reducing the solution of the transport equations to manageable form by using the regional approximation (Lampert and Mark, 1972; Lampert and Schilling, 1970), which is very useful in solving space charge injection in insulator and electrolyte problems (Lampert and Crandall, 1979). In this regional approximation the parameter space is divided into regions in which closed form solutions to the transport equations can be found. The solutions in the different regions are then connected at the boundaries to obtain a general solution. An important feature of this method is that it allows one to gain physical insight without becoming mired in endless mathematical detail. As an example of its application we shall show how to solve approximately the simplest of primary photoconduction situations, that is, an insulator between two metal blocking contacts. Actually this is a good approximation to a $p-i-n$ solar cell, since the p and n regions function mainly as blocking contacts for the photogenerated carriers. This shows the complexity in the solution of the transport equations, even in this simple case.

A calculation of the photocurrent begins with the solution of the continuity equations. These equations, along with the Poisson equation, govern the response of the photoconductor to external stimuli such as light, temperature, and applied voltage. We consider only the steady state and furthermore assume that there is a well defined mobility for each carrier.

The continuity equation for holes is

$$-(1/e) \, dJ_p/dX = G - R, \tag{32}$$

where the hole transport current is expressed by

$$J_p = e\mu_p(Fp - V_T \, dp/dX). \tag{33}$$

Here $V_T = kT/e$ is the thermal voltage. The Einstein relation $\mu_p kT/e = D$ was used to eliminate the diffusion coefficient D. The coordinate X is measured perpendicular to the plane of the insulating film. The first term in Eq. (33) is the electric-field-driven or drift current. The second term is the diffusion current. There is a similar continuity equation for the electrons, given by

$$(1/e) \, dJ_n/dX = -G + R, \tag{34}$$

where the electron transport current is

$$J_n = e\mu_n(Fn + V_T \, dn/dX). \tag{35}$$

Accompanying the continuity equations is the Poisson equation, which relates the electric field to the carrier densities and the fixed space charge Q_s. The Poisson equation is

$$\varepsilon \, dF/dX = Q_s + e(p - n), \tag{36}$$

where ε is the static dielectric constant of the insulator.

Because p, n, and F occur in these equations, they are all coupled together, making their solution formidable. Even if p and n are small enough to be neglected in the Poisson equation, the continuity equations for electrons and holes are still coupled together through the recombination rate R. The form of R given by Eq. (12) is nonlinear in n and p and thus will make these equations coupled second-order nonlinear differential equations that still appear to have no general solution.

The only hope of obtaining useful solutions to this system of equations is to realize that not all the terms are of equal importance in any given region of parameter space. In fact, except for small transition regions, only one or two terms are important over large regions of parameter space. This then permits the small terms to be neglected with respect to the dominant ones, permitting closed form solutions. This is the essence of the regional approximation. In the following we shall outline the regions of parameter space where the different physical situations predominate and then discuss each one in detail.

In the following discussion we shall represent the voltage across the insulator by V_0. If the two contacts are identical, then V_0 and the applied voltage V_A will be the same. If the two contacts are different, then there will be a built-in voltage V_b across the insulator even in the absence of an applied voltage. For this discussion we shall assume that the contacts are identical so that $V_b = 0$.

Because a-Si:H is a wide band-gap semiconductor, the density of thermally generated carriers is small enough so that the current in the absence of illumination, the dark current, is usually much smaller than the photocurrent. Thus we assume that the influence of the thermally generated carriers can be neglected in calculating the photocurrent. Furthermore, we restrict the problem to the situation in which the electric field is nearly uniform across the insulator. In fact, this may be a good approximation for high efficiency solar cell material. This requires that two conditions be met. First, the fixed space charge Q_s must be small enough not to perturb the electric field significantly in the insulator. At first this may seem to be a drastic approximation. However, measurements of the space charge in high purity glow-discharge-produced a-Si:H films show that Q_s can be less than 1×10^{14} cm^{-3}. These low space-charge densities have been obtained by DLTS measurements of the gap state density (Crandall, 1980a; Lang et al., 1982), measurements of capacitance as a function of frequency, and surface photovoltage measurements (Moore, 1983). In a 1-μm-thick insulator with a uniform space charge density of 1×10^{14} cm^{-3} and $V_0 = 1$ V, the percentage change in field across the insulator due to this space charge is about 15%. For the more usual case of a solar cell structure this change is only about 4%. Thus the uniform field approximation can be justified for these conditions.

Certainly, at sufficiently low light intensity, the density of photogenerated carriers will be so small that they will not perturb the electric field. Their effect on the field can always be estimated by solving the continuity equations with the uniform field approximation. Then with the values of p and n so obtained, Poisson's equation can be solved to find the departures of the field from uniformity. This will give a criterion for determining when the uniform field approximation will break down.

Even with the uniform field approximation, which reduces the problem to the solution of two second-order differential equations, one cannot obtain a general solution without further approximations because the recombination rate given by Eq. (12) is nonlinear in the electron and hole densities. Some workers have made a constant recombination time approximation (Dalal and Alvarez, 1981; Delahoy and Griffith, 1981; Gutkowicz-Krusin et al., 1981; Gutkowicz-Krusin, 1981; Miyamoto et al., 1981) across the entire insulator. They assumed that the hole is easily trapped, is the minority carrier, and will thus have a short recombination time. They thus approximated the recombination rate by

$$R = p/\tau_p. \tag{37}$$

This approximation, which sets the hole recombination time τ_p^R equal to a constant, the hole trapping time, cannot be correct throughout the insulator. In fact Eq. (37) is correct only if

$$p/\tau_p < n/\tau_n, \tag{38}$$

which means either that the hole density must be small or that the hole lifetime must be large. Neither of these conditions is consistent with the initial assumption of the hole being readily trapped. In fact, we shall show that if the hole mobility and lifetime are short, then $p > n$ over most of the insulator and the recombination rate over most of the insulator is controlled by the minority carrier electron, not the hole. This can readily be predicted from Eq. (12) for the recombination rate, which shows that the carrier with the longer trapping time will control the recombination kinetics over most of the insulator. It is this carrier that is the minority carrier. This situation is analogous to the situation in semiconductors in which the minority carrier is defined by doping. Here also it is the minority carrier that controls the recombination kinetics. However, in the insulator case, the minority carrier is determined not by doping but by recombination kinetics and transport. Furthermore, in this primary current mode of transport, the electrons and holes interchange their roles across the insulator.

To show how the regional approximation is applied to the solution of continuity equations, it is helpful to define certain parameters and see how their interplay determines the important terms in the continuity equations. These parameters are outlined in Table I. The lengths are the sample length L, the hole drift length l_p, the electron drift length l_n, and the mean free path of a photon l_a, equal to the reciprocal of the optical absorption coefficient α. The mean distance the charges move is proportional to the sum of the drift

TABLE I

IMPORTANT PARAMETERS

Length	Voltage	Current
L	V_0	J_d
$l_p = \mu_p \tau_p F$	$V_T = kT/e$	J
$l_n = \mu_n \tau_n F$		J_{spc}
$l_a = 1/\alpha$		
$l_c = l_p + l_n$		

length of electrons and holes and can be referred to as the collection length, which is given by $l_c = l_p + l_n$. The voltage across the insulator is V_0 and the thermal voltage V_T. The important currents are the dark current J_d, the photocurrent J, and the space-charge-limited current of photocarriers J_{spc} (Crandall, 1981a).

Because the current is a *primary* photocurrent, the *drift lengths* are the fundamental quantities rather than the *lifetimes* as for *secondary* photocurrents. Since there is no gain in a primary photocurrent, the current in the external circuit, provided that $l_c < L$, is proportional to the mean distance the charges move before trapping and eventual recombination divided by the insulator thickness.

The magnitude of the thermal voltage V_T relative to other important voltages in the insulator determines whether diffusion will be a significant factor in determining the current. For weakly absorbed light, $l_a > L$, diffusion will be important only if $V_T > V_0$ (Reichman, 1981). Since $V_T \approx$ 0.025 V at room temperature, diffusion can safely be neglected over a large range in V_0. However, for strongly absorbed light V_T must be compared with the voltage drop across the absorption length, which is $V_0 l_a / L$. For example, in the blue end of the visible light spectrum, $l_a \sim 0.5 \times 10^{-5}$ cm, so that for a 1-μm-thick a-Si:H film diffusion effects will dominate if $V_0 < 0.5$.

As long as $J_d \ll J$, the dark current can be omitted. However, there is an upper limit to the photocurrent that can flow in the insulator without perturbing the electric field. This is because the displacement of electrons to one side and holes to the other side can produce large space-charge effects (Crandall, 1981a). Thus the photocurrent must be less than the space-charge-limited current J_{spc}, given by (Crandall, 1981a)

$$J_{spc} = 4\varepsilon V_0^2 \mu^D L^{-3}, \qquad (39)$$

where μ^D is the *drift* mobility of the slowest carrier.

These ideas should become clearer as specific examples are considered. We shall give only a brief outline of the physical processes involved, keeping the mathematical details to a minimum, since they can be found in the literature.

First, we consider that $\alpha L \ll 1$ so that light is weakly absorbed in the film. In this case diffusion is unimportant, so that the current is produced only by drift of the photogenerated carriers in the field. Thus the continuity equations [Eqs. (32) and (34)] reduce to first order because of the absence of the diffusion term in the current [Eqs. (33) and (35)]. Even these equations cannot be solved in general because of their coupling to each other via the Poisson equation and because R is a nonlinear function of p and n. However, in the limit that R can be neglected with respect to the other terms, these equations can be solved along with Poisson's equation to give a description of the effect of the free-carrier space charge on the electric field (Crandall, 1981a). Because we are presently making the uniform electric field approximation, this will not be discussed at this point.

Considerable insight into the primary photoconduction process can be obtained if we neglect recombination to calculate the electron and hole densities. Then these values can be substituted into the expression for the recombination rate to show how it varies throughout the insulator. If $R \ll G$, the continuity equations can readily be solved by using the appropriate boundary conditions $p = 0$ at $X = 0$ and $n = 0$ at $X = L$. These boundary conditions are a direct consequence of the blocking nature of the contacts. In this case, the electron and hole densities increase linearly away from the boundaries $X = 0$ and $X = L$. These densities are expressed by

$$p = GX/\mu_p F \tag{40}$$

and

$$n = G(L - X)/\mu_n F. \tag{41}$$

Even though this approximation of no recombination is not expected to apply at low field, it does permit one to show readily how poor is the approximation of a constant lifetime between $X = 0$ and $X = L$.

Substitution of the values of p and n from Eqs. (40) and (41) into Eq. (12) shows that R is a strong function of position. As a function of position X, the recombination rate can be written as

$$R = GX(L - X)/[Xl_n + (L - X)l_p]. \tag{42}$$

Near $X = 0$, the recombination rate can be approximated by

$$R \sim GX/l_p, \tag{43}$$

which shows that the hole controls the recombination kinetics. In this regime the hole density is much less than the electron density, so that the hole is also the minority carrier. Near $X = L$ the electron will control the recombination kinetics. The recombination rate can be approximated by

$$R \sim (L - X)/l_n, \tag{44}$$

so that in this region the electron is the minority carrier. Equations (42)–(44) have been written in terms of the drift lengths to emphasize the fact that the drift lengths rather than the lifetimes are fundamental quantities.

We can now define two crossover points. One is the recombination rate crossover point X_R, at which the recombination changes from hole controlled to electron controlled. This point is found from the condition

$$Xl_n = (L - X)l_p,$$ (45)

which has as the solution

$$X_R \equiv Ll_p/(l_n + l_p) = Ll_p/l_c.$$ (46)

The crossover point X_R is determined by the drift length rather than by the trapping times. This is because the distribution of carriers is now determined by the interplay of the field-driven drift and trapping instead of just trapping. For this reason the drift lengths are the fundamental parameters.

The second crossover point occurs when $n = p$. At this point, X_m, the carriers interchange their roles as minority carriers. This point is given by

$$X_m = L\mu_p/(\mu_p + \mu_n).$$ (47)

For $X < X_m$ the hole is the minority carrier and for $X > X_m$ the electron is the minority carrier. The two crossover points X_m and X_R will not coincide unless $\tau_p = \tau_n$. Only when they coincide can one say that the *minority* carrier always controls recombination. For example, if $\tau_p \gg \tau_n$, then $X_m \ll X_R$ and the hole, the majority carrier for $X > X_m$, controls recombination in the interval between X_m and X_R.

If we return to the definition [Eq. (46)] of the recombination crossover point, we see the importance of the drift length in determining the recombination kinetics. In fact, it is the carrier with the *longer* drift length that determines recombination kinetics over most of the sample. This is just the opposite of the usual speculation that the carrier with the *shorter* drift length will determine the transport properties and hence the current. The reason that the carrier with the *longer* drift length determines the current is that we are dealing with a primary photoconductor, in which the charge induced in the external circuit is proportional to the distance a charge carrier moves before it is trapped and eventually recombines. If the hole, for example, has the shorter drift length, then it does not contribute as much to the current as does the electron. The hole is also more readily trapped than is the electron. Since the electron and hole recombine in pairs, there is not a continual buildup of trapped charge. There are just more trapped holes than trapped electrons.

When we consider the more general situation in which recombination cannot be neglected in calculating the carrier densities, the preceding ideas still apply. In fact, X_R is still given by Eq. (46). However, X_m must be found from the solution of a transcendental equation. To solve the continuity equations when recombination cannot be neglected, we apply the regional approximation to the recombination rate and make the lifetime constant in different parts of the insulator. As we showed above, R will be controlled by the hole for small X and by the electron for large X, with the crossover point X_R. Thus we approximate the lifetime for $X < X_R$ by the hole trapping time τ_p and for $X > X_R$ by the electron trapping time τ_n. These approximations linearize the continuity equations [Eqs. (32)–(34)], so that they may be solved by elementary means. This is an enormous simplification of the problem and permits one to obtain useful results for a variety of situations while still retaining the essential physics. Because in this review we wish to concentrate on the physical behavior, we shall reproduce here only some of the essential results and omit the mathematical details that can be found in the literature (Crandall, 1981b, 1982, 1983a).

First, we consider the situation of a weak absorption constant so that $l_a > L$. In this case diffusion can be neglected and the continuity equations can readily be solved for the electron and hole densities. The expressions for p and n in the literature (Crandall, 1982) can be rewritten in terms of l_c and X_R to produce symmetric expressions that point out the importance of the collection length l_c. They are

$$p = G\tau_p\left[1 - \exp\left(\frac{X}{X_R}\frac{L}{l_c}\right)\right], \qquad X \le X_R, \qquad (48)$$

and

$$n = G\tau_n\left[1 - \exp\left(\frac{X-L}{X_R-L}\frac{L}{l_c}\right)\right], \qquad X \ge X_R. \qquad (49)$$

Since these expressions for p and n are not simultaneously valid except at $X = X_R$, we need a further condition to obtain an expression valid for all X. This condition is that the total current J, expressed by the sum of J_p and J_n, is independent of position. For $X \le X_R$, n can be expressed in terms of p and J; for $X \ge X_R$, p can be expressed in terms of n and J. The recombination rate R can now be evaluated for all X using Eqs. (48) and (49).

The electron and hole densities throughout the insulator are plotted in Fig. 10 for the situation $l_p = l_n < L$. It is obvious that p increases linearly with X near $X = 0$, where there is little recombination because of the low hole density. Because of this weak recombination, the electron density also increases linearly with decreasing X. This region of weak recombination persists over the region from $X = 0$ to $X = l_p$ because the hole controls recombination so that it is the hole drift length that is important here. A similar situation exists near $X = L$ with the roles of electron and hole reversed. In the central region between $X = l_p$ and $X = l_n$, there is significant recombination, so that the electron and hole densities are no longer determined by drift in the field but by recombination. Since all the spatial dependence of p and n arises from the drift, they are nearly constant in this recombination region.

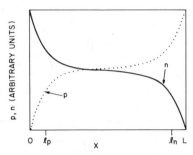

FIG. 10. Electron and hole densities as functions of distance in the insulator under the influence of an applied field. $\alpha L = 0.1$; $l_p = 0.1L$; $l_n = 0.1L$.

How the X dependence of these carrier densities affects the recombination rate is shown in Fig. 11. The solid curve in Fig. 11 is for the situation in which $l_c < L$. In this case $R \sim G$ over most of the region between $X = 0$ and $X = L$. Only within a drift length of either end of the insulator is $R \ll G$. Because of the strong recombination, the current will be much less than the saturated value eGL and will be given by the Ohm's law expression $J = eGl_c$. The other limiting situation occurs when $l_c > L$. Here we choose the case in which $l_n \gg lp$, with $l_p \ll L$, so that the electron controls recombination over most of the region. This is shown by the dashed curve in Fig. 11 for the recombination rate, where $R \ll G$ over most of the region between $X = 0$ and $X = L$.

The total current in the insulator is found by substituting Eqs. (48) and (49) into the expression for the current, to give

$$J = eF(\mu_p p + \mu_n n), \tag{50}$$

and evaluating the expression at $X = X_R$, where both expressions for n and p apply. This gives the simple expression

$$J = eGl_c[1 - \exp(-L/l_c)] \tag{51}$$

for the current, which contains only one unknown parameter, the collection length l_c. There is a reasonable physical explanation for the result that this length, the sum of the drift lengths, is sufficient to determine the current. For a primary photocurrent the charge induced in the external circuit by photoexcited electrons and holes is given by the distance that the electrons and holes move before recombining divided by the thickness of the insulator. Because both holes and electrons drift in the electric field, the charge induced in the external circuit is proportional to the sum of the drift lengths. At large l_c/L, Eq. (51) shows that the current is saturated at the value $J = eGL$, which states that all the electron–hole pairs are separated by the electric field and collected in the external circuit. In the opposite limit of

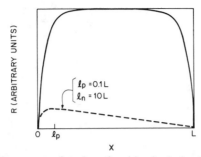

FIG. 11. Recombination rate as a function of position in the insulator. The parameters for the solid curve are the same as those in Fig. 10.

small l_c/L, there is significant recombination in the bulk, so that the current is given by the Ohm's law expression $J = eGl_c$. Of course, this picture breaks down for very short drift lengths in which the space charge due to the photogenerated carriers produces large changes in the electric field. An experimental estimate as to when space-charge effects are becoming important can be obtained from a measurement of the capacitance under illumination (Crandall, 1983b).

In the preceding discussion we have described how the current transport can be parameterized by a single quantity, the collection length l_c. By comparing this quantity with the extent of the region in which the electron–hole pairs are absorbed, we can judge whether recombination will reduce the current below its maximum value. For the case of uniformly absorbed light, as described above, the extent of the photon absorption region is the sample thickness L. However, when the light, which enters through one of the contacts, is strongly absorbed in the insulator, the region over which recombination can take place is the absorption length l_a. The reason for this is that both electrons and holes must be present for recombination to take place. In the region outside the generation region there is only one sign of carrier, the one transported out of the generation region by the field.

An example of this is shown in Fig. 12, in which the electron and hole densities are plotted as a function of X in the insulator. These densities are found from a regional approximation solution of the continuity equations for the situation in which there is little surface recombination. The details of the calculation can be found in the work of Crandall (1983a, Appendix A). What we wish to point out here is that this situation can be understood in the framework of the general solutions just obtained for weakly absorbed light. Assume that light is uniformly absorbed between $X = 0$ and $X = l_a$ and that there is no absorption between $X = l_a$ and $X = L$. Since the light is uniformly absorbed, the expressions derived above apply. The boundary condition at $X = 0$ is still $p = 0$. However, the boundary condition $n = 0$ at

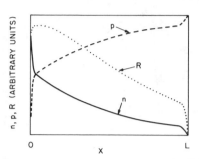

FIG. 12. Electron and hole densities and recombination rate for strongly absorbed light. The conditions are $l_p = .01L$; $l_n = 0.1L$; $\alpha L = 2$.

$L = 0$ now applies at $X = l_a$ because there is no absorption for $X > l_a$. The hole density at $X = l_a$ will be continuous across this interface. Since $n = 0$ for $X > l_a$, there can be no recombination as well as no generation. Thus the hole density will be constant in this region. In this discontinuous light absorption approximation, the current can immediately be written [using Eq. (51)], by replacing L by l_a, to give

$$J = eGl_c[1 - \exp(-l_a/l_c)]. \tag{52}$$

Note that the maximum current in the absence of recombination, $l_c \gg l_a$, is $J = eGl_a = eN_0$, where N_0 is the number of photons absorbed in the film.

The above arguments show that at large enough absorption coefficient or high enough electric field, $l_c > l_a$, so that recombination can be neglected with respect to generation. In this regime, in which there is weak recombination and strong absorption, it is no longer valid to omit the diffusion term in the continuity equations. Because l_a is short, the diffusion field, which is on the order of the thermal voltage V_T divided by the absorption length, can be on the order of the electric field across the insulator. The role of diffusion in this situation is to produce minority carrier diffusion to the contacts where surface recombination is large. This minority carrier diffusion current, which flows opposite to the electric-field-driven current, reduces the total current.

An order of magnitude expression for the reduction in the current can be obtained from the following argument. Consider the light to be incident at $X = 0$. In this case we need consider only the minority carrier holes. The field-driven hole current in the absence of diffusion is

$$J_p = eGl_a. \tag{53}$$

The hole diffusion current J_{dp} across the boundary at $X = 0$, assuming no reflection at $X = 0$, is given by the product of the gradient of the hole density and the diffusion coefficient $V_F\mu_p$. The gradient of the hole density is just the change in hole density across the absorption length $(eGl_a/\mu_p E_0)/l_a$. Thus

$$J_{dp} = -(eG/\mu_p F)V_T\mu_p. \tag{54}$$

Therefore, the total current J is reduced by this amount to

$$J = eGl_a(1 - V_T/l_a F). \tag{55}$$

Because of current continuity, only the holes need be considered. This simple expression [Eq. (55)] is equivalent to that obtained from detailed calculations (Crandall, 1983a) in the limit of small $V_T/l_a F$ and by neglecting recombination.

V. Experiments

Many details of the measurement of photoconductivity can be found in texts on the subject (Bube, 1960; Mort and Pai, 1976; Moss, 1952; Ryvkin, 1964). In this part we wish to outline only some of the experiments that may be most appropriate or particular to amorphous silicon. The discussion will be restricted to dc or quasi-dc measurements since Volume 21C contains reviews of transient current measurements that are useful in determining the drift mobilities of the carriers.

9. DETAILS

There are two basic electrode configurations appropriate to thin films of amorphous silicon. These are the planar configurations depicted in Fig. 13 and the sandwich configuration shown in Fig. 14. The planar configuration is most commonly used with ohmic contacts to measure secondary photocurrents, whereas the sandwich geometry, with blocking contacts, is appro-

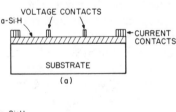

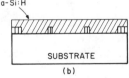

FIG. 13. (a) Electrode configuration for the planar geometry. Metal contacts deposited following deposition of a-Si:H on the substrate. (b) Insulating substrate–planar geometry. Metal deposited before a-Si:H.

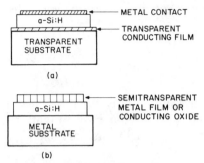

FIG. 14. (a) Electrode configuration for the sandwich geometry. A transparent conducting film, such as indium tin oxide, is applied to the transparent substrate before deposition of the a-Si:H. The metal film Schottky barrier can be platinum or palladium. (b) For this sandwich geometry the Schottky barrier is a semitransparent metal.

priate for measurements of the primary photocurrent that is typical of solar cell operation.

The large area current-source contacts in Fig. 13 should be of a metal suitably chosen to make an ohmic contact. Chromium is usually a good choice for undoped amorphous silicon. Even if the contact is not a good ohmic contact, in the sense that Ohm's law is not obeyed at low voltages, the voltage probe contacts reduce problems associated with potential drops near the current contacts. These voltage contacts should be small so as to perturb the electric field in the film as little as possible. To improve the contact between the undoped a-Si:H and the ohmic contacts in Fig. 14, it is helpful to deposit a thin n-type layer first.

In this planar geometry one is always plagued with surface problems. This is because surface charge either on the free surface or the surface adjoining the substrate can bend the bands near the surface (Solomon et al., 1978). In extreme cases, accumulation or inversion layers can be formed. These bent-band regions can dominate the measured conductivity, since they are in parallel with the bulk conductivity. Since these surface effects are discussed by Fritzsche in Chapter 10 of Volume C we shall not elaborate on them here. We mention them only as a caution to the reader that band-bending effects can cause a great deal of confusion in interpreting data and in comparing data from experiments made on different films and performed under different conditions. It has been shown that there is considerable variation in both the dark conductivity and photoconductivity with film thickness (Brodsky et al., 1980; Fritzsche, 1980; Hasegawa and Imai, 1982) because of bent bands. Since these bent-band regions can extend over distances on the order of 1 μm, one can expect considerable variation of photoconductive properties with thickness. Brodsky et al. (1980) have shown that there is considerable variation of transport properties for films thinner than $2-3$ μm.

In principle, one could use a field-effect structure to control the band bending and attempt to achieve the flat-band condition for making photoconductivity measurements. This idea was proposed by Solomon *et al.* (1978), who thought that the experiments would be difficult to control. Nevertheless, it is now possible to reliably fabricate high quality field effect transistors (FETs) of a-Si:H in order to make controlled measurements of the effects of band bending on the photocurrent (Jackson and Thompson, 1983; Jackson *et al.*, 1983a). The authors found significant changes in the spectral as well as the light intensity dependences of the photocurrent as the voltage on the FET gate electrode was varied to change the amount of band bending. These results help to clarify some of the conflicting observations in the literature.

The principal effect of band bending is to produce a high electric field near the surface that separates holes from electrons. Because carriers can recombine only when they are near each other, this increases the recombination time. Thus the photocurrent will depend on the degree of band bending. Since the band bending will be affected by the light intensity, this can give rise to a recombination time dependent on light intensity. We shall return to these effects when discussing some experimental results.

The sandwich geometry does not have the same type of band-bending problems that the planar geometry does. This is because the voltage applied across the structure controls the band bending. However, this structure has its own set of unique problems. If the contacts are ohmic, then space-charge injection can give large field distortions across the film thickness. [For a discussion of space charge in insulators see, for example, Lampert and Mark (1972) and Rose (1963).] Accompanying this space-charge-induced band bending will be a large reservoir of trapped and free charge that can alter the recombination kinetics. For example, with a voltage of, say, 1 V applied across a film thickness of 1 μm, the injected space charge Q given by the condenser relation

$$Q = CV = 4\pi\varepsilon V_a/L \tag{56}$$

is about 10^{-8} C cm^{-2}. Here ε is the dielectric constant. This corresponds to a space-charge density N_s of about 8×10^{14} cm^{-3}, which can be comparable to or greater than the space-charge density in high quality a-Si:H films. Moreover, this negative charge can affect the hole-trapping kinetics. If one assumes that these injected and trapped electrons form hole traps of density N_r, with a capture cross section σ_R of about 10^{-13} cm^2, which is reasonable for Coulomb capture, then the recombination time τ_R is

$$\tau_R = 1/N\sigma_R v \sim 10^{-15} \times 10^{13} \times 10^{-7} = 10^{-9} \quad \text{sec}, \tag{57}$$

where v is the thermal velocity of an electron. This time is much shorter than the hole trapping times that have been measured in undoped a-Si : H (Street, 1982a; Mort *et al.,* 1980b).

One uses either a dc or an ac technique to measure the photocurrent. The ac technique is the easiest to apply by using a lock-in amplifier and chopping the light. However, this technique has the disadvantage that the frequency must be lower than the response frequency of the photoconductor. Often, it is better to measure the dc current and use digital counting techniques to improve the signal-to-noise ratio. In digital counting techniques, the currents, with and without illumination, are each measured for a fixed time interval. The currents are integrated digitally so that they can be represented as numbers. These numbers are then subtracted to give the photocurrent.

10. RECOMBINATION CENTERS

There is now convincing evidence that the dangling bond is an important recombination center in a-Si : H. The dangling bond, which has an unpaired spin on a silicon atom, is characterized by an electron spin resonance (ESR) g-value of 2.0055. The dangling bond can also be occupied by two electrons having paired spins. There is evidence that the state with an unpaired spin is located about 1.25 eV below E_c and the paired spin state at about 0.85 eV below E_c (Cohen *et al.,* 1982; Jackson, 1982). The dangling bond is the dominant destroyer of the characteristic a-Si : H luminescence transition (Street, 1981). Because both states of the dangling bond lie near midgap, it will function as a recombination center rather than as a trap. Recently, two groups using different techniques have shown that the dangling bond has a significant effect on the photocurrent. In both experiments the dangling-bond density was varied by some appropriate means and measured using ESR, which is the standard technique to detect the unpaired electron spin on the dangling bond. Street (1982a) was able to deduce the cross section for capture of either electrons or holes by the dangling bond. Dersch *et al.* (1983) were able to unravel the details of the recombination process by using a combination of ESR and spin-dependent photoconductivity. They showed that the hole was the limiting factor in recombination in low-defect-density material. This result that the hole controls recombination is consistent with recent observations on solar-cell-grade devices (Faughnan *et al.,* 1984).

Street (1982a, 1983) measured the mobility–lifetime produce by using transient photocurrent techniques. He used n-type Schottky barrier devices thick enough that photogenerated electrons or holes, depending on the polarity, were unable to cross the sample before being trapped in recombination centers. Under these charge-carrier-range-limited conditions, assuming nondispersive transport and a single trapping time, the drift length and hence the mobility–lifetime product could be accurately measured. The different dangling-bond densities in different samples were measured by using ESR. He concluded that a dangling bond would trap either a hole or an electron with about the same probability. This is a reasonable result if the dangling bond is neutral, so that it gives an ESR signal. Indeed, this seemed to be the experimental condition. The capture cross section was on the order of 1×10^{-15} cm^{-3}, which is a reasonable value for a neutral trap.

On the other hand, Dersch *et al.* (1983) used a single sample and varied the dangling-bond density in a controlled way. Electron bombardment was used to produce dangling-bond densities greater than 1×10^{18} cm^{-3}. The density was then reduced by subsequent isochronal anneals at various temperatures. The data showing the temperature dependence of the presumably *n*-type photoconductivity at different spin concentrations is shown in Fig. 15. The dark conductivity is also shown. Since this does not change with spin density, the Fermi level is pinned at an energy of about 0.85 eV below E_c. This is important since it ensures that changes in the photocurrent with Fermi level commonly observed (Rehm *et al.*, 1977; Vanier *et al.*, 1981) will not obscure the dependence of the photocurrent on the dangling-bond density. The authors also observed that the photocurrent varied linearly with light intensity. This is good evidence for monomolecular recombination resulting from a single recombination center. Furthermore, band-bending effects do not appear to be a problem insofar as the lifetime does not depend on light intensity.

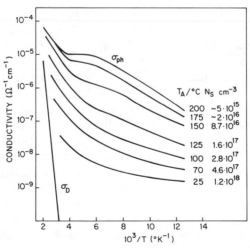

FIG. 15. Dark conductivity σ_D and photoconductivity σ_{ph} for different spin densities N_s in electron irradiated a-Si:H. [After Dersch *et al.* (1983).]

To unravel the details of the recombination process, Dersch *et al.* (1983) measured the dependence of the photoconductivity on magnetic field. A resonant decrease in the photocurrent was detected at a magnetic field corresponding to that for the ESR resonance of the dangling-bond defect. Thus it was clear that the dangling bond was involved in the recombination process. They also observed two other characteristic resonances in the photoconductivity: one corresponding to an electron localized in conduction-band tail states and the other associated with holes localized in valence-band tail states. The implication of this is clear: it confirms a recombination model in which trapped holes and electrons recombine via dangling-bond defect states. Capture by free carriers is not the dominant process. The authors were able to further clarify the situation by showing that at high defect density the rate-limiting step in the recombination process was the tunneling of band-tail electrons to the neutral dangling bond, whereas at low defect density, the rate-limiting step was the diffusion of holes to the dangling bond. Essentially similar conclusions that recombination proceeds via band-tail states have been reached by Street *et al.* (1982). Other authors (Dersch *et al.*, 1983; Faughnan *et al.*, 1984; Kirby and Paul, 1982) have concluded that the electron is trapped more readily than the hole. The difference between their conclusions and that of Street *et al.* (1982) may be due to different material. In fact, recent experiments by Street (1983) and Kirby *et al.* (1983) show that hole and electron trapping depend strongly on doping. Holes are trapped more readily than electrons in *n*-type material, and electrons are trapped more readily than holes in *p*-type material.

This qualitative picture makes the peak in the temperature dependence of the photocurrent shown for low defect density in Fig. 15 understandable. At high temperature, thermal excitation of the hole into extended or high mobility states increases its trapping by the dangling bond. With decreasing temperature, this mechanism is reduced and the electron lifetime increases, resulting in a rise in the photocurrent. This increase in the photocurrent with decreasing temperature is often observed (Paul and Anderson, 1981; Spear *et al.*, 1974; Vanier *et al.*, 1981) and may have a similar origin. However, there may be other defects determining the recombination process, so that the model involving the dangling bond may not be appropriate. Nevertheless, the idea that the electron and hole recombine from tail states should still apply. The quenching of the photocurrent with infrared radiation (Persans, 1980; Vanier *et al.*, 1981) is also consistent with this model that the hole limits recombination. The infrared radiation would excite holes from low mobility states to the mobility edge, where they could be trapped by dangling bonds. However, it is always possible that some other defect could be responsible for the recombination.

It is not certain that the dangling bond is the universal recombination center. Certainly, if the dangling-bond density is low enough, other defects can be more important recombination centers. Phosphorus doping generally produces a center with a larger cross section for hole capture than for electron capture (Lang *et al.,* 1982; Wronski *et al.,* 1982). However, a new defect may not be introduced, because the dangling-bond density is found to increase with phosphorus doping (Cohen *et al.,* 1982; Jackson and Amer, 1982). The Fermi level also moves toward the conduction band with phosphorus doping. Both of these effects increase the amount of hole trapping, the former by increasing the number of dangling bonds and the latter by increasing the number of occupied dangling bonds. It is important, if at all possible, to pin the Fermi level when making studies of recombination centers. As we shall discuss in Section 11, the movement of the Fermi level has a profound effect on the photoconductivity. Boron doping may also introduce dangling bonds rather than a new defect (Jackson and Amer, 1982).

11. DEMARCATION LEVELS

As an example of the use of demarcation levels, we shall discuss two sets of experimental data that can be understood in terms of the movement of the demarcation levels with external influences. The first is the common observation that the photocurrent varies sublinearly with light intensity (Hoheisel *et al.,* 1981; Paul and Anderson, 1981; Williams and Crandall, 1979; Wronski and Daniel, 1981). The second is the variation of the photocurrent with activation energy of the dark conductivity (Hoheisel *et al.,* 1981; Paul and Anderson, 1981; Rehm *et al.,* 1977; Vanier *et al.,* 1981). We must keep in mind, however, that band-bending effects can influence the experimental data.

In general, the photocurrent can be written as a function of the light intensity f as

$$J_n = cf/N_r = cf^\gamma, \qquad (58)$$

where c is a constant, N_r the density of recombination centers, and γ a fraction to account for nonlinear variations of current with light intensity.

A typical example of this variation of the photocurrent with light intensity is shown in Fig. 16. For these data, a Schottky-barrier device was used so that band-bending effects could be minimized (Williams and Crandall, 1979). The data in Fig. 16 show that $\gamma = \frac{3}{4}$, a commonly observed value. Rose (1951, 1963) was the first to give a physical explanation of these fractional power law variations. His model is based on the fact that the demarcation level moves with light intensity. As the demarcation level sweeps toward the band edge with increasing light intensity, it converts states that were originally shallow traps into recombination centers. Thus the density of recombi-

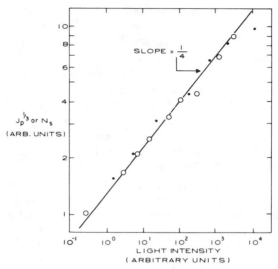

FIG. 16. Dependence of photocurrent J_p (●) and space-charge density N_s (○) on light intensity for a Schottky-barrier device. N_s is measured in reverse bias, J_p is measured in forward bias, so that the current is a secondary photocurrent. [After Williams and Crandall (1979).]

nation centers increases with light intensity, producing a sublinear variation of photocurrent with light intensity. The actual value of the exponent γ depends on the distribution of states in the gap. He postulated a distribution of states that increased exponentially toward the band edge to explain values of γ in the range 0.5–1. Of course, we now know that the shallow states in a-Si:H have an exponential distribution. Using Rose's model, with $\gamma = \frac{3}{4}$, gives a characteristic energy for the exponential distribution of 0.08 eV. This value is larger than the value determined for the shallow states measured by drift mobility by a factor of 2 to 3 (Tiedje *et al.*, 1980). Since one expects that the states involved in recombination will be farther removed from the band edges and therefore have a broader distribution, this result is reasonable.

Another feature shown in Fig. 16 is the variation of the density of the photogenerated space charge N_s with light intensity. The space charge, which is positive, is thought to be a measure of the number of trapped holes. Since these presumably act as the recombination centers for the n-type photoconductivity, N_s should be proportional to N_r. That they are indeed proportional is demonstrated by Fig. 16. The data show that N_s varies as $f^{1/4}$ and the photocurrent as $f^{3/4}$. From Eq. (58), we see that this shows that N_r varies as $f^{1/4}$, as does N_s.

Unfortunately, the measurements were not made at lower temperature to see whether γ would approach unity as predicted by the model (Rose, 1963). Others have made measurements of γ as a function of temperature and other parameters. The results do not, however, lead to a consistent picture for a-Si:H. This is due, in part, to the large variation in a-Si:H among laboratories and even in the same laboratory. Band-bending effects also contribute to variations among samples. Nevertheless, we shall describe some of the findings.

Wronski and Daniel (1981) found two values of γ for their material, namely, 0.83 and 0.5, with the lower value occurring at higher light intensity. The exponent γ did not vary with temperature, but the transition point between the two γ values did shift to lower light intensity with lower temperature.

On the other hand, Hoheisel *et al.* (1981) found that γ did increase from a value near 0.5 at room temperature to nearly unity at low temperature. Their studies were carried out on samples with different doping concentrations, which varied from n-type through fully compensated to p-type. From the behavior of the photoconductivity with doping, they concluded that the photocurrent was carried by electrons in the n-type samples and by holes in the p-type samples. This belief was based on the result that the room temperature photocurrent increased with decreasing activation energy of the dark current, regardless of conductivity type. The minimum in the photoconductivity was for an activation energy of 0.8 eV, corresponding to midgap. Similar results for ion-implanted a-Si:H were obtained by Le-Comber *et al.* (1980). They also found a minimum in the photoconductivity for closely compensated samples. Thermopower measurements on photoexcited p-type a-Si:H made by Dresner (1983) show the photocarriers to be holes, supporting the contention of Hoheisel *et al.* (1981).

On the other hand, Paul and Anderson (1981) found a complex variation of γ with temperature in different samples of sputtered a-Si:H. The exponent decreased, increased, or was relatively insensitive to temperature, depending on the sample. They did, however, find that the photocurrent decreased with increasing activation of the dark current, in agreement with Hoheisel *et al.* (1981).

The decrease in photoconductivity with increasing activation energy of the dark current is not observed under all conditions, as was pointed out by Vanier *et al.* (1981). Their measurements of the room temperature photoconductivity for a large number of films is shown in Fig. 17, in which the photoconductivity is plotted as a function of activation energy of the dark conductivity. Their data are in substantial agreement with those mentioned above; however, they observed a new feature for activation energies less than 0.4 eV, a region not covered by the others. Here the photoconductivity decreases with decreasing activation energy of the dark conductivity. A reasonable explanation for this is that as the Fermi level rises toward the band edge, more and more band-tail states are occupied by electrons and

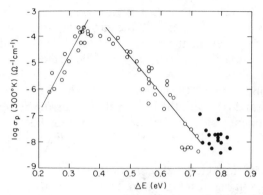

FIG. 17. Logarithm of the photoconductivity versus the activation energy of the dark conductivity. ○, sample that exhibited a temperature dependence similar to that shown in Fig. 15; ●, sample that exhibited a more complicated temperature dependence. [After Vanier *et al.* (1981).]

can thus trap holes to become effective recombination centers. To apply the same type of argument to the lower activation energy region necessitates a change in the nature of the states below the Fermi level. They must be poor recombination centers when occupied in the dark. This actually is the model of Anderson and Spear (1977). They proposed that occupied and unoccupied dangling-bond pairs would overlap at midgap, producing positively charged states above midgap and negatively charged states below midgap. The positive states would be effective traps for electrons when the Fermi level was at midgap. As the Fermi level rises and these states become occupied and therefore neutral, their effectiveness as recombination centers would be reduced. A similar situation occurs as the Fermi level moves from midgap toward the valence band, with the roles of electrons and holes interchanged.

It is noteworthy that the photoconductivity is a maximum when the activation energy of the dark conductivity is in the vicinity of 0.4 eV. There is a minimum in the density of gap states at this energy (Lang *et al.*, 1982). It is now believed that this energy represents the transition from band-taillike states to dangling-bondlike states. The band of states beginning below 0.4 eV has been shown (Cohen *et al.*, 1982), by a combination of spin resonance spectroscopy and DLTS, to be associated with the dangling bond.

12. SUB-BAND-GAP PHOTOCURRENT

The earliest measurements of steady-state secondary photocurrent by Spear and co-workers (Loveland *et al.*, 1973/1974) on a-Si:H showed considerable photoconductivity for photon energies well below the optical band gap, which is between 1.5 and 1.7 eV, depending on sample preparation conditions. A complete discussion of the meaning and determination of this optical gap can be found in Chapter 2 by Cody.

Figure 18 is typical of the spectral dependence of the photocurrent found for a variety of samples. For these measurements, using planar geometry on films varying in thickness between 0.1 and 3.9 μm, the assumption is that the holes are mostly trapped so that the photocurrent is due predominantly to electrons. Under these conditions, the photocurrent is

$$J_n = eN_0(1 - R_e)[1 - \exp(-\alpha L)]\eta_p\mu_n\tau_n^R F/L, \qquad (59)$$

where R_e is the reflectivity and η_p the photogeneration efficiency. The ordinate in Fig. 18 represents the number of electrons in the external circuit per photon entering the sample. The features in Fig. 18 have generally been observed by other workers (Crandall, 1980c; Evangelisti *et al.*, 1983; Jackson *et al.*, 1983b; Moddel *et al.*, 1980; Suzuki *et al.*, 1977; von Roedern and Moddel, 1980; Welsch *et al.*, 1981). The photoconductivity decreases rapidly for photon energies below about 1.5 eV, the typical energy for the optical gap, and disappears below about 0.6–0.5 eV. Most of the curves exhibit a shoulder between 1 and 1.3 eV. The photoconductivity generally follows the shape of the optical absorption curve, at least for high absorption coefficients for which optical transmission measurements are reliable. This

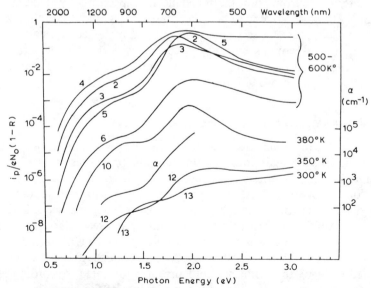

FIG. 18. Wavelength dependence of photocurrent i_p for a variety of samples; the indicated temperature refers to the substrate temperature during deposition. [After Loveland *et al.* (1973/1974). Copyright North-Holland Publ. Co., Amsterdam, 1973/1974.]

is in distinct contrast to those materials such as a-Se that exhibit significant geminate recombination at room temperature.

However, there is considerable controversy over interpretation of the results. Band-bending effects, mentioned previously, certainly have an influence on the data. Further complications arise because of the slow response time of the photocurrent at low energies. Since the response time is wavelength dependent, this can skew the data when a fixed chopping frequency is used (Crandall, 1980d; Evangelisti *et al.*, 1983; Moddel *et al.*, 1980; Persans, 1980). For this reason, dc techniques are appropriate for this region of the spectrum. Another problem is that the photocarrier lifetime can be wavelength dependent. We expect this because the important optical transitions are from states above the valence band to E_c. As the excitation energy is changed, the hole is produced in different regions of the valence-band tail. It stands to reason that this will affect the recombination kinetics. A further complication is that the photocurrent does not vary linearly with light intensity. Various schemes have been used to take this into account in extracting the absorption coefficient from the data. Still another approach is to measure the primary photocurrent. This is not easy, especially if low barrier-height metals are used as the blocking contact. Nevertheless, when all the above precautions are observed, the absorption coefficient, determined from photocurrent measurements, is in qualitative agreement with that determined from photothermal deflection spectroscopy (Jackson *et al.*, 1983b). The close agreement between these two techniques implies that the photocarriers must be separated without geminate recombination being important. This would seem to be in conflict with the data of Carasco and Spear (1983) that indicated geminate recombination at room temperature for below band-gap excitation.

13. PRIMARY AND SECONDARY CURRENTS

Most of the previous discussion has centered around the use of secondary photocurrents to obtain information about the properties of the photoconductor. To date, the measurement of primary currents has been limited to the domain of solar cells, since they operate in a primary photocurrent mode. However, the solar cell structure, particularly the Schottky-barrier device, is well suited for study of both primary and secondary currents. An example of these two currents is given in Fig. 19, which is a current–voltage curve for a Schottky-barrier structure. At large enough reverse bias (negative applied voltage), the primary photocurrent is saturated. According to Eq. (51) this current is eGL, and thus is a measure of the electron–hole pair generation rate G. Therefore it is not necessary to have the light system

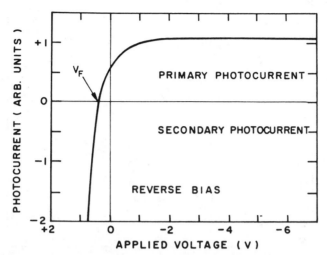

Fig. 19. Example of primary and secondary photocurrents in an a-Si:H Schottky-barrier solar cell. Measurements are at room temperature with weakly absorbed light.

calibrated or to know the absorption constant, both of which are necessary in an ordinary secondary photocurrent measurement. At the voltage V_F in the figure, the photocurrent reverses sign, changing from a primary to a secondary current. At this voltage, the flat-band voltage, the potential drop across the photoconducting region is zero. The flat-band voltage V_F is not zero in the figure because there is a built-in voltage in this structure produced by the difference in work functions of the two dissimilar contact materials. In this example, platinum metal forms the Schottky-barrier contact and stainless steel makes the quasi-ohmic contact. For voltages more positive than V_F the photocurrent is a secondary current. Referring to Eq. (26), we see that this current is now a direct measure of the mobility–lifetime products of the electrons and holes because G is known from the primary photocurrent measurement.

Figure 20 is an example of the measurement of both primary and secondary photocurrents to determine the photocurrent gain g. According to Eq. (26), Eq. (27), and Eq. (51), the gain is just the ratio of the secondary photocurrent to the saturated primary current. The large values of gain for the electron photocurrent shown in the figure are an indication that holes are trapped in the insulator. If there were no hole trapping, then the gain would saturate at the value unity (Rose, 1951). This is the same type of a-Si:H as that used for the data in Fig. 16, which shows a large amount of space charge due to hole trapping. Because holes are trapped more readily

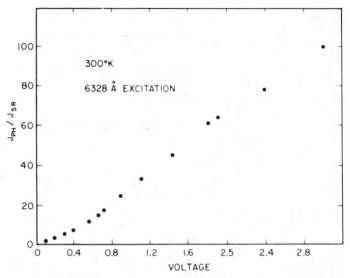

FIG. 20. Ratio of forward-bias photocurrent to saturated reverse-bias photocurrent as a function of voltage ($V - V_F$) across the i layer of an a-Si:H Schottky-barrier solar cell. [After Williams and Crandall (1979).]

than electrons, $l_n \gg l_p$ and g is a measure of l_n and hence of the electron mobility–lifetime product.

In high quality solar cell a-Si:H it is possible to reduce hole trapping sufficiently so that the gain will saturate at values near unity. This is seen in Fig. 21, which shows the primary and secondary photocurrents in a high efficiency $p-i-n$ solar cell at different temperatures. At room temperature, the forward-bias secondary current saturates at a value near that of the reverse-bias primary photocurrent. This can happen only if the holes are extracted from the i layer. The data for lower temperature show the secondary current exceeding that of the primary current. Here there is a small amount of hole trapping so that the gain can exceed unity.

The current–voltage curves shown for primary photocurrent in Figs. 19 and 21 can be fit exceedingly well by Eq. (51), which assumes a uniform electric field. At high enough light intensities this will no longer be the case because of the photogenerated space charge that is produced by the displacement of electrons and holes by the field. The presence of this space charge has been observed by making measurements of the capacitance under illumination. This photocapacitance has a characteristic voltage dependence that distinguishes it from other sources of capacitance, such as fixed space charge. Measurements of this photocapacitance, which is due to

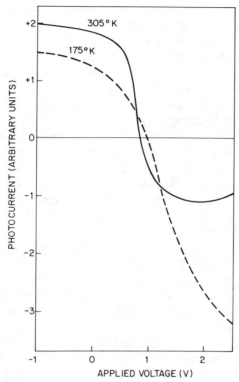

FIG. 21. Primary and secondary photocurrents in a $p-i-n$ solar cell. The cell was illuminated at two different temperatures with weakly absorbed light. The chopped light frequency was 10 Hz.

mobile charge, have been used to determine the drift velocity of the space-charge packet (Crandall, 1983b).

VI. Conclusion

The purpose of this chapter has been to outline the important concepts involved in photoconductivity in a-Si:H. A thorough understanding of the mechanism of photoconduction requires a detailed knowledge of the physical properties of the material. Therefore we have presented only a brief review, partly because of a lack of knowledge and partly because the field is undergoing such rapid expansion that there is still controversy surrounding some of the work, which, no doubt, will be resolved in the future. Much uncertainty stems from the poor characterization of the material. As the major defects such as dangling bonds are reduced trace impurities become important. Reported midgap densities of states in the low 1×10^{14} cm^{-3} range indicate a significant improvement in purity for an amorphous material. The fact that thin films are usually produced with significant structural and chemical inhomogeneities also has a significant bearing on the analysis of its properties. In any complete study of photoconductivity this must be taken into account.

The analytical discussion has centered on steady-state nondispersive transport. This may be appropriate in some situations; however, recombination is, in the final analysis, a random walk process. Thus it would be appropriate to describe recombination in terms of the detailed diffusion of the electron and hole to each other to accomplish a recombination event.

It is comforting that there is substantial agreement that the dangling bond is an important, if not the most important, recombination center. Much more work must be done in this area to determine unambiguously the trapping parameters of this important defect. It would be useful to be able to change the charge state of the dangling bond and study its trapping parameters. The spin-dependent photoconductivity technique is an important tool for probing details of the recombination. There should also be some effort to determine what other defects can be important recombination centers. This seems to be a distinct possibility now that a-Si:H can be produced with such a low density of states near midgap, where recombination takes place.

The question as to whether geminate recombination is important at room temperature should be answered. There is still not a satisfactory explanation for the results of some experiments that indicate a significant amount of geminate recombination at room temperature. Perhaps this could be resolved if the different experiments were to be performed on the same sample.

The theory of two-carrier primary photocurrents should be investigated in detail. This mode of photoconduction is becoming increasingly important because of its use in device operation. Along with this is the photogenerated space charge problem that occurs in both primary and secondary modes of photoconduction, in the former because of the transport itself and in the latter because of band bending at the surfaces. In fact, the band bending and subsequent charge separation are probably the most significant source of discrepencies among various experiments. Care should be exercised to eliminate this source of error, perhaps by means of FET geometry.

Because of the low density of midgap states now obtainable, it is important that one take pains to reduce the extrinsic impurity concentration in a-Si:H. This may require increased cleanliness and care in gas handling. However, a small number of impurities, on the order of 1×10^{14} cm^{-3}, acting as recombination centers can have a deleterious effect on the photocurrent.

REFERENCES

Ackley, D. E., Tauc, J., and Paul, W. (1979). *Phys. Rev. Lett.* **43**, 715.

Allan, D., LeComber, P. G., and Spear, W. E. (1977). *In Proc. Int. Conf. Amorphous and Liquid Semicond., 7th* (W. E. Spear, ed.), p. 323. Edinburgh University (CICL), Edinburgh.

Anderson, D. A., and Spear, W. E. (1977). *Philos. Mag.* **36**, 695.

Bell, R. O. (1980). *Appl. Phys. Lett.* **36**, 936.

Brodsky, M. H., and Title, R. S. (1969). *Phys. Rev. Lett.* **23**, 787.

Brodsky, M. H., Evangelisti, F., Fischer, R., Johnson, R. W., Reuter, W., and Solomon, I. (1980). *Solar Cells* **2**, 401.

Bube, R. H. (1960). "Photoconductivity in Solids." Wiley, New York.

Biegelsen, D. K. (1980). *Solar Cells* **2**, 421.

Carasco, F., and Spear, W. E. (1983). *Philos. Mag. B* **47**, 495.

Chen, I., and Lee, S. (1982). *J. Appl. Phys.* **53**, 1045.

Cody, G., Tiedje, T., Abeles, B., Brook, B., and Goldstein, Y. (1981). *Phys. Rev. Lett.* **47**, 1480.

Cohen, J. D., Harbison, J. P., and Wecht, K. W. (1982). *Phys. Rev. Lett.* **48**, 109.

Coppage, F. N., and Kepler, R. G. (1967). *Mol. Cryst.* **2**, 231.

Crandall, R. S. (1965). *Phys. Rev. A* **138**, 1242.

Crandall, R. S. (1980a). *J. Electron. Mater.* **9**, 713.

Crandall, R. S. (1980b). *Appl. Phys. Lett.* **36**, 307.

Crandall, R. S. (1980c). *Phys. Rev. Lett.* **44**, 749.

Crandall, R. S. (1980d). *Solar Cells* **2**, 319.

Crandall, R. S. (1981a). *RCA Rev.* **42**, 458.

Crandall, R. S. (1981b). *RCA Rev.* **42**, 441.

Crandall, R. S. (1982). *J. Appl. Phys.* **53**, 3350.

Crandall, R. S. (1983a). *J. Appl. Phys.* **54**, 7176.

Crandall, R. S. (1983b). *Appl. Phys. Lett.* **42**, 451.

Crandall, R. S., Williams, R., and Tompkins, B. E. (1979). *J. Appl. Phys.* **50**, 5506.

Dalal, V., and Alvarez, F. (1981). *J. Phys. Colloq. Orsay Fr.* **42**, C4-491.

Datta, T., and Silver, M. (1981). *Solid State Commun.* **38**, 1067.

Delahoy, A., and Griffith, R. W. (1981). *J. Appl. Phys.* **52**, 6337.

Dersch, H., Schweitzer, L., Stuke, J. (1983). *Phys. Rev. B.* **28**, 4678.

Döhler, G. H. (1979). *Phys. Rev. B* **19**, 2083.

Döhler, G. H. (1981). *J. Phys. Colloq. Orsay Fr.* **42**, C4-115. *Supplement 10.*

Dolezalek, F. K., and Spear, W. E. (1975). *J. Phys. Chem. Solids,* **36**, 819.

Dresner, J. (1983). *J. Non-Cryst. Solids.* **58**, 353.

Enck, R. G., and Pfister, G. (1976). *In* "Photoconductivity and Related Phenomena" (J. Mort and D. M. Pai, eds.), Chap. 7. Elsevier, Amsterdam.

Evangelisti, F., Fiorini, P., Fortunato, G., Frovd, A., Giovanella, C., and Peruzzi, R., (1983). *J. Non-Cryst. Solids.* **55**, 191.

Faughnan, B., Moore, A., and Crandall, R. (1984). *Appl. Phys. Lett.* To be published.

Fritzsche, H. (1980). *Solar Cells* **2**, 289.

Gudden, B., and Pohl, R. W. (1921). *Z. Phys.* **6**, 248.

Gutkowicz-Krusin, D. (1981). *J. Appl. Phys.* **52**, 5370.

Gutkowicz-Krusin, D., Wronski, C. R., and Tiedje, T. (1981). *Appl. Phys. Lett.* **38**, 87.

Hasegawa, S., and Imai, Y. (1982). *Philos. Mag. B* **46**, 235.

Hoheisel, B., Fischer, R., and Stuke, J. (1981). *J. Phys. Colloq. Orsay Fr.* **42**, C4-819.

Jackson, W. B. (1982). *Solid State Commun.* **44**, 477.

Jackson, W. B., and Amer, N. M. (1982). *Phys. Rev. B* **25**, 5559.

Jackson, W. B., and Thompson, M. J. (1983). *Physica Amsterdam* **117 & 118B**, 883.

Jackson, W. B., Street, R. A., and Thompson, M. J. (1983a). *Solid State Commun.* **47**, 435.

Jackson, W. D., Nemanich, R. J., and Amer, N. M. (1983b). *Phys. Rev. B.* To be published.
Kirby, P. K., and Paul, W. (1982). *Phys. Rev. B* **25**, 5373.
Kirby, P. B., Paul, W., Lee, C., Lin, S., von Roedern, B., Weisfield, R. L. (1983). *Phys. Rev. B* **28**, 3635.
Lampert, M. A., and Crandall, R. S. (1979). *Chem. Phys. Lett.* **68**, 473.
Lampert, M. A., and Mark, P. (1972). "Current Injection in Solids." Academic Press, New York.
Lampert, M. A., and Schilling, R. B. (1970). *In* "Semiconductors and Semimetals" (R. K. Willardson and A. C. Beer, eds.), Vol. 6, pp. 1–95. Academic Press, New York.
Lang, D. V., Cohen, J. D., and Harbison, J. P. (1982). *Phys. Rev. B* **25**, 5321.
Langevin, P. (1903). *An. Chem. Phys.* **28**, 287, 433.
Lax, M. (1960). *Phys. Rev.* **119**, 1502.
LeComber, P. G., and Spear, W. E. (1970). *Phys. Rev. Lett.* **25**, 509.
LeComber, P. G., Spear, W. E., Müller, G., and Kalbitzer, S. (1980). *J. Non-Cryst. Solids* **35/36**, 327.
Loveland, R. J., Spear, W. E., Al-Sharbaty, A. (1973/1974). *J. Non-Cryst. Solids* **13**, 55.
Madan, A., Czubatyj, W., Adler, D., and Silver, M. (1980). *Philos. Mag. B* **42**, 257.
Meyer, J. W. (1968). *In* "Semiconductor Detectors" (G. Bertolini and A. Coch, eds.), Chap. 5. North-Holland Publ., Amsterdam.
Miyamoto, H., Konagai, M., and Takahashi, K. (1981). *Jpn. J. Appl. Phys.* **20**, 1691.
Moddel, G., Anderson, D. A., and Paul, W. (1980). *Phys. Rev. B* **22**, 1918.
Moore, A. M. (1977). *Appl. Phys. Lett.* **31**, 762.
Moore, A. M. (1980). *Appl. Phys. Lett.* **37**, 327.
Moore, A. M. (1983). *J. Appl. Phys.* **54**, 222.
Mort, J. (1981). *J. Phys. Colloq. Orsay Fr.* **42**, C4-433.
Mort, J., and Pai, D. M. (eds.) (1976). "Photoconductivity and Related Phenomena." Elsevier, Amsterdam.
Mort, J., Chen, I., Troup, A., Morgan, M., Knights, J. C., and Lujan, R. (1980a). *Phys. Rev. Lett.* **45**, 1348.
Mort, J., Grammatica, S., Knights, J. C., and Lujan, R. (1980b). *Solar Cells* **2**, 451.
Mort, J., Troup, A., Morgan, M., Grammatica, S., Knights, J. C., and Lujan, R. (1981). *Appl. Phys. Lett.* **38**, 277.
Moss, T. S. (1952). "Photoconductivity in the Elements." Academic Press, New York.
Mott, N. F. (1968). *J. Non-Cryst. Solids* **1**, 1.
Mott, N. F., and Gurney, R. W. (1946). "Electronic Processes in Ionic Crystals." Oxford Univ. Press, London and New York.
Noolandi, J., Hong, K. M., and Street, R. A. (1980). *J. Non-Cryst. Solids* **35/36**, 669.
Okamoto, H., Yamaguchi, T., and Hamakawa, Y. (1980). *J. Phys. Soc. Jpn.* **49**, *Suppl. A*, 1213.
Onsager, L. (1938). *Phys. Rev.* **54**, 554.
Orenstein, J., and Kastner, M. (1981). *Phys. Rev. Lett.* **46**, 1421.
Paul, W., and Anderson, D. (1981). *Solar Energy Mat.* **5**, 229.
Persans, P. D. (1980). *Solid State Commun.* **36**, 851.
Rehm, W., Fischer, R., Stuke, J., and Wagner, H. (1977). *Phys. Status Solidi B* **79**, 539.
Reichman, J. (1981). *Appl. Phys. Lett.* **38**, 251.
Rose, A. (1951). *RCA Rev.* **12**, 362.
Rose, A. (1963). "Concepts in Photoconductivity and Allied Problems." Wiley (Interscience), New York.
Ryvkin, S. H. (1964). "Photoelectric Effects in Semiconductors." Consultants Bureau, New York.
Shockley, W., and Read, W. T. (1952). *Phys. Rev.* **87**, 835.

Silver, M., Cohen, L., and Adler, D. (1981). *Phys. Rev. B* **24**, 4855.

Solomon, I., Dietl, T., and Kaplan, D. (1978). *J. Phys. Paris* **39**, 1241.

Spear, W. E. (1980). *EEC Photovoltaic Solar Energy Conf., 3rd, Cannes* (W. Palz, ed.), p. 302. Reidel Publ. Co., London.

Spear, W. E., Loveland, R. J., and Al-Sharbaty, A. (1974). *J. Non-Cryst. Solids* **15**, 410.

Street, R. A. (1981). *Adv. Phys.* **30**, 593.

Street, R. A. (1982a). *Appl. Phys. Lett.* **41**, 1060.

Street, R. A. (1982b). *Philos. Mag. B* **46**, 273.

Street, R. A. (1983). *Appl. Phys. Lett.* **43**, 672.

Street, R. A., Biegelsen, D. K., and Zesch, J. (1982). *Phys. Rev. B* **25**, 4334.

Suzuki, T., Hirose, M., Ogose, S., and Osaka, Y. (1977). *Phys. Status Solidi* **42**, 337.

Swartz, G. (1982). *J. Appl. Phys.* **53**, 712.

Tiedje, T., and Rose, A. (1981). *Solid State Commun.* **37**, 49.

Tiedje, T., Abeles, B., Morel, D. L., Moustakas, T. D., and Wronski, C. R. (1980). *Appl. Phys. Lett.* **36**, 695.

Tiedje, T., Cebulka, J. M., Morel, D. L., and Abeles, B. (1981). *Phys. Rev. Lett.* **46**, 1425.

Vanier, P. E., Delahoy, A. E., and Griffith, R. W. (1981). *AIP Conf. Proc.* **73**, 227.

Vardeny, Z., Strait, J., Pfost, D., Tauc, J., and Abeles, B. (1982). *Phys. Rev. Lett.* **48**, 1132.

von Roedern, B., and Moddel, G. (1980). *Solid State Commun.* **35**, 467.

Welsch, H. M., Fuhs, W., Greeb, K. H., and Mell, H. (1981). *J. Phys. Colloq. Orsay Fr.* **42**, C4-567.

Williams, R., and Crandall, R. S. (1979). *RCA Rev.* **40**, 371.

Wronski, C. R., and Daniel, R. E. (1981). *Phys. Rev. B* **23**, 794.

Wronski, C. R., Abeles, B., Tiedje, T., and Cody, G. D. (1982). *Solid State Commun.* **44**, 1423.

CHAPTER 9

Time-Resolved Spectroscopy of Electronic Relaxation Processes

J. Tauc

DIVISION OF ENGINEERING AND DEPARTMENT OF PHYSICS
BROWN UNIVERSITY
PROVIDENCE, RHODE ISLAND

I. Introduction

Relaxation processes of photogenerated carriers, which follow the generation of an electron–hole pair, are hot carrier thermalization, trapping, and recombination. The methods usually used for studying these processes are time-resolved luminescence and photoconductivity, which are discussed elsewhere in this volume. In this chapter we shall summarize the results obtained by measuring the time evolution of the change in absorption in a sample produced by illuminating the sample with a light pulse. This method has not yet been developed as much as the other two methods, but its advantages and limitations are understood. It should be considered an additional tool for studying electronic processes in amorphous semiconductors, a tool that often provides information not obtainable by other methods. For example, it has been successfully applied to the study of relaxations in the range from a fraction of a picosecond to a few picoseconds, a time domain so far not explored by other methods.

Two different techniques have been used in different time domains. For the long time domain (from nanoseconds up) semiconductor detectors are available and time-resolved measurements are feasible. There are no detectors fast enough in the subpicosecond and picosecond time domains. The

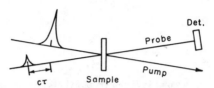

FIG. 1. The pump and probe technique.

detectors available for the region of tens and hundreds of picoseconds are not very sensitive, and there are difficulties with removing spurious responses associated with the circuit. Therefore, from 0.5 psec up to 1.5 nsec we have used a correlation technique, the pump and probe method (Fig. 1) (Ackley *et al.*, 1979). The pump pulse generates carriers whose time-resolved absorption is studied by measuring the transmission with the probe pulse as a function of the delay τ between the probe and pump pulses. The detector integrates over pulse duration and does not have to be very fast (Part II).

The processes studied by time-resolved absorption spectroscopy are shown in Fig. 2a. An electron–hole pair is produced by absorbing a photon of energy $\hbar\omega_p$. Both hot electron and hole then relax to lower energy states, which are shown in the figure only for the electron. The first process is thermalization (the loss of the excess energy ΔE of a hot carrier by interaction with phonons), at the completion of which the carrier is at the bottom of the band. In an amorphous semiconductor, the bottom of the band is defined as the *mobility edge*. The thermalized carrier is trapped in states just below the mobility edge to which we shall refer as *shallow traps* or the *band tail*. As time proceeds, the carriers go into deeper and deeper states in the

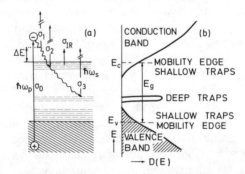

FIG. 2. Electronic relaxation processes in amorphous semiconductors. (a) Generation of a hot electron–hole pair by absorption of a pump photon $\hbar\omega_p$ followed by their thermalization into shallow traps and then into deep traps. The processes are shown for the electron only, with cross sections σ_1, σ_2, and σ_3 for the probe photon absorption. σ_{IR} is the cross section for the midgap PA absorption (Part V). (b) Density of states $D(E)$ in an amorphous semiconductor.

tail, and simultaneously or subsequently into deep traps and/or recombination centers. In recombination centers they recombine with the carrier of opposite sign and the relaxation process is completed.

In photoconductivity (PC) experiments one obtains information about carriers above the mobility edge (the product of their concentration and mobility). In photoluminescence (PL) experiments, one can directly follow the radiative part of the recombination process. Photoinduced absorption (PA) depends on the product of carrier concentration and the absorption cross section.

When a photon is absorbed, the optical absorption is usually decreased ($\Delta\alpha < 0$) by the generation of the electron–hole pair because there are fewer initial and final states available for probe absorption. Therefore the transmission through the sample is increased (we have photoinduced transmission or "bleaching"); this effect is observed in crystalline semiconductors. We shall characterize this process by a cross section σ_0 whose definition and meaning are discussed by Vardeny and Tauc (1984). However, in amorphous semiconductors immediately after excitation we usually observe induced absorption ($\Delta\alpha > 0$). This means that σ_1 (Fig. 2a) is larger than σ_0. The reason is that hot carriers in amorphous semiconductors have a higher optical absorption cross section than hot carriers in crystals because the k-vector conservation is relaxed. The absorption cross section of thermalized carriers in shallow traps is σ_2 (we shall neglect the dependence of σ_2 on energy) and in deep traps σ_3. It is observed that the optical cross section in amorphous semiconductors is smaller for deeper states ($\sigma_1 > \sigma_2 > \sigma_3$). This trend is plausible since in the absence of k-vector conservation the absorption cross section is primarily determined by the density of final states, which is larger at higher energies (Fig. 2b).

The PA response is proportional to the product of carrier concentration and $\sigma_1 - \sigma_0$ (for hot carriers), $\sigma_2 - \sigma_0$ (carriers in shallow traps), and $\sigma_3 - \sigma_0$ (carriers in deep traps). If $\sigma_i > \sigma_0$ ($i = 1, 2,$ or 3) we observe induced absorption ($\Delta\alpha > 0$); if $\sigma_i < \sigma_0$ we observe bleaching ($\Delta\alpha < 0$).

We note an important limitation of the PA technique: In general the PA response is a superposition of the responses of electrons and holes; however, it appears that it often happens that the response of one or the other carrier prevails, as we shall see later.

The processes described in Fig. 2a occur in a-Si:H at short times (< 1 nsec) before the recombination becomes important. Therefore the total carrier concentration is constant and the PA response corresponds to the time evolution of the distribution of carriers over states with different σ.

Our measurements with the pump and probe technique have been done using radiation with photon energy in the absorption edge of a-Si:H, $\hbar\omega_p = 2$ eV for both pump and probe. In this case, the heat produced by the dissipation of carrier energy increases the temperature of the material. This increase shifts the absorption edge and produces a change in absorption coefficient. The time constants of this effect are longer than 1 nsec, and therefore thermal effects are not seen in the picosecond studies. However, at times exceeding 1 nsec, thermal effects may become appreciable. We avoided this problem by using a probe with energies below the absorption edge, where the absorption coefficient α and $d\alpha/dT$ are very small. In this case, the probe measured transitions of carriers trapped in shallow traps into the states close to the bottom of the same band (Fig. 2a) rather than deep into the band as is the case with 2 eV photons. These transitions form a band in the infrared region (Part V) whose total oscillator strength decreases as the carriers disappear by recombination.

In this chapter, we shall discuss effects observed on undoped and doped a-Si : H films. After describing the pump and probe technique in Part II, we shall discuss hot carrier thermalization (Part III), trapping (Part IV), the infrared PA absorption band (Part V), and recombination (Part VI). In this chapter, we do not cover chalcogenide glasses; picosecond effects in chalcogenides are discussed by Vardeny and Tauc (1984).

II. Picosecond Pump and Probe Technique

The equipment used for studying relaxation in the time domain from 0.5 psec to 1.5 nsec by the pump and probe technique (Fig. 1) is shown in Fig. 3. The subpicosecond pulses were produced by a passively mode-locked dye laser, using the design by Ippen and Shank (1977a). It produced linearly polarized, transform-limited light pulses at $\hbar\omega_p = 2$ eV with pulse duration $t_p = 0.6$ to 1 psec, energy 1 to 2 nJ and repetition rate 10^4 to 10^6 sec^{-1}. The laser beam is divided into the pump (about 97% of the intensity) and probe. The pump is chopped by a mechanical chopper and focused on a spot on the sample of about 25 μm in diameter. The probe passes through a polarization rotator so that its polarization relative to the pump can be changed. Then it is delayed by time τ by passing through a translation stage, focused on the same spot on the sample as the pump is, and finally is focused on a detector (usually a Si photodiode). The signal from the detector is amplified by a lock-in amplifier and transformed into a multichannel analyzer that integrates the responses over several scanning times of the translation stage.

An essential advantage of this system is that the peak pulse power (about 0.3 GW cm^{-2}) and energy are relatively small and therefore a superposition of many nonlinear effects occurring at high-power excitations is avoided. Also, the samples may be damaged less by photoinduced structural changes. An acceptable signal-to-noise ratio is obtained by integration over many pulses, which is feasible since the pulse repetition rate is high.

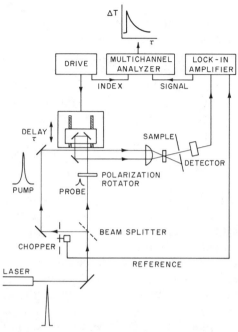

FIG. 3. The experimental setup for studying picosecond relaxation of induced absorption.

A limitation of this system is that it is not tunable and the pump and probe have the same photon energies (= 2 eV). Since the pump has to be absorbed in the sample and the probe has to get through, one must work in some part of the absorption edge of the semiconductor. This means that the probe pulse is absorbed in the material too. This is an undesirable feature because it reduces the signal-to-noise ratio. Another disadvantage is the thermal effect, as discussed in Part I. These difficulties are absent when the probe has a different wavelength. This can be done by using two synchronously pumped dye lasers fed from the same actively mode-locked lasers (Ar⁺ or YAG) and tuned to different wavelengths. This is a preferable method for measuring beyond 20 or 30 psec, but for very fast relaxations the method is not applicable because it is difficult (perhaps even not feasible) to reduce the jitter between the pulses to values below 10 psec.

A more powerful solution of the tunability of the probe is a system in which the intensity of the pulse from the mode-locked dye laser is amplified into the gigawatt range. This pulse can generate a broad spectrum of subpicosecond duration when it is focused in a nonlinear optical medium, e.g., in water (Shank *et al.*, 1979; Shank *et al.*, 1982). A disadvantage of this system is that the pulse repetition rate is only 10 Hz.

The pump and probe pulses can be considered as delta functions for delays beyond ~ 10 psec, but in order to obtain meaningful information from measurements when the pulses overlap, one must take into account the actual shapes of the pulses and deconvolute the data. The aim of the measurements is to determine the response function $A(t)$, which is a characteristic of the material. We define it as the change $\Delta\alpha(t)$ of the absorption coefficient produced by a delta function excitation. The observed change ΔI_2 of the probe intensity I_2 is related to $A(t)$ by the relation (Vardeny and Tauc, 1981b)

$$\Delta I_2(\tau) = ad \int d\tau' \, A(\tau' + \tau) \int dt' \, I_1(\tau' + t')I_1(t'),\qquad(1)$$

where we assume that the probe beam I_2 is a delayed replica of the pump beam I_1: $I_2(t) = aI_1(t - \tau)$ (the constant $a \ll 1$); d is the sample thickness. The second integral is the intensity autocorrelation function. It can be determined experimentally in the setup shown in Fig. 3 by replacing the sample by a nonlinear crystal (KDP) and measuring the second harmonic generation as a function of τ. In general, ΔI_2 contains, in addition to the part represented by Eq. (1), another part produced by the interference of the coherent pump and probe beams in the material [the so called coherent artifact (Ippen and Shank (1977b)]. Vardeny and Tauc (1981b) have shown that in amorphous semiconductors the coherent artifact is usually zero if the pump and probe polarizations are perpendicular. It is also absent if the pump and probe have different wavelengths.

Another technique for studying picosecond relaxations is transient grating spectroscopy (Fayer, 1982). It is a similar correlation technique, but instead of probing directly the decay of absorption, one probes the decay of a grating produced by the pump beam. The pump beam is split into two parts that combine at the sample. The interference of those two beams with parallel polarizations produces a grating associated with periodically excited photocarriers that deflects the probe in a certain direction. The detector placed in this direction sees a signal only when the probe beam is deflected; by changing the delay between the probe and the pump, one can follow the decay of the grating. The gratings expected in a-Si:H are concentration gratings. The photocarriers change the absorption constant and the index of refraction; if one of the effects prevails, we speak of "amplitude gratings" or "phase gratings," but both effects are usually present. The grating disappears by recombination (or trapping in states of low absorption cross sections) but also by diffusion of carriers, which erases the spatial concentration modulation. By varying the angle between the pump beams one changes the pitch of the grating, i.e., the distance over which the carriers have to diffuse in order to erase the grating. From the dependence of the decay time on the grating pitch one can determine the recombination (or trapping) time and the diffusion constant. This is very useful in crystalline semiconductors in which the carriers have high mobilities. In amorphous semiconductors diffusion over distances on the order of 1 μm is usually negligible in the picosecond and subnanosecond time domains. The transient grating method was recently applied to a-Si : H by Komuro et al. (1983).

III. Hot Carrier Thermalization

Hot carrier relaxation in semiconductors occurs on a time scale of a picosecond, and can be directly studied by subpicosecond laser spectroscopy. Since the thermalization time t_0 is often close to the pulse duration t_p, one can obtain meaningful data only after removing the contribution of the coherent artifact. The coherent artifact is a field enhancement due to the interaction of the electric fields of two coherent waves in a nonlinear medium. During the overlap of the pulses around $\tau = 0$, this interaction gives an additional contribution to $\Delta\alpha$. In amorphous semiconductors the simplest way to avoid the coherent artifact is to use perpendicular polarizations of the probe and pump.

In crystalline semiconductors, thermalization has been studied by measuring the response associated with band filling (bleaching), for example, in GaAs (Shank et al., 1979; von der Linde and Lambrich, 1979) and Ge (Elci et al., 1977). Because of the k-vector conservation, σ_0 is larger than σ_1 (see Fig. 1). In amorphous semiconductors, on the other hand, σ_1 is larger than σ_0 and the absorption by hot carriers prevails over the band-filling effect.

Theories dealing with hot carrier absorption in crystals were derived by Seeger (1973) and Elci et al. (1977), but their approach is not applicable to amorphous solids. We shall discuss a simple model applicable to this case. In the absence of k-vector conservation and because of the lack of knowledge about the matrix elements, we make the simplest possible assumption, i.e., that the absorption cross section of the excited carriers is proportional to the density of final states. The contribution of hot electrons to σ_1 is proportional to $D_c(\Delta E + \hbar\omega_s)$, where $D_c(E)$ is the density of the conduction-band states, ΔE is the electron excess energy, and $\hbar\omega_s$ is the probe photon energy (Fig. 2). If we assume that in the range of interest $D_c(E) \propto E^r$ (E being measured from the bottom of the conduction band) and $\Delta E/\hbar\omega_s \ll 1$, we obtain

$$\sigma_1(\Delta E) = a_1(\Delta E + \hbar\omega_s)^r = a_1(\hbar\omega_s)^r (1 + r\Delta E/\hbar\omega_s) = a\Delta E + \sigma_s, \quad (2)$$

where $a = a_1 r(\hbar\omega_s)^{r-1}$ and σ_s is the absorption cross section of an electron at the bottom of the conduction band.

The measured $\Delta\alpha$ depends on the hot carrier distribution $f(\Delta E)$ as

$$\Delta\alpha \sim \int f(\Delta E)\sigma(\Delta E)\, d(\Delta E) = a\, \Delta E + \Delta\alpha_s, \qquad (3)$$

where $\sigma(\Delta E) = \sigma_1(\Delta E) - \sigma_0$ (σ_0 is defined in Part I). In this model, measuring the time evolution of $\Delta\alpha$ gives the time evolution of the electron excess energy. For comparison with experiment, the hole contribution must be added.

In Fig. 4 fast decays of $\Delta\alpha(\tau)$ are shown in nonhydrogenated, hydrogenated, and hydrogenated fluorinated amorphous silicon (Vardeny and Tauc, 1981a). Curves (1) and (2) correspond to parallel and perpendicular polarizations. In a-Si and a-Si:H a maximum in $\Delta\alpha$ around $\tau = 0$ decays to a steady value $\Delta\alpha_s$ at longer times. The depolarization factor $\rho = \Delta\alpha_{s\perp}/\Delta\alpha_{s\parallel}$ is seen to be 1 beyond 4 psec and shows that there is no polarization memory (the response does not depend on the relative polarizations of the probe and pump). Curve (3) is $\Delta\alpha_\parallel - \Delta\alpha_\perp$ and gives directly the coherent artifact present with parallel polarizations. For a-Si (Fig. 4a) a peak in $\Delta\alpha_\perp$ around $\tau = 0$ still exists and this rules out the possibility that $A(t)$ is a step function and that the observed peak around $\tau = 0$ is caused by the coherent artifact. The response function $A(t)$ must therefore contain a fast decay component. The dotted curves in Fig. 4 were obtained from Eq. (1) by convoluting the measured pulse shapes with response functions $A(t)$ of the form shown in Fig. 5; the fit is seen to be good.

A link between these data and the interpretation as hot carrier absorption is provided by the plot in Fig. 6 (Vardeny and Tauc, 1981a). Here the relative strength of absorption at $\tau = 0$ measured by $[\Delta\alpha(0) - \Delta\alpha_s]/\Delta\alpha_s$ is plotted versus the hydrogen content C_H. It is seen that the response is weaker when C_H is larger. Larger C_H gives a larger energy gap E_g, which in turn leads to smaller excess energy $\hbar\omega_p - E_g$. The fluorinated samples show the same trend, since fluorination gives a larger E_g for the same C_H. These data show that the decay mechanism that we have to explain has the property that the response is weaker when the excess energy ΔE is smaller and is a linear function of ΔE. The hot carrier absorption discussed above has this property and appears to be a plausible interpretation of the data shown in Fig. 4.

In this interpretation, t_0 (shown in Fig. 5) is the average time the hot carrier needs to reach the band bottom (thermalization time). Another parameter can be estimated from the data: the excess energy dissipation rate $R = \Delta E(0)/t_0$. The carrier excess energy at $t = 0$ depends on how the total excess energy $\hbar\omega_p - E_g$ is divided between the electron and the hole. We assume that the excess energy is divided equally between them; then $E(0) = \frac{1}{2}(\hbar\omega_p - E_g)/t_0$. There is also an uncertainty in defining the band gap E_g in an amorphous semiconductor; we took it as the photon energy at

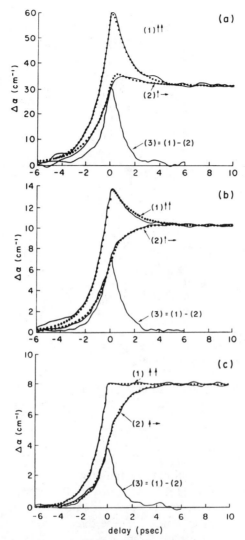

FIG. 4. Photoinduced absorption $\Delta\alpha(t)$ in (a) a-Si at $T = 80°$K, (b) a-Si:H ($C_H = 10$ at. %) at $T = 80°$K, and (c) a-Si:H:F ($C_H = 14$ at. %, $C_F = 16$ at. %) at $T = 300°$K. [After Vardeny and Tauc (1981a).]

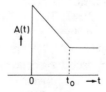

FIG. 5. Response function $A(t)$ with which the data in Fig. 4 were fitted.

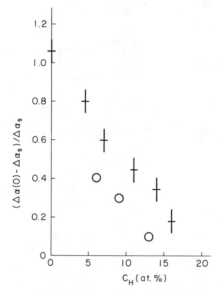

FIG. 6. The relative strengths of the induced absorption at $t = 0$ as a function of hydrogen content in a-Si:H (crosses) and a-Si:H:F (circles). $C_F = 16$ at. %. Data taken at room temperature. [From Vardeny *et al.* (1981a).]

which the absorption coefficient α is 10^3 cm^{-1} (Freeman and Paul, 1979). For samples in Fig. 4 we found

$$\text{a-Si:} \quad t_0 = 0.7 \quad \text{psec,} \quad R = 0.5 \quad \text{eV psec}^{-1},$$

$$\text{a-Si:H } (C_H = 10\%): \quad t_0 = 1.2 \quad \text{psec,} \quad R = 0.1 \quad \text{eV psec}^{-1}.$$

We also obtained an estimate of the absorption cross section. For the saturation region in a-Si:H (Fig. 4) $\sigma_s \simeq 4 \times 10^{-17}$ cm^2.

For comparison with theory we must take into account the fact that we are dealing with carrier distributions. These distributions are different if the average excess energy dissipation rate $R = d\,\overline{\Delta E}/dt$ is larger or smaller than the carrier–carrier dissipation rate R_c, which depends on the carrier density and excess energy (Conwell, 1967). In the conditions of our experiments, R_c was about 0.3 eV psec^{-1}. If $R < R_c$, the hot carrier temperature T_e can be defined; the hot carrier distribution is $f(\Delta E) = \exp(-\Delta E/kT_e)$ and $\overline{\Delta E} = 3kT_e/2$. This is the case for a-Si:H. The experimental response function shows that within the precision of the experiment, T_e decays linearly with t. On the other hand, in a-Si $R > R_c$ and T_e cannot be defined. A method for dealing with this case was proposed by Vardeny and Tauc (1981c).

A theory of the dissipation process must explain the rate and the approximate linearity of the decay for small ΔE. We originally proposed Fröhlich coupling (polar optical scattering) as the dominating dissipation process. M. Cardona (1982, private communication) showed that this was unlikely and suggested electron interaction with longitudinal optical phonons in nonpolar materials [optical deformation potential scattering (Seeger, 1982)]. The optical deformation potential constant D necessary to explain the rate was estimated to be roughly 5×10^8 eV cm^{-1} (i.e., of the same order as in crystals), but no detailed analysis has been made. The higher rate of R observed in a-Si is probably due to an enhancement of electron–phonon interaction by disorder. According to J. C. Phillips (1981, private communication), internal surfaces provide particularly suitable conditions for the enhancement.

IV. Carrier Trapping

1. UNDOPED a-Si:H

When the thermalization process is completed, the electron and the hole have diffused a distance apart that is called the thermalization radius $r_0 = \sqrt{D_{hot}t_0}$. Here D_{hot} is the average hot carrier diffusion constant. The subsequent processes depend on whether r_0 is smaller or larger than the Coulomb capture (Onsager) radius $r_c = e^2/4\pi\varepsilon kT$, where ε is the dielectric constant of the medium. If $r_0 < r_c$, a fraction of carriers recombine by geminate recombination; when $r_0 > r_c$, the thermalized electron and hole move independently of each other.

Geminate recombination was shown to occur in chalcogenide glasses As$_2$S$_{3-x}$Se$_x$ $(0.25 < x < 1)$ (Ackley *et al.*, 1979). For these glasses the pump photon energy is below the absorption edge E_g, i.e., in the Urbach tail. In this case the photocarriers have very small initial excess energies (actually with our definitions, the energies are small and negative) and their mobilities are very small. Consequently, r_0 is very small (3 to 9 Å), not only smaller than the Onsager radius (which is about 80 Å at 300°K) but short enough for geminate recombination by tunneling to occur. This was in fact observed.

In a-Si:H, the Onsager radius is about 50 Å at 300°K and 170 Å at 80°K. When hot carriers are excited with 2-eV photons and we assume that their mobility is 10 cm^2 V^{-1} sec^{-1}, for a-Si:H ($C_H = 10\%$) with $E_g = 1.7$ eV, t_0 is 1.2 psec, and the thermalization radius $r_0 = 55$ Å. At low temperatures, geminate recombination should occur since r_0 (expected to be little temperature dependent) is much smaller than the Onsager radius. In this case it is unlikely that geminate recombination would occur by direct tunneling of carriers just after thermalization, as in the previous case when the excitation was in the Urbach tail. In a-Si:H at low T before the geminate recombination occurs, the electron and hole would first have to move toward each other by some kind of transport (such as multiple trapping or hopping).

At room temperature, the Onsager criterion does not give a clear answer because r_0 can only be estimated to be close to r_c, but one cannot be sure whether in reality it is larger or smaller. This is an important question for solar cells made of a-Si : H because when $r_0 < r_c$, the fraction of geminately recombining carriers $1 - \exp(-r_c/r_0)$ is lost for energy conversion. Mort *et al.* (1981) interpret their xerographic and delayed-collection-field data as evidence that at 300°K about 50% of photocarriers recombine geminately even for photon energies larger than the gap. This contradicts the results obtained by a more straightforward method, namely, measurements of collector efficiencies of a-Si : H solar cells that gave a much smaller upper limit on geminately recombining pairs (Abeles *et al.*, 1980).

We shall show that our results on the PA response in a-Si : H beyond the thermalization region are consistent with a negligible or small contribution of geminate recombination. In the region from 5 psec up to 1.5 nsec, we used the same setup as that shown in Fig. 3 but with a longer translation stage.

In a-Si : H, we found that the saturation shown in Fig. 4 is actually a slow decrease of $\Delta\alpha$ as seen in Figs. 7 and 8 (Vardeny *et al.*, 1983). In these measurements, the thermalization process is not resolved. Close to $t = 0$ some broadening occurs due to the finite pulse widths, but the onset is instantaneous within our resolution. Then a decay follows that is faster in the sputtered (SP) sample, reaching a saturation value in about 1 nsec. In the high quality glow-discharge (GD) sample, the decay is slower and does not

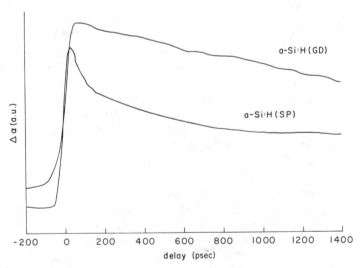

FIG. 7. Induced absorption decays in glow-discharge (GD) and sputtered (SP) a-Si : H at 300°K. [From Vardeny *et al.* (1983).]

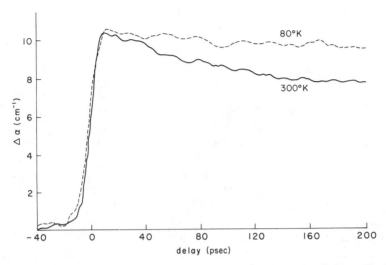

FIG. 8. Temperature dependence of the decay in a-Si : H (SP). $C_H = 16$ at. %. [From Vardeny *et al.* (1983).]

saturate in our time range. The interpretation of this observation is that carriers are trapped in deeper and deeper states in which their absorption cross section is smaller and smaller. Their trapping rate is higher if the concentration of traps is larger. (The SP sample in Fig. 7 was of poorer quality and had more states in the gap than the GD sample.) When all carriers are trapped, a new saturation occurs.

The curves in Fig. 7 could also be interpreted as responses associated with geminate recombination. In this picture, the SP curve shows that 45% of carriers disappear by geminate recombination. An objection against this interpretation is that it does not explain the observed dependence of the decay rate on the trap concentration. Vardeny *et al.* (1983) measured the decay rates in different amorphous silicon samples (a-Si : H : F, a-Si (SP), and a-Si prepared by evaporation in ultrahigh vacuum) and found that they were correlated with the density of uncompensated spins. Such a relation is difficult to understand on the basis of geminate recombination.

The temperature dependence of the decay shown in Fig. 8 is also incompatible with geminate recombination, which should remove a larger fraction of carriers at lower temperature (since r_c is larger). The observed temperature dependence actually contradicts the Onsager model because at low temperatures $r_c \gg r_0$ if it is assumed that the hot carrier mobility is 10 cm² V⁻¹ sec⁻¹ or smaller. The discrepancy would be resolved if the carrier mobility were much larger. This tentative interpretation is in agreement with the results obtained from reverse recovery experiments on *p–i–n* diodes by Silver *et al.* (1982), which were interpreted as evidence for high carrier mobilities (larger than 100 cm^2 V^{-1} sec^{-1}).

Recently, Komuro *et al.* (1983) applied the transient grating method to study decays associated with photogenerated carriers in a-Si : H; their result in the range 100–400 psec is qualitatively similar to curve SP in Fig. 7.

Picosecond PC response in a-Si : H was recently reviewed by Johnson (1984). The curve shown in Fig. 9 was measured by the electronic correlation technique invented by Auston *et al.* (1980). In this case, the true response is convoluted with the known response of another photoconductor that is fast. If we compare the rate of the PC decay (Fig. 9) with the PA decay (Fig. 7) in similar materials we see that the PC decay is much faster. This is easily explained. The PA response depends on the distribution of carriers in the states in the gap whose absorption cross sections $\sigma(E)$ decrease when the energy E moves deeper into the gap. The PC response depends on the mobility $\mu(E)$ of carriers in these states, which decreases much more sharply with E than does $\sigma(E)$ (the deeper states are more localized).

The mobilities of trapped carriers are negligible and only carriers in the band contribute to PC. This explains why the PC decay becomes faster at low T (thermal excitation of trapped carriers is less probable), although the PA response is slower. (A significant decrease of σ requires a deeper trap, and at low T it takes a longer time for the carrier to reach it.) For both PA and PC

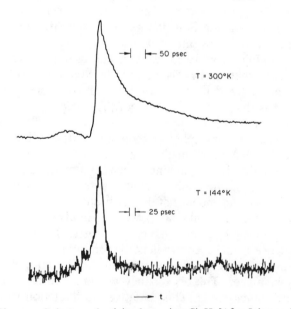

FIG. 9. Picosecond photoconductivity decays in a-Si : H. [After Johnson (1984).]

a higher defect concentration increases the decay rate (the decays are much faster in a-Si than in a-Si:H).

2. Doped a-Si:H

If geminate recombination can be neglected, the PA and PC decays give information about carrier transport in the picosecond time domains. However, a detailed comparison with theory is difficult because of the unknown trap distributions and unknown dependencies of σ on E. It was found that the situation is much simpler in a-Si:H doped with phosphorus, which, besides being an n-type dopant, also introduces deep traps confined to a relatively narrow range of E (Street *et al.*, 1981, and others.) We analyzed our data with two cross sections assumed to be independent of energy, σ_2 for shallow traps and σ_3 for deep traps, as shown schematically in Fig. 2. However, as we shall see, the results in a-Si:H:P show that the carriers producing the observed effects are holes rather than electrons.

As shown in Fig. 10 (Vardeny *et al.*, 1982), the decay rate is much more enhanced in phosphorus doped a-Si:H than in undoped a-Si:H. It is even more so in boron doped and compensated samples. Close to $t = 0$ one observes induced absorption that sharply decreases and eventually changes its sign, becoming induced transmission. The decay is faster at high temperature and higher P concentrations.

One can understand the sign reversal of $\Delta\alpha(t)$ if one assumes that the absorption cross section of a carrier in the deep trap σ_3 is smaller than σ_0. We shall present arguments showing that the species we actually see in a-Si:H:P are holes. If the concentration of holes in the shallow traps (near the valence-band top) with absorption cross section σ_2 is $N(t)$ and the recombination time is much longer than the time of the experiment, then

$$\Delta\alpha(t) = (\sigma_2 - \sigma_0)N(t) + (\sigma_3 - \sigma_0)[N(0) - N(t)], \qquad (4)$$

so that from the measurement of $\Delta\alpha(t)$ we can determine $N(t)$ and consequently obtain information about the transport of holes from the shallow traps into the deep traps.

At times close to $t = 0$ one has to deconvolute the data of Fig. 10 because Eq. (4) actually holds for $A(t)$. We could fit the decays with $\sigma_2 - \sigma_0 = 4 \times 10^{-17}$ cm^2, $(\sigma_2 - \sigma_3)/(\sigma_0 - \sigma_3) = 1.8$, and

$$N(t) = N(0)/[1 + (t/\tau)^{T/T_0}] \qquad (5)$$

with $T_0 = 350°$K and τ shown in Fig. 11. Equation (5) was proposed by Orenstein and Kastner (1981b) for nongeminate recombination of carriers in the multiple trapping (MT) model (see also Parts V and VI). It is applicable to our case of deep trapping as long as the traps are not saturated. Their recombination time constant is in our case the trapping time constant

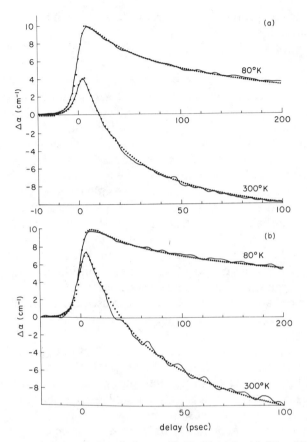

FIG. 10. Decays in P-doped glow-discharge a-Si:H with (a) 5×10^{-3} P and (b) 10^{-3} P. Solid lines: experimental; dotted lines: calculated. [From Vardeny *et al.* (1982).]

τ, and it holds that

$$\tau = \nu_0^{-1} \exp(E_a/kT), \tag{6}$$

$$E_a = kT_0 \ln(b_t N_t/b_{dt} N_{dt}), \tag{7}$$

where N_t and N_{dt} are the densities of shallow and deep traps, and b_t and b_{dt} are the corresponding capture coefficients. Figure 11 shows that τ is indeed activated. The intercept of the two lines gives $\nu_0 = 2 \times 10^{11}$ sec^{-1} (ν_0 is the carrier attempt frequency to escape from shallow traps). The difference in the activation energies E_a for the two doping levels shown in Fig. 11 is 12 meV, which is in agreement with Eq. (7) if we use the observation of Street *et al.* (1981) that changing the nominal doping from 1×10^{-3} to 5×10^{-3} changes the density of deep traps only be a factor of 1.5.

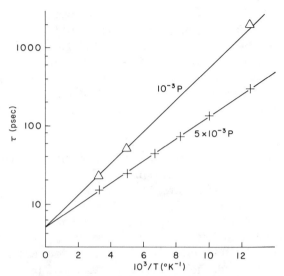

FIG. 11. Temperature dependence of the trapping time τ for the two samples whose PA decays is shown in Fig. 9. [From Vardeny et al. (1982).]

From our data we can obtain information about the deep traps. Since the measured activation energy is close to kT_0 ($= 30$ meV), it follows from Eq. (7) that the ratio $b_t N_t / b_{dt} N_{dt} \simeq 3$. If we use the estimated values of $N_t \simeq 10^{20}$ cm^{-3} (Tiedje et al., 1981) and $N_{dt} \simeq 10^{18}$ cm^{-3} (Street et al., 1981), we find that the b_{dt}/b_t ratio has to be large ($\simeq 10^2$). This means that the capture cross section of deep traps is substantially larger than that of shallow traps. Since an estimate of the latter is 10^{-15} cm^2 (Tiedje et al., 1981), the capture cross section of deep traps should be of order 10^{-13} cm^2. Such large capture cross sections are expected for charged traps (Rose, 1978). The energy E_{dt} of deep traps (measured from the band edge) was obtained from the measurement of the photoinduced midgap absorption band in this material (Vardeny et al., 1982): $E_{dt} \simeq 0.4$ eV.

The drift mobility of carriers in thermal equilibrium with deep traps with concentration N_{dt} can be estimated from the equation

$$\mu = \mu_0 (N_0/N_{dt}) \exp(E_d/kT), \qquad (8)$$

where μ_0 is the band mobility and N_0 is the effective density of states at the band edge. By taking $\mu_0 \simeq 10$ cm^2 V^{-1} sec^{-1}, $N_0 \simeq 10^{20}$ cm^{-3}, and $N_{dt} \simeq 10^{18}$ cm^{-3}, we obtain $\mu \simeq 2 \times 10^{-5}$ cm^2 V^{-1} sec^{-1}. This value is several orders lower than the drift mobility of about 4×10^{-2} cm^2 V^{-1} sec^{-1} obtained by PC measurements in a-Si:H:P at $t = 3$ μsec (Hvam and Brodsky, 1981). We conclude that the carriers trapped in deep traps cannot be the carriers responsible for the PC response. Since PC in P-doped a-Si:H:P is due to electrons, the induced transmission is due to trapped holes.

Another observation is consistent with this assignment. The transient photoconductivity measurements by Hvam and Brodsky (1981) showed that the majority carrier (electron) lifetime increased by orders of magnitude on doping a-Si:H with P. This can be understood if the defect states produced by P doping are hole traps that are negatively charged when empty.

The most important result of this study is that the detailed agreement between experiment and theory conclusively shows that the hole transport in a-Si:H is dispersive (Scher and Montroll, 1975), starting at times shorter than 5 psec. The dispersion is correctly described by the MT model (Tiedje and Rose, 1981; Orenstein and Kastner, 1981a). This means that the hole transport is dominated by the interaction with shallow traps whose energy distribution is exponential; the width of this distribution is kT_0. Moreover, in this model the time and energy scales are related by the equation $E = kT$ ln $v_0 t$. Since the MT model holds at times shorter than 5 psec, the distribution of shallow traps has to be exponential, starting very close to the valence-band top (10^{-2} eV or less).

This conclusion explains the result of Johnson *et al.* (1981) that the initial mobility in a-Si:H determined from their picosecond photoconductivity measurements is activated with an activation energy of 58 meV. With their time resolution, the initial mobility is the mobility during the first 25 psec, when the carriers, according to our results, are definitely subjected to multiple trappings. In this case the effective mobility associated with carriers occupying deeper and deeper traps is activated (Tiedje and Rose, 1981; Orenstein and Kastner, 1981a), and the activation energy depends on the time scale (a small activation energy is expected in the picosecond range, as observed).

We shall now discuss an example of a temperature-independent decay. Figure 12 shows the PA decay in an apparently lower quality sputtered P-doped a-Si:H at 80 and 300°K. The decays are similar to those in Fig. 11 except for the temperature dependence: there is practically no change in their shape or in the decay time constant with temperature. The probable reason for this is that the energy distribution of the shallow traps is nonexponential and broad. The MT model as used above is applicable only for exponential distributions. The time dependence of $N(t)$ deduced from the data in Fig. 11 can be described by the equation

$$N(t) = N(0)/[1 + (t/\tau)^\alpha], \tag{9}$$

which is similar to Eq. (5) except that the T/T_0 factor is replaced by a temperature-independent parameter α (which in this sample is $\simeq 0.7$).

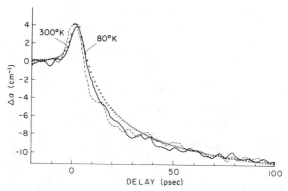

FIG. 12. Decays in P-doped sputtered a-Si: H. Solid and dashed lines: experimental; dotted: calculated.

Following the current usage, the letter α is used for denoting the absorption coefficient and also for the exponent in Eq. (9) (and in Eqs. (18) and (19) below). In the latter meaning, α is usually referred to as the "dispersion parameter." This parameter was introduced by Scher and Montroll in 1975 in their theory of dispersive transport [as is apparent from Eq. (5), in the MT model $\alpha = T/T_0$]. The trapping time τ is also temperature independent (= 8 psec). It appears plausible to assume that because of the high density of states in the gap, the carrier transport occurs by nonactivated tunneling between traps. So far there is no theoretical work justifying Eq. (9); however, its long-time limit $N(t) \rightarrow t^{-\alpha}$ follows from various theoretical approaches to dispersive transport (Scher, 1981).

V. Photoinduced Midgap Absorption

In studies of decay processes at times longer than 1 nsec it is necessary to avoid thermal effects (Part I). It is therefore convenient to use PA in the infrared region. O'Connor and Tauc (1979) observed that illumination with photon energies above the gap produces a transient absorption band in a-Si: H and similar amorphous semiconductors. The steady-state spectrum is shown in Fig. 13. In a-Si: H, its low energy end can be represented by the function (O'Connor and Tauc, 1980b, 1982)

$$\Delta\alpha(\omega) = (\hbar\omega - E_0)^{1/2}/\hbar\omega. \tag{10}$$

Its origin was ascribed to transitions of carriers trapped in the band tails into the same band, as shown schematically in Fig. 2. However, for reasons to be discussed later, it is likely that the midgap PA band in a-Si: H is due to trapped holes for which the model shown in Fig. 14 is applicable. We assume

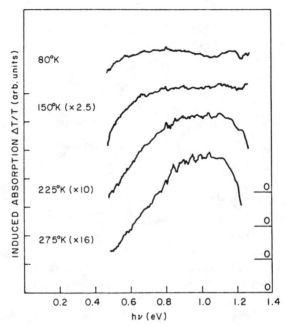

FIG. 13. Midgap PA spectra in a-Si : H (SP). [From O'Connor and Tauc (1980a). Copyright North-Holland Publ. Co., Amsterdam, 1980.]

that the valence band has a tail of localized states that have an exponential energy distribution

$$D_t(E) = (N_t/kT_0) \exp(-E/kT_0), \qquad (11)$$

where N_t is the total concentration of these states (shallow traps); in this case, the energy E is measured from the top of the valence band E_v. In the steady state the tail will be filled following the Fermi distribution with a quasi-Fermi level F_p. States with $E > F_p$ will be filled with holes, states with $E < F_p$ will be empty of holes. The change occurs sharply at F_p (within kT).

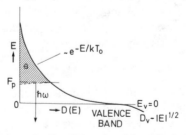

FIG. 14. State density close to the top of the valence band in the MT model for holes. The transition $\hbar\omega$ shows the origin of the PA band.

The combined sharpness of the Fermi distribution and of the energy distribution $D_t(E)$ have as a consequence that practically all holes are concentrated within a narrow energy interval at F_p. Since in amorphous semiconductors no k-vector selection rule applies, the absorption produced by hole transitions from the tail into the valence band is proportional to the convolution of the state densities (assuming a constant matrix element). In the convolution, we approximate the initial state density by a delta function at $E = F_p$ and assume that the density of the valence-band states is proportional to $|E|^{1/2}$:

$$\hbar\omega \,\Delta\alpha \propto \int \delta(E - F_p)D_v(\hbar\omega - E) \, dE \propto D_v(\hbar\omega - F_p) \sim (\hbar\omega - F_p)^{1/2}. \quad (12)$$

From studying the PA band and its temperature dependence, information about the shallow state distribution can be obtained (Tauc, 1982). In this chapter, we shall discuss only time-dependent properties, and in this section, we shall show that in high quality a-Si:H the observed time evolution of the shape of the spectrum is in agreement with the MT model; in Part VI we shall discuss the decay of the whole PA band, from which information about recombination was obtained.

Tiedje and Rose (1981) and, independently, Orenstein and Kastner (1981a) proposed a simple multiple trapping model for explaining the dispersive transport of carriers in amorphous semiconductors. The idea of dispersive transport was originated by Scher and Montroll (1975) to explain the results of the time-of-flight measurements in highly disordered materials and has since been widely applied to analyze a variety of effects associated with transport of excess carriers in such materials. The multiple trapping model is a simple particular case of the general theory. It connects the dispersive behavior to energy distributions and to other parameters by using a direct appeal to intuition.

We shall apply the MT model to discuss the time evolution of the PA spectra in a-Si:H. The energy distribution shown in Fig. 15 is the same as that in the steady-state case in Fig. 14. However, in time-resolved spectros-

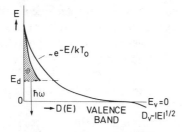

FIG. 15. Filling of the tail states in the transient regime (before the steady state shown in Fig. 14 is attained).

copy one deals with situations before the steady state is attained and the Fermi distribution realized. At a time t the tail is partly filled with holes but only down to an energy E_d ("demarcation energy") below which the states are empty of holes. More accurately, the holes in the states with energies below E_d are in thermal equilibrium with the holes in the valence bands. The demarcation energy plays a role similar to that of F_p in the steady state. The main difference is that in the steady state the tail states above E_d ($\equiv F_p$) are completely filled with holes (saturated). The definition of E_d is based on the formula for the thermal release time t of a hole from a trap at energy E_d

$$t = v_0^{-1} \exp(E_d/kT), \tag{13}$$

where v_0 is an "attempt to escape" frequency (as in Part IV). Because of the exponential dependence of t on energy, at a certain time t the holes in the states with E above E_d will never be reexcited into the valence band, whereas the hole distribution below E_d corresponds to thermal equilibrium established by hole trapping and thermal reexcitation. However, the holes in the valence band fall not only into the states below E_d but also into the deep states ($E > E_d$) from which there is no return. Therefore, as time progresses the demarcation energy E_d moves away from the band edge according to

$$E_d = kT \ln v_0 t, \tag{14}$$

and the partial filling of the states above E_d with holes increases. This may eventually lead to trap saturation. In this case the demarcation energy E_d stops moving with time, i.e., steady state is reached and E_d becomes identical with F_p. If there is no recombination, trap saturation will always be reached. However, recombination may prevent this from happening.

It follows from these considerations that the spectrum at a time t should have the shape given by Eq. (10), but E_0 should be replaced by the time-dependent $E_d(t)$ from Eq. (14).

Ray *et al.* (1981a) measured the time-dependent PA spectrum using a dye laser pumped with a nitrogen laser (photon energy 2.1 eV, pulse duration 10 nsec, pulse energy 60 μJ, repetition rate 20 Hz). The spectra measured at different times with a boxcar integrator after the pump pulse are shown in Fig. 16 for a glow-discharge a-Si:H sample. The shape of the spectra could be described by Eq. (10) and from the fit with the data the temperature and the time dependence of E_0 were determined (Fig. 17). The linearity of the plot shows that $E_0(T,t)$ is determined within the accuracy of the measurement by Eq. (14) (recall that $E_0 \equiv E_d$). A single adjustable parameter $v_0 = 10^{13} \sec^{-1}$ describes the shapes of all the spectra measured at different T and t.

However, this simple situation was found only in a-Si:H prepared by glow discharge. The time evolution of the spectra in sputtered a-Si:H

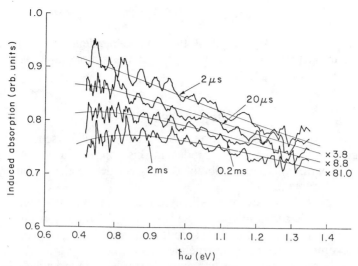

FIG. 16. Midgap PA spectra measured at different delays after the excitation pulse. a-Si : H ($C_H = 16$ at. %) at $T = 220°K$. [From Ray *et al.* (1981b).]

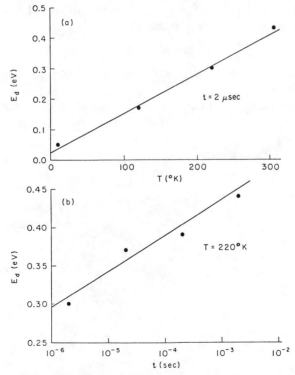

FIG. 17. Temperature and time dependencies of the demarcation energy E_d determined from the data in Fig. 16. [From Ray *et al.* (1981b).]

showed the same basic trend of the time dependence (shift to higher frequencies with increasing t and T), but the simple analysis was not applicable (Ray *et al.,* 1981b).

VI. Recombination

The theory of the recombination processes in amorphous semiconductors is discussed in Chapter 8 by Crandall in this volume. In this chapter we present experimental results obtained from PA decays that we analyze in terms of a simple model.

We studied the decay of $P = \int \Delta\alpha(\omega)\, d\omega$, the total PA band optical strength, which, in the model discussed in Part V, is proportional to the total concentration of holes N. If the absorption cross section of holes in the band tail is assumed to be energy independent, the changes in the hole distribution should not matter. In the first work on PA decay (Vardeny *et al.,* 1980, 1981) (Fig. 18), it was found that the initial decay rate depended on the level of illumination, suggesting bimolecular recombination. At long times it was observed that $N \propto P \propto t^{-\alpha}$ with $\alpha < 1$ rather than with $\alpha = 1$ as follows from the bimolecular recombination equation

$$dN/dt = -bN^2. \tag{15}$$

It was realized (Vardeny *et al.,* 1980) that $\alpha < 1$ indicated that dispersive transport was involved.

A simple model for recombination in the presence of dispersive transport,

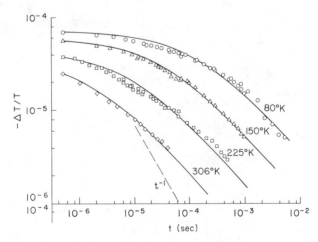

FIG. 18. Decay of photoinduced absorption (total midgap PA band) in sputtered a-Si:H. [From Vardeny *et al.* (1980).]

based on the MT model, was proposed by Orenstein and Kastner (1981b). A free carrier in the band can either be trapped in the tail or recombine at a recombination center. Since the carrier can be trapped only in an empty trap with an energy below E_d (above E_d it is immediately thermally reexcited) and since E_d sinks with time deeper and deeper into the tail, the density of states available for trapping decreases with time while the density of available recombination centers N_r remains constant. A recombination time τ_r can be defined as the instant at which the capture rate by recombination centers becomes equal to the trapping rate by the tail states. If the trapping coefficient is b_t (independent of trap energy) and the capture coefficient of the recombination center b_r, and if we assume that the traps are far from saturation, then the two rates are equal at a time τ_r determined by

$$b_t \int_{E_d(\tau_r)}^{\infty} D_t(E)\, dE = b_r N_r. \tag{16}$$

If we substitute for $D_t(E)$ from Eq. (11) and E_d from Eq. (14) we obtain

$$\tau_r = (1/\nu_0)(b_t N_t/b_r N_r)^{T_0/T}. \tag{17}$$

This is equivalent to Eqs. (6) and (7) written in a different form. For monomolecular recombination in the case $b_r/b_t > 1$, which is apparently justified in the conditions of our experiments, the decay of N is given by

$$N(t) = N(0)/[1 + (t/\tau_r)^\alpha], \tag{18}$$

where $\alpha = T/T_0$. This equation holds also for bimolecular recombination; however, N_r has to be replaced by $N(0)$. At long times $t \gg \tau_r$ Eq. (18) gives $N(t) \propto t^{-\alpha}$. For bimolecular recombination it follows from Eqs. (17) and (18) that $N(t)$ becomes independent of the initial concentration $N(0)$ as

$$N(t) \rightarrow (b_t N_t/b_r)(\nu t)^{-\alpha}. \tag{19}$$

The results shown in Fig. 18 were obtained on a sputtered sample for which α (equal to the slope of the log–log plots at long times) was temperature independent. Apparently, the MT model does not apply in this case, but Eq. (18) is still valid.

The results shown in Fig. 18 were obtained by using a cavity-dumped rhodamine 6G dye laser with the following parameters of the pump: photon energy 2.1 eV, duration 0.2 μsec, energy 60 nJ, and repetition rate 100 Hz.

Later (Ray et al., 1981b), the same effect was measured with a pump produced by a pulsed nitrogen dye laser (photon energy 2.1 eV, pulse duration 10 nsec, pulse energy 60 μJ, repetition rate 10 to 20 Hz). A typical result obtained on a sample with a low density of gap states is shown in Fig. 19 (Kirby et al., 1982). The sample was prepared by a sputtering process that yields high quality a-Si: H (Paul and Anderson, 1981). In the time domain

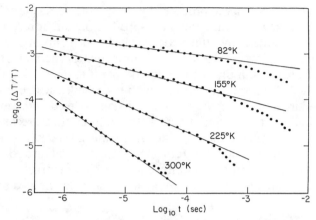

FIG. 19. PA decay in high quality sputtered a-Si : H. [Reprinted with permission from *Solid State Communications,* Vol. 42, by P. B. Kirby, W. Paul, S. Ray, and J. Tauc, Comparison of the dispersion parameters from times of flight and photoinduced midgap absorption measurements on sputtered a-Si : H. Copyright 1982, Pergamon Press, Ltd.]

shown in Fig. 19 the sample is already in the region where $t \gg \tau_r$ and the slope of the log–log plot gives the dispersion parameter α [Eq. (18)]. In this case α was temperature dependent and equal to T/T_0, where $T_0 \simeq 300°$K.

Wake and Amer (1983) found that when the measurements were extended to shorter times (down to 10 nsec) the log–log plot was still linear. They used a Nd : YAG dye laser (4 nsec pulses, 10 Hz the repetition rate) and studied a-Si : H samples with different concentrations of dangling bonds (Fig. 20). They found the dispersion parameter α to be temperature depen-

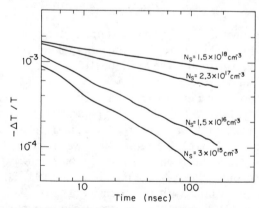

FIG. 20. PA decays in a-Si : H samples at $T = 295°$K with different densities of dangling bonds N_s. [From Wake and Amer (1983).]

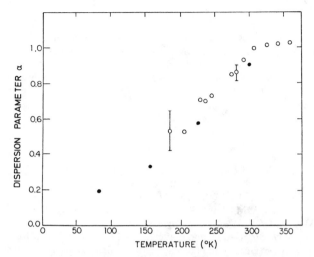

FIG. 21. Temperature dependence of the dispersion parameter α obtained from PA decays shown in Fig. 19 (filled circles) and time-of-flight measurement (open circles). [After Kirby *et al.* (1982).]

dent $(= T/T_0)$. The value of T_0 was larger when the dangling-bond density was larger. This correlated very well with the broadening of the absorption edge that they measured on the same samples. The MT model explains these results if the presence of dangling bonds increases the width of the valence-band tail.

Wake and Amer's results are in agreement with the assignment of the midgap PA band to holes trapped in the valence-band tail, as we proposed (O'Connor and Tauc, 1980b). The correlation between the values of T_0 measured from the PA decay and the tail width observed optically was interpreted as evidence that the species whose transport dominates the recombination process are holes, except in samples with a low dangling-bond density in which the electron transport appears to be significant. The work of Kirby *et al.* (1982) indicates that in samples with low densities of dangling bonds the electrons indeed dominate the transport. The open circles shown in Fig. 21 were obtained from measuring the α of electrons by the time-of-flight method (Scher and Montroll, 1975; Tiedje and Rose, 1981) as a function of temperature. It is apparent that the agreement is good with the values of α obtained by the PA decay (filled circles).

The studies of the midgap PA band in a-Si:H have not yet been extensive enough to really resolve all the questions that arose during the first stage of research. The current work promises to shed more light on this method and develop it into a reliable new tool.

ACKNOWLEDGMENTS

The work reviewed in this paper was done at Brown University with the support of the National Science Foundation (grants DMR 79-09819 and DMR 82-09148) and the NSF Materials Research Laboratory. My collaborators include my former and present students D. E. Ackley, P. O'Connor, S. Ray, J. Strait, and D. Pfost; research associate Z. Vardeny; and research engineer T. R. Kirst. I thank Nabil M. Amer for showing me some results of his work before publication.

REFERENCES

Abeles, B., Wronski, C. R., Tiedje, T., and Cody, G. D. (1980). *Solid State Commun.* **36,** 537.
Ackley, D. E., Tauc, J., and Paul, W. (1979). *Phys. Rev. Lett.* **43,** 715.
Ackley, D. E., Tauc, J., and Paul, W. (1980). *J. Non-Cryst. Solids* **35/36,** 957.
Auston, D. H., Johnson, A. M., Smith, R. R., and Bean, J. C. (1980). *Appl. Phys. Lett.* **37,** 371.
Conwell, E. M. (1967). *Solid State Phys. Suppl.* **9.**
Elci, A., Scully, M. O., Smirl, A. L., and Matter, J. C. (1977). *Phys. Rev. B* **16,** 191.
Fayer, M. D. (1982). *In* "Picosecond Phenomena III" (K. B. Eisenthal, R. M. Hochstrasser, W. Kaiser, and A. Laubereau, eds.), p. 82. Springer-Verlag, Berlin and New York.
Freeman, E. C., and Paul, W. (1979). *Phys. Rev. B* **20,** 716.
Hvam, T. M., and Brodsky, M. H. (1981). *Phys. Rev. Lett.* **46,** 371.
Ippen, E. P., and Shank, C. V. (1977a). *Appl. Phys. Lett.* **27,** 488.
Ippen, E. P., and Shank, C. V. (1977b). *In* "Ultrashort Light Pulses" (S. L. Shapiro, ed.), p. 83. Springer-Verlag, Berlin and New York.
Johnson, A. M. (1984). *In* "Semiconductors Probed by Ultrafast Laser Spectroscopy" (R. R. Alfano, ed.), Vol. 2. Academic Press, New York. To be published.
Johnson, A. M., Auston, D. H., Smith, P. R., Bean, T. C., Harbison, J. P., and Adams, A. C. (1981). *Phys. Rev. B* **23,** 6816.
Kirby, P. B., Paul, W., Ray, S., and Tauc, J. (1982). *Solid State Commun.* **42,** 533.
Komuro, S., Aoyagi, Y., Segawa, Y., Namba, S., Masuyama, A., Okamoto, H., and Hamakawa, Y. (1983). *Appl. Phys. Lett.* **42,** 79.
Mort, J., Chen, I., Morgan, M., and Grammatica, S. (1981). *Solid State Commun.* **39,** 1329.
O'Connor, P., and Tauc, J. (1979). *Phys. Rev. Lett.* **43,** 311.
O'Connor, P., and Tauc, J. (1980a). *J. Non-Cryst. Solids* **35/36,** 699.
O'Connor, P., and Tauc, J. (1980b). *Solid State Commun.* **36,** 947.
O'Connor, P., and Tauc, J. (1982). *Phys. Rev. B* **25,** 2748.
Orenstein, J., and Kastner, M. (1981a). *Phys. Rev. Lett.* **46,** 1421.
Orenstein, J., and Kastner, M. (1981b). *Solid State Commun.* **40,** 85.
Paul, W., and Anderson, D. Z. (1981). *Solar Energy Mat.* **5,** 229.
Ray, S., Vardeny, Z., and Tauc, J. (1981a). *J. Phys. Colloq. Orsay Fr.* **42,** C4-555.
Ray, S., Vardeny, Z., and Tauc, J. (1981b). *AIP Conf. Proc.* **73,** 253.
Rose, A. (1978). "Concepts in Photoconductivity and Allied Problems." Krieger Publ. Co., Huntington, New York.
Scher, H. (1981). *J. Phys. Colloq. Orsay Fr.* **42,** C4-547.
Scher, H., and Montroll, E. W. (1975). *Phys. Rev. B* **12,** 2245.
Seeger, K. (1982). "Semiconductor Physics." Springer-Verlag, Berlin and New York.
Shank, C. V., Fork, R. L., Leheny, R. F., and Shah, J. (1979). *Phys. Rev. Lett.* **42,** 112.
Shank, C. V., Fork, R. L., and Yen, R. T. (1982). *In* "Picosecond Phenomena" (K. B. Eisenthal, R. M. Hochstrasser, W. Kaiser, and A. Laubereau, eds.), Vol. 3, p. 2. Springer-Verlag, Berlin and New York.

Silver, M., Giles, N. C., Snow, E., Cannella, V., and Adler, D. (1982). *Appl. Phys. Lett.* **41**, 935.
Street, R. A., Biegelsen, D. K., and Knights, T. C. (1981). *Phys. Rev. B* **24**, 969.
Tauc, J. (1982). *Solar Energy Mat.* **8**, 259.
Tiedje, T., and Rose, A. (1981). *Solid State Commun.* **37**, 49.
Tiedje, T., Cebulka, T. M., Morel, D. L., and Abeles, B. (1981). *Phys. Rev. Lett.* **46**, 1425.
Vardeny, Z., and Tauc, J. (1981a). *Phys. Rev. Lett.* **46**, 1223. [Erratum: *Phys. Rev. Lett.* **47**, 700 (1981).]
Vardeny, Z., and Tauc, J. (1981b). *Opt. Commun.* **39**, 396.
Vardeny, Z., and Tauc, J. (1981c). *J. Phys. Colloq. Orsay Fr.* **42**, C4-477.
Vardeny, Z., and Tauc, J. (1984). *In* "Semiconductors Probed by Ultrafast Laser Spectroscopy" (R. R. Alfano, ed.), Vol. 2. Academic Press, New York. To be published.
Vardeny, Z., O'Connor, P., Ray, S., and Tauc, J. (1980). *Phys. Rev. Lett.* **44**, 1267.
Vardeny, Z., O'Connor, P., Ray, S., and Tauc, J. (1981). *Phys. Rev. Lett.* **46**, 1108.
Vardeny, Z., Strait, J., Pfost, D., Tauc, J., and Abeles, B. (1982). *Phys. Rev. Lett.* **48**, 1132.
Vardeny, Z., Strait, J., and Tauc, J. (1983). *Appl. Phys. Lett.,* **42**, 580.
von der Linde, D., and Lambrich, R. (1979). *Phys. Rev. Lett.* **42**, 1090.
Wake, D. R., and Amer, N. M. (1983). *Phys. Rev. B. Rapid Commun.* **27**, 2598.

SEMICONDUCTORS AND SEMIMETALS, VOL. 21, PART B

CHAPTER 10

IR-Induced Quenching and Enhancement of Photoconductivity and Photoluminescence

P. E. Vanier

METALLURGY AND MATERIALS SCIENCE DIVISION
BROOKHAVEN NATIONAL LABORATORY
UPTON, NEW YORK

I. Introduction

Optoelectronic experiments on hydrogenated amorphous silicon (a-Si:H) can be divided into four broad classes, briefly described by the optical excitation conditions as follows.

(a) *Dark Equilibrium.* In measurements such as dark conductivity, thermopower, and electron spin resonance, the free carriers and those trapped in localized states are in thermal equilibrium.

(b) *Steady State Illumination.* In the measurement of optical absorption, steady-state photoconductivity, photoluminescence, and light-induced ESR, excess electrons and holes are generated in the bands by a beam of intrinsic light and become trapped at localized states in the gap. The occupation of a given state is then determined by the relative rates of emission and capture for electrons and holes interacting with that state.

329

(c) *Transient (Pulsed) Excitation.* The decay of photocurrent or photoluminescence after a short pulse of light can be used to determine the rates
at which the excited carriers return to equilibrium conditions.

(d) *Infrared (IR) Modulation.* An illuminated sample in a steady state
can be perturbed by a second beam of photons of energy less than that of the
band gap. This category of experiment includes photoinduced infrared
absorption and infrared modulation of either photoconductivity or photoluminescence.

This chapter is mainly concerned with the last two techniques of class (d),
in which sub-band-gap optical excitations can change the occupations of the
trapping centers from one quasi-equilibrium condition to another, thereby
influencing the rate of carrier recombination. By varying the wavelength of
the IR beam, it is possible to measure a spectrum of sub-band-gap excitations that produce a change in the overall recombination rate. Both quenching and enhancement of steady-state signals can be observed, depending on
the chosen excitation conditions. By chopping the IR beam at various
frequencies, one can explore the rates at which the trapped carriers redistrib-
ute themselves in response to the perturbation. Since the steady state itself is
dependent on temperature, both the spectra and the chopping frequency
response must also be temperature dependent.

The principal result of these experiments has been to confirm a previous
deduction from single-beam photoconductivity experiments, i.e., that recombination in a-Si:H is controlled by more than one species of defect,
each with a distinctly different set of carrier capture-rate constants. The net
transfer of minority carriers from one species of center to another can result
in a remarkable change in recombination rate. This transfer depends on the
position of the Fermi level E_F, the intensities of both intrinsic and extrinsic
excitation, and the sample temperature.

The modulation of photoluminescence by infrared excitation is a much
smaller effect because only a small fraction of the electron–hole pairs
interacts with the defects. The liberation of trapped holes by an IR pulse
causes a transient enhancement of the luminescence, but since the majority
of these holes can tunnel to defects rather than find unpaired electrons and
form new pairs, the steady-state signal is quenched, i.e., IR excitation may
also favor recombination via nonradiative recombination centers.

II. Historical Background

1. PHOTOCONDUCTIVITY IN OTHER MATERIALS

The physical principles involved in carrier recombination in semiconductors have been understood for a long time (Shockley and Read, 1952; Bube,

1960; Rose, 1963). The materials being studied at the time were crystalline, and the models used included abrupt band edges and two or three discrete levels in the gap. With these models it was possible to explain a wide variety of phenomena, such as sublinear, linear, and supralinear dependence of photocurrent on light intensity. Thermal quenching and infrared quenching of photoconductivity were observed in crystalline CdSe : I : Cu (Bube, 1957) and Ge : Mn (Newman et al., 1956). Early theories of amorphous materials (Mott, 1967; Cohen et al., 1969) predicted smoothly varying densities of states that would be unlikely to produce such interesting behavior.

The generalization of Shockley–Read statistics to semiconductors containing an arbitrary distribution of traps (Simmons and Taylor, 1971, 1972; Taylor and Simmons, 1972) helped to explain the single broad peak in the temperature dependence of photoconductivity of chalcogenide glasses (Arnoldussen and Bube, 1972). However, the data were more closely fit by a model containing relatively narrow trap bands (Simmons and Taylor, 1974) than by one with slowly varying trap distributions (Simmons and Taylor, 1973). The same authors were then able to estimate a number of trapping parameters from their best fits (Taylor and Simmons, 1974). Although these authors made it clear that there could be more than one species of trap, each of which would have a different occupation function and a different fraction of the recombination traffic, they usually simplified the problem by assuming that only one species was present. More recently, the application of Shockley–Read statistics to a semiconductor containing several different discrete traps with differing capture cross sections was carried out (Hoffmann and Stöckmann, 1979) for the case of CdS. The authors were able to fit experimental plots of photoconductivity over 15 orders of magnitude of light intensity and extract the densities, energies, and occupations of three species of traps, along with some of the capture cross sections.

2. PHOTOCONDUCTIVITY IN a-Si : H

Early studies of the temperature dependence of photoconductivity in a-Si : H (Spear et al., 1974) revealed even sharper transitions from one recombination mechanism to another than had been seen in the chalcogenide glasses. These results were a strong indication that the recombination was controlled by relatively narrow bands of traps. The analysis was based on a picture of the density of states obtained by field effect measurements, which indicated bands of traps labeled x and y at energies given by $E_c - E_y = 0.4$ eV and $E_x - E_v = 0.4$ eV, where E_c, E_v are the conduction- and valence-band edges. This group also explained some abrupt changes of slope in Arrhenius plots of photoconductivity as transitions from conduction in extended states to conduction by hopping through traps (Spear et al., 1974; Anderson and Spear, 1977). The density of states derived from field effect measurements was later found to be strongly affected by interface states (Goodman and Fritzsche, 1980), raising serious questions about the detailed interpretation of the photoconductivity data of Spear *et al.*

The detection of an electron spin resonance (ESR) signal by changes in photoconductivity showed that electron lifetimes are controlled by a fast recombination center that has an ESR signal at $g = 2.0055$ (Solomon *et al.*, 1977). These centers are widely accepted to be singly occupied (neutral) dangling bonds (Biegelsen, 1980), sometimes referred to as T_3^0 centers (Kastner *et al.*, 1976). The photoconductivity of a-Si:H samples has been shown to decrease rapidly with increasing spin density (Voget-Grote *et al.*, 1980). Further evidence of the influence of dangling bonds on photoconductivity came from the changes associated with dehydrogenation and rehydrogenation (Staebler and Pankove, 1980), a comparison of ESR spin density with the density of recombination centers obtained from Schottky diode measurements (Tiedje *et al.*, 1981), and the passivation of dangling bonds by hydrogen in sputtered a-Si:H (Paul, 1981).

However, the dangling bond by itself was not sufficient to explain the complicated photoconductive behavior of a-Si:H. Measurements of the decay of photoconductivity in a-Si:H from a steady state after abrupt termination of the illumination enabled workers at RCA to determine electron lifetimes as a function of photogeneration rate (Wronski and Daniel, 1981). From these results several deductions were made. The kinetics of trapping and recombination were found to be consistent with free-carrier transport over the whole temperature range (120 to 350°K). The electron lifetime, which varied between 10^{-6} and 10^{-3} sec, was shown to depend on two types of recombination centers located at or below midgap. One of these was a fast center (presumably the dangling bond) and the other was a "sensitizing" slow center. No evidence was found for the previously mentioned E_y peak in the density of states.

While the paper by Wronski and Daniel (1981) was in press, films of a-Si:H prepared at Brookhaven National Laboratory were found to exhibit clearly the photoconductive behavior characteristic of at least two competing recombination centers at different energies in the gap with very different electron capture cross sections (Griffith *et al.*, 1980; Vanier *et al.*, 1981a,b). These features included thermal quenching of photoconductivity σ_p at low temperatures, supralinear dependence of σ_p on light intensity, and infrared quenching of σ_p. Meanwhile, two-beam modulated absorption experiments (Olivier *et al.*, 1980; O'Connor and Tauc, 1980, 1982) showed that illumination of a-Si:H or a-Si:Ge:H alloys with high intensity visible light caused an increase in infrared absorption. This light-induced absorption was interpreted as excitations from the valence band into gap states that were occupied in the dark but vacant under visible illumination. The spectrum of these excitations showed a threshold that shifted with temperature, reflecting the movement of the quasi-Fermi level for holes E_{F_p}. At about the same time, room temperature, two-beam modulation experiments (Persans, 1980) revealed that either quenching or enhancement of the photoconductivity could be produced, depending on the photon energy of the IR excitation. This study was later extended to low temperatures (Persans and Fritzsche, 1981) and the chopping frequency was varied (Persans, 1982a). Some of those results will be summarized in Part IV.

Infrared quenching of photoconductivity is not just a transient effect, nor is it detectable only in modulation experiments as a small perturbation from the single-beam photocurrent. Under steady-state conditions, dc measurements by Vanier and Griffith (1980, 1982) showed that the photocurrent can be quenched by a factor of 20 by the IR beam. In their latter paper it was also shown that the energy threshold of infrared quenching can shift with temperature.

3. LUMINESCENCE QUENCHING AND ENHANCEMENT

The extensive literature on luminescence in a-Si:H has been reviewed (Street, 1981a) and is covered by Street in Chapter 7 of this volume. High quality undoped a-Si:H at temperatures below 100°K exhibits a single broad luminescence band with a peak at an energy of about 1.3 eV. Optical excitation produces electron – hole (e – h) pairs that are very rapidly trapped in band-tail states. Radiative recombination then takes place at rates dependent on the distance between the electron trap and the hole trap. At low excitation rates, the density of pairs is so low that there is no interaction between pairs, and the recombination is geminate. Some pairs are created close enough to defects in the material that nonradiative tunneling to the defects can take place. The luminescence efficiency Y_L is given by

$$Y_L = P_R/(P_R + P_{NR}),\qquad(1)$$

where P_R, P_{NR} are the radiative and nonradiative recombination rates. If $P_{NR} \ll P_R$, this becomes

$$Y_L \approx 1 - P_{NR}/P_R.\qquad(2)$$

At temperatures above 100°K, the e–h pairs break up, and the luminescence is thermally quenched. The free carriers then give rise to photoconductivity. There is much evidence that the dangling bonds that limit the photoconductivity by providing fast recombination centers also reduce the luminescence efficiency (Street et al., 1978). Since the occupancy of these defects can be changed by IR excitation, it is possible to modulate the photoluminescence with a chopped IR beam (Hirsch and Vanier, 1982). Only the fraction of e–h pairs that interacts with nearby defects can be affected, so the ratio of the modulation signal to the total luminescence signal cannot be greater than P_{NR}/P_R. The effect is therefore not expected to be as dramatic as that for the quenching and enhancement of photoconductivity.

III. Experimental Details

4. SAMPLES

Infrared quenching effects have been reported for only a relatively small number of samples of a-Si:H, all of which were produced by rf glow discharge at substrate temperatures in the range 200–300°C. The samples were not heavily doped and had Fermi levels not far from midgap. For photoconductivity, the usual substrate is Corning 7059 glass or quartz. Molybdenum contacts can be evaporated by electron beam onto the substrate prior to or after the a-Si:H deposition, using a mask to define a gap of 1–2 mm. An underlayer of chromium about 200 Å thick helps to improve the adhesion of the molybdenum to the substrate. For these high resistance samples, only two contacts are necessary, since the voltage drops in the contacts and connecting wires are negligible.

For photoluminescence, sapphire substrates have been used because of their good thermal conductivity at liquid helium temperatures. In order to minimize the interference fringes in the IR excitation spectrum, the substrates can be roughened by sandblasting prior to the a-Si:H deposition. This procedure probably also enhances the absorption of IR radiation by means of light trapping (Tiedje et al., 1982).

5. CRYOGENICS

Samples must be mounted in an evacuated Dewar in order to remove surface moisture and to provide thermal isolation in temperature-dependent measurements. A simple liquid nitrogen cold finger with a heater is adequate for experiments in which the temperature is varied continuously from 80 to 450°K. Thermocouples or platinum thin film resistance thermometers can be used to sense the temperature. For experiments having long duration at constant temperature, such as spectral or chopping frequency response measurements, a regulated cryogenic system with either continuous flow or closed cycle refrigeration is required. Commercial systems of this type operating with helium rather than nitrogen can be used for both photoconductivity and photoluminescence experiments, since temperatures down to ~ 10°K can be attained. However, most of these systems are not designed for use above 300°K.

6. OPTICS

Two light sources are needed for IR-modulated experiments. The visible (or pumping) light can be of fixed wavelength, e.g., a small He–Ne laser or a tungsten–halogen lamp with band-pass filters. A second source similar to the latter can be chopped mechanically to provide the modulating IR beam for measurements of temperature dependence and chopping frequency dependence. For IR excitation spectra a small monochromator is necessary. If it is of the grating type, care must be taken to insert appropriate long-pass filters to separate overlapping orders of diffraction. Thick slices of Si and Ge are useful for this purpose since they transmit IR but absorb visible light. Calibration of the IR intensity incident on the sample as a function of wavelength is essential since the lamp output and monochromator efficiency both vary enormously over the range of wavelengths of interest. This can be achieved by means of a silicon photodetector, calibrated in the visible range, combined with a thermal detector. (The latter, which could be pyroelectric, thermoelectric, or thermomechanical, is assumed to have a response independent of wavelength.) For wavelengths up to 3 μm, quartz windows and lenses can be used. For longer wavelengths, with photon energies less than 0.4 eV, focusing mirrors and special IR transmitting windows are required, and an IR source such as a hot carbon rod could be helpful.

In photoluminescence experiments, an additional monochromator is required to analyze the emission from the sample and to reject the scattered light. A schematic of the IR-modulated luminescence experiment is shown in Fig. 1. A cooled S1 photomultiplier detects the main luminescence band of a-Si:H at 1.3–1.4 eV with high sensitivity, but does not respond to the scattered (modulated) IR with photon energies less than 1.0 eV. Since the luminescence is detectable only at low temperatures, one can check at room temperature for a spurious modulated signal originating from scattered light.

7. ELECTRONICS

A schematic diagram of a typical apparatus for dual-beam photoconductivity is shown in Fig. 2. A regulated dc power supply is used to apply a constant voltage (e.g., 10 V) across the sample. Dark currents and direct photocurrents are measured with an electrometer, such as the Keithley Model 616 operated in the fast mode. Typical dark currents at room temperature for a film 2 cm wide and 0.5 μm thick are about 10^{-11} A. Modulated photocurrents are measured with a lock-in amplifier using either a current-sensitive preamplifier or a suitable load resistor across the inputs of a voltage-sensitive preamplifier. If measurements are made as a function

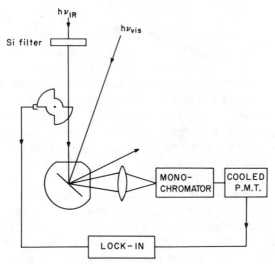

FIG. 1. Diagram of IR modulated luminescence experiment. Specularly reflected beams are not collected.

of chopping frequency to determine the dynamical behavior of the modulation effects, a frequency-tracking lock-in, such as the EG & G-PAR Model 186A or the Ithaco Model 393, is extremely useful. The readout of most lock-in amplifiers is calibrated to give the rms value of the fundamental Fourier component of the signal. For a 50% duty cycle square wave, this output must be multiplied by $\pi/\sqrt{2}$ ($=2.22$) to obtain the equivalent dc

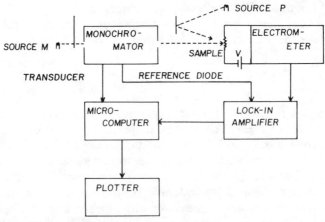

FIG. 2. Diagram of IR modulated photoconductivity experiment. P, pump beam; M, modulated beam. [From Persans (1982a).]

photocurrent. Interfacing the experiment to a desk-top computer is quite important, since a large number of data points must be recorded and normalized appropriately.

IV. Experimental Results and Interpretation

8. TRANSITIONS

Figure 3 shows a schematic of the gap states in a-Si : H that are required to explain most of the effects discussed in this chapter. Shallow electron traps are labeled t and shallow hole traps are labeled u. These are drawn as discrete levels, but represent some sort of average behavior of the distributions of band-tail states. Fast recombination centers near midgap are denoted by r, and are identified with neutral (singly occupied) dangling bonds. The fact that the dangling bond can have three charge states (T_3^+, T_3^0, or T_3^-) is not included in Fig. 3, because this causes further complexity. In the n-type samples to be discussed, it is assumed that the density of T_3^+ is negligible, and one need only consider a one-electron state that is either vacant (T_3^0) or filled (T_3^-). States s are "safe" hole traps with relatively small capture cross sections for electrons. States s and u may be just different parts of a continuous band-tail distribution or they may be structurally distinguishable species.

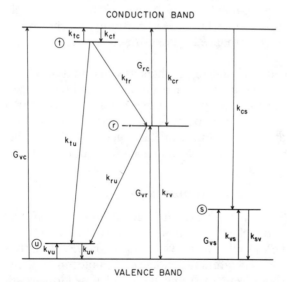

FIG. 3. Energy level diagram for gap states in a-Si : H. r, recombination centers; s, safe hole traps; t, shallow electron traps; u, shallow hole traps; G_{ij}, optical generation rates; k_{ij}, emission and capture rate coefficients.

Four possible optical excitation processes are shown, with generation rates G_{ij} that depend on the intensity of the incident light as a function of wavelength. The spontaneous capture and thermal emission processes are labeled by rate constants k_{ij}, which refer to electron transitions from level i to level j (i,j = c,r,s,t,u,v). In principle, one can follow the approach of Simmons and Taylor and write down rate equations for all these processes. Then if all of the rate constants are known, the occupation of each species of gap state and the densities of free carriers can be self-consistently calculated as a function of G_{ij} and temperature. This has not yet been done, but the concepts of quasi-Fermi levels and demarcation lines described by Crandall in Chapter 8 can still be used to give qualitative explanations of the observations. By using various approximations, estimates of some rate constants can be obtained.

Note that each species of defect i will have its own occupation function determined by its rate constants k_{ij} and k_{ji}. Thus each species will have its own quasi-Fermi levels and demarcation lines. The shifts of these levels by the addition of an IR beam can be taken into account by adding terms containing G_{vi} and G_{ic} to the rate equations.

The transition t → u represents the radiative recombination giving rise to the 1.3 eV luminescence at low temperatures. This process competes with t → r and r → u transitions involving the fast recombination center. One or both of these may be responsible for the weak 0.8 eV luminescence that scales with dangling-bond density (Street, 1980). Alternatively, the 0.8 eV defect luminescence may be a t → s transition (not shown in Fig. 3), since the conditions that produce high densities of r states may also increase the number of s states.

It was argued by Crandall in Chapter 8 that in a-Si:H the densities of carriers in extended states are small compared to those trapped in the band tails and therefore that the processes t → r and r → u, rather than c → r and r → v, dominate the recombination at or below room temperature. However, at elevated temperatures in the range 100–200°C the free carrier densities become high enough that recombination from extended states could be important.

9. SINGLE-BEAM PHOTOCONDUCTIVITY

Not all a-Si:H samples exhibit large IR enhancement and quenching effects. If E_F is too close to the conduction band or the valence band, the r states are either all filled or all vacant, and the IR has only a very small influence. The temperature dependence of photoconductivity $\sigma_p(T)$ measured with a single beam of visible light can be used to select those samples most likely to show big effects. Only when E_F is in such a position that the r states are partially occupied does there exist a delicate balance between the population of the r and s states. Then the IR excitation can tip the balance, and cause an appreciable change in the recombination rate. Figure 4 shows a series of logarithmic plots of σ_p versus $1000/T$ for six samples with different positions of the dark Fermi level E_F. Relative to the conduction band E_c, this level was determined from the dark conductivity σ_d by the relation

$$E_c - E_F = \Delta E = kT \ln(\sigma_0/\sigma_d), \tag{3}$$

where the minimum metallic conductivity was assumed to be $\sigma_0 = 200$ Ω^{-1} cm^{-1} (Mott and Davis, 1979). This method of estimating E_F has been found to be more reliable than taking the slope of Arrhenius plots of σ_d (Vanier *et al.*, 1981a,b) because of statistical shifts of E_F with temperature (Redfield, 1982). Other values of σ_0 as large as 5000 Ω^{-1} cm^{-1} have been used in this way (Overhof and Beyer, 1981), but ΔE is insensitive to this value, and is shifted by less than 0.1 eV if the larger value is chosen.

Clearly, the position of E_F has a very strong effect on both the magnitude of σ_p at room temperature ($1000/T = 3.3°K^{-1}$) and the shape of the $\sigma_p(T)$ curve. In Fig. 4 sample (a), with $\Delta E = 0.31$ eV, is lightly doped *n*-type and exhibits a simple maximum near room temperature. The shape of this curve is as one might expect for a smoothly varying density of states. Below room temperature, the signal is sublinear with respect to light intensity (i.e., bimolecular) and decreases exponentially with decreasing temperature. The activation energy is 0.11 eV over more than three decades. The physical process involved is most likely a successively deeper trapping of the carriers

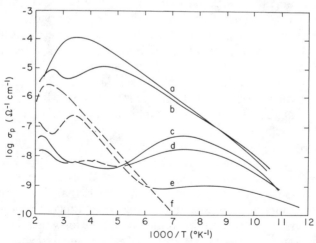

FIG. 4. Temperature dependence of photoconductivity of six a-Si : H samples with different Fermi level positions. $E_c - E_F = 0.31$ eV (a), 0.55 eV (b), 0.65 eV (c), 0.80 eV (d), 1.00 eV (e), 1.29 eV (f). Solid lines, *n*-type photoconduction; broken lines, *p*-type photoconduction. Photon flux $= 10^{14}$ cm^{-2} sec^{-1}.

with decreasing temperature (i.e., a thermally activated mobility). At temperatures higher than that of the peak in σ_p, the magnitude of σ_d exceeds σ_p, indicating that the density of photogenerated carriers is less than the density of thermal carriers. Around that point, σ_p becomes linear with light intensity because the photoelectrons recombine mainly with thermally generated holes rather than with photogenerated holes. Only small modulation effects have been reported for samples of this type (Persans, 1982b). Since nearly all of the r and s states are occupied, the only IR transition available is r $\rightarrow$ c, and this empties fast recombination centers at the same time that it creates free electrons. The net effect is about 10^{-5} times that of the steady-state signal.

Samples (b) ($\Delta E = 0.55$ eV) and (c) ($\Delta E = 0.65$ eV) of Fig. 4 show the appearance and growth of a valley in $\sigma_p(T)$ near room temperature as E_F becomes lower. This is interpreted as an increase in recombination rate through the r centers as more of them become vacant. All samples in Fig. 4 should have about the same density of dangling bonds since they were deposited under similar conditions and none was doped with more than 1 ppm PH_3 or B_2H_6. For samples with lower Fermi levels a larger fraction of these defects is T_3^0 rather than T_3^- at room temperature and is therefore able to capture electrons. Raising the temperature first causes thermal excitations v $\rightarrow$ r that reduce the recombination rate and then excites v $\rightarrow$ c transitions that increase the recombination rate, hence we see the high temperature peak.

The low temperature peak in curves (b) and (c) of Fig. 4 is the feature that allows one to predict the infrared quenching effect. The photoconductivity rises with decreasing temperature because holes become trapped or "frozen" in s states for which the recombination rate constant k_{ts} is very small. By charge neutrality, most of the r states become occupied, and the electron lifetime increases. Thus the recombination is limited by the holes as mentioned in Chapter 8 by Crandall. Excitations v $\rightarrow$ s create free holes that can then participate in r $\rightarrow$ v or r $\rightarrow$ u recombination transitions, thus accounting for IR quenching.

Curves (d) ($\Delta E = 0.08$ eV) and (e) ($\Delta E = 1.00$ eV) of Fig. 4 show the appearance of a third peak in $\sigma_p(T)$ between the high and low temperature peaks. Comparison of the shape of this peak with the single peak in curve (f) for a diborane-doped film ($\Delta E = 1.29$ eV) shows that this feature arises from hole conduction. This assertion has been confirmed by photothermopower measurements on a sample similar to sample (e), which was p-type at room temperature and n-type at 200°C (Delahoy and Vanier, 1982). The added complication of bipolar conduction may introduce further difficulties in the interpretation of IR modulation effects, because, although a v $\rightarrow$ s transition may quench the electron current, it must also enhance the hole current. The key to sorting out these effects lies in the response time (or frequency dependence) measurements, since the former process is slow and indirect whereas the latter is instantaneous.

An interesting feature in Fig. 4 is that curves (a)–(e) all seem to converge in the low temperature limit. In other words, regardless of the dark Fermi level position, all samples that are at least slightly n-type—when cooled to 90°K under illumination—will reach a steady state in which nearly all the r states are filled. For this to happen, the number of holes p_s trapped in s states must be approximately equal to the number of dangling bonds that convert from T_3^0 in the dark to T_3^- in the light. In this regime, the intensity exponent γ was found by Vanier $et~al.$ (1981a) to be 0.68, which suggests that the process t → u may contribute to the overall recombination rate. However, it could still be that the dominant recombination pathway is through the small number of r states p_r left unoccupied. If the r → v process is extremely fast, p_r will increase with light intensity, and σ_p will be sublinear.

Rough estimates of the capture coefficients k_{cr} and k_{cs} can be made by assuming that the density N_r of dangling bonds in these samples is the same as that measured in similar material (8×10^{15} cm^{-3}) by Pontuschka $et~al.$ (1982). The bulk generation rate is given by

$$G_{vc} = f(1 - R)[1 - \exp(-\alpha d)]/d \tag{4}$$

(Wronski $et~al.$, 1980). Using measured values of the photon flux ($f = 10^{14}$ cm^{-2} sec^{-1}), the reflectance ($R = 0.3$), the absorption coefficient at a wavelength of 600 nm ($\alpha = 24{,}900$ cm^{-1}), and the sample thickness ($d = 0.4~\mu$m), one obtains $G_{vc} = 10^{18}$ cm^{-3} sec^{-1}. For sample (c) in Fig. 4, under the steady-state illumination at room temperature, when virtually all the dangling bonds are active recombination centers (i.e., neutral),

$$G_{vc} \approx k_{cr}n_c N_r + k_{tr}n_t N_r, \tag{5}$$

where n_c and n_t are the densities of free and trapped electrons. Neglecting the localized-to-localized transitions, one obtains

$$k_{cr} \approx G_{vc}/n_c N_r. \tag{6}$$

The density of free electrons n_c is estimated by using

$$n_c = \sigma_p/e\mu_n = 1.4 \times 10^9 \quad \text{cm}^{-3}, \tag{7}$$

where the free-electron mobility $\mu_n = 13$ cm^2 V^{-1} sec^{-1} is assumed (Tiedje $et~al.$, 1981) and $\sigma_p = 3 \times 10^{-9}$ Ω^{-1} cm^{-1} is the photoconductivity of sample (c) at room temperature. Substituting in Eq. (6), one finds $k_{cr} = 9 \times 10^{-8}$ cm^3 sec^{-1}, which corresponds to a capture cross section $S_r = 9 \times 10^{-15}$

cm² (assuming a thermal velocity of 10^7 cm sec⁻¹). If a significant fraction of the recombination traffic goes by t → r transitions at room temperature, the above value for S_r is an overestimate.

Comparing the room temperature values of σ_p for samples (a) and (c), one finds that when the r centers are nearly filled, the recombination rate drops by a factor of 10^4. Since there must be at least as many s states as r states, the former must have capture rate coefficients $k_{cs} < k_{cr} \times 10^{-4}$. Thus the electron capture cross section for the safe traps is $S_s < 9 \times 10^{-19}$ cm². Values for capture cross sections in a-Si : H given in the literature are shown in Table I.

10. DUAL-BEAM PHOTOCONDUCTIVITY

a. Temperature Dependence

For samples similar to (c) in Fig. 4, lowering the temperature from 300 to 90°K causes the quenching effect on the photocurrent to become progressively stronger. This is depicted in Fig. 5, in which the low temperature peak is shown on a linear scale for single- and dual-beam dc experiments. In both cases the visible beam flux was 10^{14} cm⁻² sec⁻¹ with a photon energy of 2.07 eV. The broadband IR beam was estimated to have a photon flux of 2×10^{17} cm⁻² sec⁻¹ with photon energies in the range 0.4–1.2 eV. The solid curve shows the quenching ratio $Q(T)$ defined by

$$Q = \sigma \text{(visible + IR)}/\sigma \text{(visible)}. \tag{8}$$

This quantity decreases from 1 at 225°K to 0.05 at 90°K with an activation energy of 0.06 eV. Assuming that at these temperatures most of the electrons are trapped in t, and recombination proceeds from t to r and from t to s,

$$Q = \frac{k_{tr}p_r + k_{ts}p_s}{k_{tr}(p_r + \Delta p) + k_{ts}s(p_s - \Delta p)}, \tag{9}$$

TABLE I

CAPTURE CROSS SECTIONS FOR r AND s CENTERS

References	Carrier type	S_r (cm²)	S_s (cm²)
Wronski and Daniel, 1981	n	—	10^{-19}
Moustakas et al., 1981	n,p	6×10^{-15}	—
Persans, 1982	n	—	5×10^{-20}
Street, 1982	$\begin{cases} n \\ p \end{cases}$	4×10^{-15} 2×10^{-15}	— —
Abeles et al., 1982	p	1.3×10^{-15}	—
This work	n	$<9 \times 10^{-15}$	$<9 \times 10^{-19}$

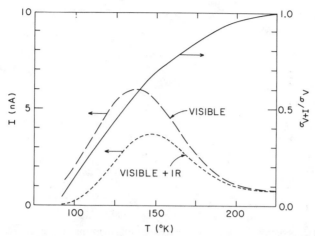

FIG. 5. Temperature dependence of dc photocurrent with one beam (visible) and two beams (visible + IR). The solid curve gives the quenching ratio σ_{v+I}/σ_v. Quenching effect of IR increases with decreasing temperature. [From Vanier and Griffith (1982).]

where p_r and p_s are the densities of holes in states r and s and Δp is the number of holes transferred from s to r by the IR excitation. Rearranging Eq. (9),

$$\frac{1}{Q} = 1 + \frac{(k_{tr} - k_{ts})\,\Delta p}{k_{tr}p_r + k_{ts}p_s}. \tag{10}$$

Since $k_{ts} \ll k_{tr}$,

$$\frac{1}{Q} - 1 = \frac{k_{tr}\,\Delta p}{k_{tr}p_r + k_{ts}p_s}. \tag{11}$$

An Arrhenius plot of the left-hand side of Eq. (11) is given in Fig. 6, showing a reasonably straight line over the whole temperature range in which IR quenching is observed. At room temperature,

$$k_{tr}p_r \gg k_{ts}p_s. \tag{12}$$

As the temperature decreases, thermal quenching indicates that p_s increases and p_r decreases. If at some point $k_{ts}p_s$ were to become equal to $k_{tr}p_r$, one would expect a sudden change in slope of the plot in Fig. 6. Since this is not found, the recombination through s states must be negligible, and Eq. (11) becomes

$$Q^{-1} - 1 = \Delta p/p_r. \tag{13}$$

It is not possible at this point to attribute the temperature dependence in Fig.

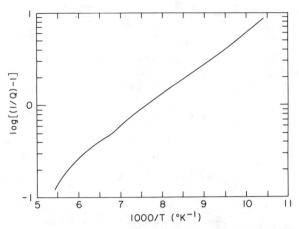

FIG. 6. Temperature dependence of the photoconductivity quenching ratio Q. The quantity plotted vertically gives the relative change in the density of holes trapped in r states after IR excitation. Slope corresponds to an activation energy of 0.06 eV. [From Vanier and Griffith (1982).]

6 to either Δp or p_r alone. The temperature dependence of the light induced IR absorption causes Δp to increase with decreasing temperature, whereas the trapping of holes in shallow valence-band tail states causes p_r to decrease (by charge neutrality).

b. Spectra

Examples of normalized photocurrent spectra of the IR quenching transitions at two different temperatures are shown in Fig. 7. These measurements of modulated photocurrent ΔI were carried out with a visible light flux of $f_v = 6 \times 10^{16}$ photons cm^{-2} sec^{-1} and peak IR flux of $f_{IR} = 6 \times 10^{15}$ photons cm^{-2} sec^{-1}. A quenching effect was observed (using 6 Hz modulation) for a broad range of photon energies between 0.4 and 1.6 eV. Above 1.6 eV (not shown) the chopped beam made a positive contribution to the photocurrent, due to band-to-band excitation. Without assuming a particular functional form for the density of states in either the valence band or the s states, an empirical threshold energy was obtained by a linear extrapolation. This threshold E_t was found to shift monotonically with temperature as shown in Fig. 8. The direction of the shift agrees with the expected shift of the quasi-Fermi level for holes trapped in the s states E_{Fp}^s, which should move toward the valence band with decreasing temperature. There is therefore qualitative agreement with photoinduced absorption (PA) measurements (O'Connor and Tauc, 1980). In those experiments, the transitions were assumed to be between a parabolic valence band and a narrow band of trap states bounded below by E_{Fp}^s and above by an exponentially

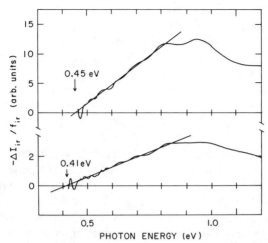

FIG. 7. Normalized spectra of low frequency in-phase IR modulation of photocurrent. The upper curve is for $T = 160°$K, the lower for $T = 100°$K. Quenching effect is plotted upward. [From Vanier and Griffith (1982).]

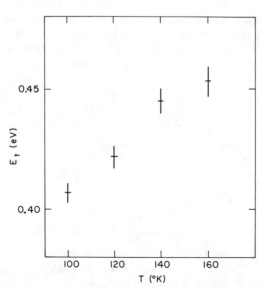

FIG. 8. Shift in threshold of quenching spectrum with temperature. [From Vanier and Griffith (1982).]

decreasing density of states. That model predicts that a plot of $(\Delta\alpha\, E_{IR})^2$ versus E_{IR} should give a straight line, where $\Delta\alpha$ is the photoinduced change in the absorption coefficient and E_{IR} is the IR photon energy.

Although reasonably good straight lines were obtained for PA, this type of plot did not work well for the photoconductivity IR quenching spectrum. Persans (1982a) found that for the normalized modulated photoconductance P_m given by

$$P_m = \Delta I\, l/f_{IR}\, Vew, \qquad (14)$$

where l is the contact separation, w the sample width, and V the applied voltage, a better straight line was obtained by plotting $(P_m E_{IR})^{3/2}$ versus E_{IR}, which implies that the density of s states can be approximated by a step function. These plots are shown in Fig. 9 for an undoped sample at three different temperatures and for two lightly doped samples. In this case, the threshold was at 0.60 ± 0.05 eV, independent of temperature and dopant. The implication was made that the s states are not distributed throughout the gap but have an optical low-energy limit at 0.60 eV above the valence-band edge. It is also possible that E_{Fp}^s became pinned by a high density of states and that the region below 0.60 eV was not accessible with the light intensity used. Clearly, the threshold of the IR quenching transitions depends on the method of extracting a value from the raw data and varies from the samples of one laboratory to those of another.

At temperatures above about 200°K, additional features in the low-frequency IR modulation spectrum were observed by Persans (1982a), particularly at high intensities of visible light. The broad low-temperature quenching peak already described is shown in Fig. 10a, whereas in Fig. 10b a positive modulation peak is observed for $T = 220$°K. The origin of this

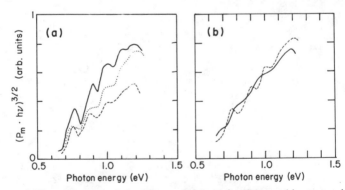

FIG. 9. A different method of obtaining a threshold for IR quenching, assuming a step function distribution of s states. Both (a) undoped (Si-221-A) at $T = 100$°K (solid curve), 148°K (dotted curve), and 172°K (dashed curve) [from Persans (1982a)]; and (b) lightly doped (Si–B–62–A, dashed curve; Si–P–22–A, solid curve) samples showed IR quenching [from Persans, personal communications (1982)].

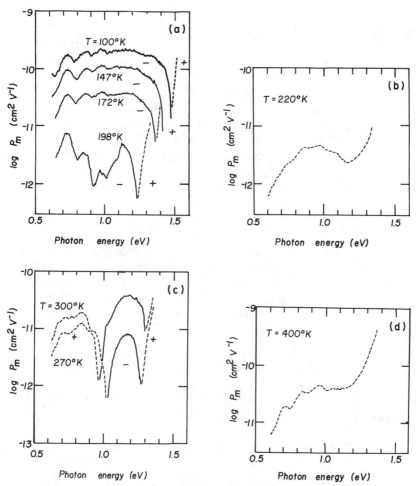

FIG. 10. Spectra of the in-phase component of dual-beam modulated photoconductance measured at 5 Hz chopping frequency with a pump beam intensity of 2×10^{16} cm^{-2} sec^{-1}. The structure with ~0.13 eV spacing is due to interference fringes in the optical absorptance. Both quenching and enhancement processes are seen, depending on temperature and photon energy. [From Persans (1982a).]

signal is somewhat uncertain. Persans (1982a) suggests that the peak might consist of (1) a positive component due to an excitation of an electron from a center 0.65 eV below the conduction-band edge to the conduction band and (2) a negative component due to an excitation of an electron from the valence-band edge into a correlated center 0.95 eV above the valence band. Another possibility, suggested by the peaked nature of the signal, is a localized-to-localized state transition.

At temperatures in the range 270–300°K (see Fig. 10c) a third process appeared in the form of a negative modulation in a relatively narrow band in the energy range 1.0–1.3 eV. This signal was observed only at high pump beam intensity and was quite sample dependent. It was attributed to another set of centers near midgap with a small capture cross section for electrons. These defects were labeled x centers and, though not included in Fig. 3, they are shown schematically in Fig. 11. Since high light intensities were used, it is possible that these states are actually light-induced donorlike states as discussed by Cohen *et al.* (1982) and Vanier (1982, 1983).

At higher temperatures, such as 400°K, the second quenching effect disappears and the positive modulation remains (see Fig. 10d). Either the x states are no longer enclosed by the quasi-Fermi levels, or they are removed by annealing.

c. Chopping Frequency Dependence

The spectra shown in the previous section all correspond to the in-phase component of a signal obtained at low chopping frequencies (5–6 Hz). The time required for the occupation of gap states to reach a steady state is in the range of 1–100 msec, so the response varies considerably with chopping frequency in the range 2–100 Hz. For a single relaxation time τ, the signal can be fitted to an expression of the form

$$g_m(\omega) = \frac{\Delta I(\omega)}{V} = g_1(1 - i\omega\tau)/(1 + \omega^2\tau^2), \qquad (15)$$

where g_1 is the zero frequency asymptotic value and ω is 2π times the frequency. Figure 12 shows the frequency response of the in-phase (0°) and

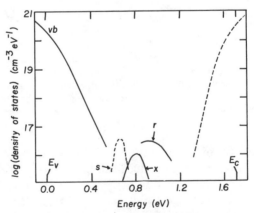

FIG. 11. Schematic picture of density of states of a-Si:H consistent with data in Fig. 10. [From Persans (1982a).]

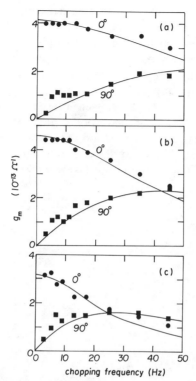

FIG. 12. In-phase and quadrature dual-beam modulated conductance g_m as a function of chopping frequency at $T = 148°K$ using visible modulating light with three different pump beam intensities given by g_p: (a) 3.1×10^{-9}, (b) 3.4×10^{-10}, and (c) 2.3×10^{-11} Ω^{-1}. Enhancement response times were in the range 3–6 msec. [From Persans (1982a).]

quadrature (90°) signals obtained when the modulating beam consisted of visible light with a photon energy of 1.9 eV. As one might expect, only enhancement signals were observed. The sample temperature was $T = 148°K$, and three different pump beam intensities were used: (a) $g_p = 3.1 \times 10^{-9}$ Ω^{-1}, (b) $g_p = 3.4 \times 10^{-10}$ Ω^{-1}, and (c) $g_p = 2.3 \times 10^{-11}$ Ω^{-1}. As pointed out by Persans (1982a), the measurement of g_1 is a differential method for obtaining the intensity dependence exponent γ:

$$g_1(1.9 \text{ eV}) = \gamma f_p^{(\gamma - 1)} f_m \tag{16}$$

where f_p and f_m are the pump beam and modulation-beam fluxes. He obtained a value of $\gamma = 0.95$ from the increase in g in going from Fig. 12a to 12b and a supralinear value $\gamma = 1.15$ from the decrease in g_1 in going from Fig. 12b to 12c. This differential method of detecting supralinearity is undoubtedly more sensitive than would be merely plotting the intensity dependence of σ_p. The response times varied from 2.7 to 6.4 msec, with higher pump intensities giving faster responses.

For the same sample at the same temperature, the response to an IR modulation beam (photon energy = 0.9 eV) was completely different, as shown in Fig. 13. In this case, the frequency response data could not be fit by a curve calculated using a single relaxation time, but was found to consist of a faster positive component and a slower negative component. Reasonable fits to the data were obtained if the response times of the faster component were taken to be independent of the modulation beam energy. The negative component was then found to have a relaxation time in the range 10–50 msec, depending on the pump beam intensity. For each intensity,

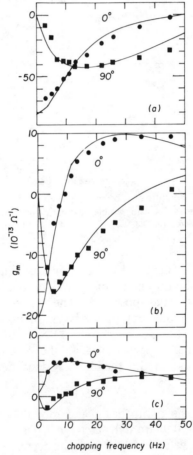

FIG. 13. As for Fig. 12, but with IR modulation (photon energy = 0.9 eV). Response time of quenching effect: (a) 11, (b) 27, and (c) 52 msec. [From Persans (1982a).]

this response time was independent of temperature between 100 and 150°K. The IR quenching process does not seem to depend on a thermally activated step and is rather slow. When the hole is freed from the s state, it is not immediately captured from the extended state into an r state; nor does it diffuse to an r state by a series of trapping events, each followed by thermal reemission; rather, the transport must be by tunneling from one localized state to another.

11. MODULATED PHOTOLUMINESCENCE

a. Temperature Dependence

Photoluminescence is observed in a-Si : H at temperatures below about 150°K, partially overlapping the range covered by photoconductivity measurements and continuing down to liquid helium temperatures. A typical temperature dependence of the intensity I emitted at 1.4 eV is shown by the broken line in Fig. 14. The thermal quenching above 30°K has been attributed to the dissociation of geminate e–h pairs (Tsang and Street, 1979; Austin et al., 1979; Paesler and Paul, 1980). The slight decrease in I below 30°K could be due to the Auger-like interaction between geminate pairs that limits I at high excitation intensities (Street, 1981b). This effect is small since the excitation intensity used here was low (10^{13} photon cm^{-2} sec^{-1}).

At a chopping frequency of 200 Hz, the in-phase modulated fraction of the luminescence $\Delta I/I$ rises steeply with decreasing temperature according to the solid line (positive signal) in Fig. 13. At 15°K, the IR induced enhancement exceeds 10% of the single-beam emission intensity. For com-

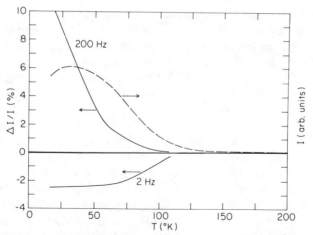

FIG. 14. Temperature dependence of IR enhancement (200 Hz) and quenching (2 Hz) effects on photoluminescence. The broken line shows the single-beam signal.

parison on the same figure is shown the net quenching effect obtained with identical illumination with a chopping frequency of 2 Hz (solid line, negative signal). This low-frequency response is essentially constant at about -2.5% for temperatures below $70°K$ and diminishes to zero between 70 and $120°K$. In order for the net quenching to be constant in the temperature region in which the fast process is rising steeply with decreasing temperature, the actual magnitude of the slow quenching process must also be rising steeply and always be slightly greater than the fast process. Thus the enhancement and quenching seem to be correlated effects rather than independent processes.

In order to explain the fast enhancement, one can consider several possible IR induced transitions based on Fig. 3. In the first place, one can rule out transitions $t \to c$ or $v \to u$, since these would break up geminate pairs and cause fast quenching of the luminescence. The $v \to r$ excitation would tend to enhance the $t \to u$ emission by preventing the $t \to r$ recombination. However, it has been argued in Section 9 that the r states are almost all occupied at these low temperatures under illumination, so this mechanism does not by itself seem likely to produce a 10% enhancement in I. As mentioned previously, an $r \to c$ transition creates free electrons (not $e-h$ pairs), and the resulting vacant r state can participate in $t \to r$ recombination. A two-photon process consisting of the steps $r \to c$ followed by $v \to r$ would create additional $e-h$ pairs and enhance the luminescence, but the probability of these events must be negligibly small. Furthermore, any excitation of this sort would generate the $e-h$ pairs very close to the fast recombination centers, so their most likely fate would be nonradiative recombination at the defects.

The simplest explanation of the data in Fig. 14 starts with the excitation $v \to s$. Recall that below $T = 140°K$ holes accumulate in the s states while the r states become occupied. With visible excitation only, these holes are unable to contribute to the luminescence at 1.4 eV since the s states are $0.4-0.8$ eV above the valence band. Switching on the IR beam generates a large number of free holes by $v \to s$ transitions. Some of these holes are rapidly trapped into u states, where they have a second chance to contribute to the 1.4 eV luminescence. The fast enhancement of the emission is therefore an example of nongeminate radiative recombination in a-Si:H. The steep increase in the fast enhancement with decreasing temperature reflects an increase in the density of holes in shallow u traps at short times. This density is not controlled by the thermal reemission rate constant k_{vu}, because the activation energy observed is very small (5 meV in the temperature range $25-50°K$). It is more likely to be controlled by the relative rates of the processes $t \to u$ and $r \to u$ compared with $u \to v$, which are only weakly temperature dependent.

After a few milliseconds of IR excitation, the excess holes in v and u must decrease in number because of r → u and r → v tunneling transitions whose rates are determined partly by the spatial separation of the r and s defects. When a steady state is established, the number of vacant r states and the nonradiative recombination rate through them have both inreased. A net quenching of the luminescence results. This signal (measured at 2 Hz) is temperature independent below about 70°K because both the r → u and the t → u transitions are enhanced by an amount proportional to the number of holes Δp_u transferred to u by the IR excitation.

b. Spectrum

The low frequency quenching modulation of photoluminescence is produced by a broad spectrum of IR excitations, as shown in Fig. 15. The resemblance to the spectra in Fig. 7 is quite striking, allowing for the fact that the quenching signal is plotted positively in Fig. 7 but negatively in Fig. 15. This result confirms that the IR modulation of luminescence involves excitations to the same gap states detected by photoconductivity IR quenching. The threshold at 0.44 eV is quite close to those measured by photoconductivity, although the measurement conditions are somewhat different (lower temperature and lower pump intensity). The positive signal above 1.07 eV may not be a real modulation of the luminescence. At about this photon energy the scattered modulated IR beam starts to be detected by the photomultiplier.

c. Frequency Dependence

The transition from the slow quenching signal to the fast enhancement signal as the chopping frequency is varied from 2 to 200 Hz is demonstrated

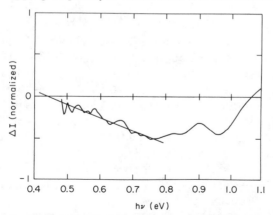

FIG. 15. Normalized spectrum of IR quenching of photoluminescence at $T = 80°K$ and $f = 2$ Hz. [Hirsch and Vanier, unpublished data (1982).]

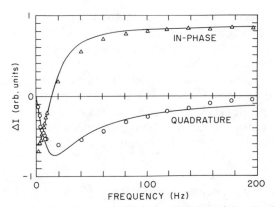

FIG. 16. In-phase and quadrature dual-beam modulation of photoluminescence. Response time of quenching effect $\tau_q = 9$ msec, $T = 80°K$. [Hirsch and Vanier, unpublished data (1982).]

in Fig. 16. The curves fitted to the data points were calculated with the assumption that the fast process would be instantaneous and the slow process would respond according to Eq. (15). The response time of the quenching effect τ_q obtained from such fits varied slightly from 9 msec at $T = 80°K$ to 6 msec at $T = 120°K$ under constant excitation conditions. If $1/\tau_q$ is taken as the rate of a particular relaxation step, an activation energy of 8 meV is obtained. This value is of the same order of magnitude as that obtained from the temperature dependence of the enhancement effect in the range 25–50°K. Again, the weak temperature dependence indicates a tunneling process rather than thermal emission into a band.

V. Suggested Future Work

Many questions remain about the gap states in a-Si:H that can continue to be explored by IR modulation techniques. In the first place, only a few samples have been examined in any detail. How is the density of each type of state influenced by deposition conditions? Do the same transitions occur in sputtered or chemical vapor deposition (CVD) a-Si:H as in glow-discharge material? What is the spatial distribution of each species of defect? What is the chemical or configurational identity of the species other than the dangling bond? Which species are induced by light exposure and related to the Staebler–Wronski effect? Which species are related to impurities? What influence do these states, with differing energies and capture cross sections, have on the performance of devices?

One fruitful approach to some of these problems might be to combine IR modulation with spin-dependent photoconductivity, light-induced electron spin resonance, or optically detected magnetic resonance. These techniques have already identified a hole trap with a g-factor of 2.01 that may be the same as that for the s states described here (Morigaki *et al.*, 1981; Depinna *et al.*, 1982; Street *et al.*, 1981; Voget-Grote *et al.*, 1980; Dersch *et al.*, 1983). Alternatively, there may be new resonance signals that appear under high intensity IR excitation.

To improve the signal-to-noise ratio, it would be useful to have an IR source that is brighter than a tungsten-halogen lamp. This may soon be possible with the use of new synchrotron light sources, or IR lasers. Pulsed IR sources of this sort might make it easier to extract recombination rate constants.

The use of higher intensities of visible pumping light introduces the complication of light-induced states, at the same time pushing the quasi-Fermi levels further apart and exploring a wider fraction of the gap. The limit in this direction is set by the thermal conductivities of the sample and substrate, which determine whether the sample can be kept at a low temperature. The temperature dependence of IR-modulated luminescence develops complicated oscillations at high pump intensities, and deserves further study.

ACKNOWLEDGMENTS

The author is grateful to Dr. F. J. Kampas for many useful discussions, to Dr. M. D. Hirsch for providing some of the modulated luminescence data, to Dr. J. I. Pankove for suggesting the luminescence experiments, and to Dr. P. D. Persans and Taylor and Francis, Ltd., for allowing the reproduction of some figures. This work was supported in part by the U.S. Department of Energy, Division of Materials Sciences, Office of Basic Energy Sciences, under Contract No. DE-AC02-76CH00016.

REFERENCES

Abeles, B., Wronski, C. R., Goldstein, Y., and Cody, G. D. (1982). *Solid State Commun.* **41,** 251.
Anderson, D. A., and Spear, W. E. (1977). *Philos. Mag.* **36,** 695.
Arnoldussen, T. C., and Bube, R. H. (1972). *J. Non-Cryst. Solids* **8–10,** 933.
Austin, I. G., Nashashibi, T. S., Searle, T. M., LeComber, P. G., and Spear, W. E. (1979). *J. Non-Cryst. Solids* **32,** 373.
Biegelsen, D. (1980). *Solar Cells* **2,** 421.
Bube, R. H. (1957). *J. Phys. Chem. Solids* **1,** 234.
Bube, R. H. (1960). "Photoconductivity of Solids." Wiley, New York.
Cohen, M. H., Fritzsche, H., and Ovshinsky, S. R. (1969). *Phys. Rev. Lett.* **22,** 1065.
Cohen, J. D., Harbison, J. P., and Wecht, K. W. (1982). *Phys. Rev. Lett.* **48,** 109.
Delahoy, A. E., and Vanier, P. E. (1982). *Bull. Am. Phys. Soc.* **27,** 268.
Depinna, S., Cavenett, B. C., Austin, I. G., Searle, T. M., Thompson, M. J., Allison, J., and LeComber, P. G. (1982). *Philos. Mag. B* **46,** 473.
Dersch, H., Schweitzer, L., and Stuke, J. (1983). *Phys. Rev. B* **28,** 4678.

Goodman, N., and Fritzsche, H. (1980). *Philos. Mag. B* **42**, 149.

Griffith, R. W., Kampas, F. J., Vanier, P. E., and Hirsch, M. D. (1980). *J. Non-Cryst. Solids* **35/36**, 391.

Hirsch, M. D., and Vanier, P. E. (1982). *Bull. Am. Phys. Soc.* **27**, 246.

Hoffmann, H.-J., and Stöckmann, F. (1979). *Festkörperprobleme* **19**, 271.

Kastner, M., Adler, D., and Fritzsche, H. (1976). *Phys. Rev. Lett.* **37**, 1504.

Morigaki, K., Sano, Y., and Hirabayashi, I. (1981). *Solid State Commun.* **39**, 947.

Mott, N. F. (1967). *Adv. Phys.* **16**, 49.

Mott, N. F., and Davis, E. A. (1979). *In* "Electronic Processes in Noncrystalline Materials," 2nd Ed., (W. Marshall and D. H. Wilkinson, eds.) p. 31. Oxford Univ. Press (Clarendon), London and New York.

Moustakas, T. D., Wronski, C. R., and Tiedje, T. (1981). *Appl. Phys. Lett.* **39**, 721.

Newman, R., Woodbury, H. H., and Tyler, W. W. (1956). *Phys. Rev.* **102**, 613.

O'Connor, P., and Tauc, J. (1980). *Solid State Commun.* **36**, 947.

O'Connor, P., and Tauc, J. (1982). *Phys. Rev. B* **25**, 2748.

Olivier, M., Penzin, J., and Chenevas-Paule, A. (1980). *J. Non-Cryst. Solids* **35/36**, 693.

Overhof, H., and Beyer, W. (1981). *Phys. Status Solidi B* **107**, 207.

Paesler, M. A., and Paul, W. (1980). *Philos. Mag. B* **41**, 393.

Paul, W. (1981). *In* "Fundamental Physics of Amorphous Semiconductors" (F. Yonezawa ed.), p. 72. Springer-Verlag, Berlin and New York.

Persans, P. D. (1980). *Solid State Commun.* **36**, 851.

Persans, P. D. (1982a). *Philos. Mag. B* **46**, 435.

Persans, P. D. (1982b). *Bull. Am. Phys. Soc.* **27**, 337.

Persans, P. D., and Fritzsche, H. (1981). *J. Phys. Colloq. Orsay, Fr.* **42**, C4-597.

Pontuschka, W. M., Carlos, W. W., Taylor, P. C., and Griffith, R. W. (1982). *Phys. Rev. B* **25**, 4362.

Redfield, D. (1982). *Proc. IEEE Photovoltaic Specialists Conf., 16th, San Diego*, p. 1327. IEEE, New York.

Rose, A. (1963). "Concepts in Photoconductivity and Allied Problems." Wiley, New York.

Shockley, W., and Read, W. T. (1952). *Phys. Rev.* **87**, 835.

Simmons, J. G., and Taylor, G. W. (1971). *Phys. Rev. B* **4**, 502.

Simmons, J. G., and Taylor, G. W. (1972). *J. Non-Cryst. Solids* **8–10**, 947.

Simmons, J. G., and Taylor, G. W. (1973). *J. Phys. C* **6**, 3706.

Simmons, J. G., and Taylor, G. W. (1974). *J. Phys. C* **7**, 3051.

Solomon, I., Biegelsen, D., and Knights, J. C. (1977). *Solid State Commun.* **22**, 505.

Spear, W. E., Loveland, R. J., and Al-Sharbaty, A. (1974). *J. Non-Cryst. Solids* **15**, 410.

Staebler, D. L., and Pankove, J. I. (1980). *Appl. Phys. Lett.* **37**, 609.

Street, R. A. (1980). *Phys. Rev. B* **21**, 5775.

Street, R. A. (1981a). *Adv. Phys.* **30**, 593.

Street, R. A. (1981b). *Phys. Rev. B* **23**, 861.

Street, R. A. (1982). *Appl. Phys. Lett.* **41**, 1060.

Street, R. A., Knights, J. C., and Biegelsen, D. K. (1978). *Phys. Rev. B* **18**, 1880.

Street, R. A., Biegelsen, D. K., and Knights, J. C. (1981). *Phys. Rev. B* **24**, 969.

Taylor, G. W., and Simmons, J. G. (1972). *J. Non-Cryst. Solids* **8–10**, 940.

Taylor, G. W., and Simmons, J. G. (1974). *J. Phys. C* **7**, 3067.

Tiedje, T., Moustakas, T. D., and Cebulka, J. M. (1981). *Phys. Rev. B* **23**, 5634.

Tiedje, T., Abeles, B., Cebulka, J. M., and Pelz, J. (1982). *Proc. IEEE Photovoltaic Specialists Conf., 16th, San Diego*, p. 1423. IEEE, New York.

Tsang, C., and Street, R. A. (1979). *Phys. Rev. B* **19**, 3027.

Vanier, P. E. (1982). *Appl. Phys. Lett.* **41**, 986.

Vanier, P. E. (1983). *Solar Cells.* **9,** 85.
Vanier, P. E., and Griffith, R. W. (1980). *Bull. Am. Phys. Soc.* **25,** 330.
Vanier, P. E., and Griffith, R. W. (1982). *J. Appl. Phys.* **53,** 3098.
Vanier, P. E., Delahoy, A. E., and Griffith, R. W. (1981a). *J. Appl. Phys.* **52,** 5135.
Vanier, P. E., Delahoy, A. E., and Griffith, R. W. (1981b). *AIP Conf. Proc.* **73,** 227.
Voget-Grote, U., Kümmerle, W., Fischer, R., and Stuke, J. (1980). *Philos. Mag. B* **41,** 127.
Wronski, C. R., and Daniel, R. E. (1981). *Phys. Rev. B* **23,** 794.
Wronski, C. R., Abeles, B., Cody, G. D., Morel, D., and Tiedje, T. (1980). *Proc. Photovoltaic Specialists Conf., 14th, San Diego,* p. 1057. IEEE, New York.

CHAPTER 11

Irradiation-Induced Metastable Effects

H. Schade

RCA/DAVID SARNOFF RESEARCH CENTER
PRINCETON, NEW JERSEY

I. Introduction

The structure of a-Si:H contains various types of bonds with a range of bond energies and bond angles, voids, internal surfaces, and various degrees of internal stresses. Such materials are potentially affected more readily by the interaction with various kinds of radiation than defect-free crystalline materials. Just as the optical and electrical properties of a-Si:H are very sensitive to the electron density of states that correspond to a particular bonding network, irradiation-induced bonding changes must equally strongly affect these properties. Examples of irradiation-induced effects that have been reported specifically for a-Si:H are changes of the dark conductivity, photoconductivity, photoluminescence, spin density, and infrared absorption, all as a result of exposure to light. Furthermore, changes in some of these properties have also been reported for irradiation with electrons, ions, and x rays. All these irradiation-induced effects are typically found to be reversible by an annealing treatment at about 200°C. The structural changes due to irradiation thus are metastable and hence lead to metastable electronic states and metastable optical and electrical effects. The use of irradiation-induced changes allows us to study in a given network the effects of certain types of states that are controllable by the type and the amount of the irradiation.

359

In this chapter, we shall present mainly optical irradiation-induced effects, namely, changes in photoluminescence, photoconductivity, and infrared absorption. In order to further understand the observed effects we shall draw on other, nonoptical, metastable effects that are observed, e.g., with electron spin resonance (ESR) or deep level transient spectroscopy.

II. Experimental Observations

1. PHOTOLUMINESCENCE

Photoluminescence (PL) has been extensively applied to obtain information on recombination mechanisms and the role of defects and impurities in both undoped and doped a-Si: H. For a review of this field we refer the reader to Chapter 7 by Street. Here we shall restrict our discussion only to *changes* in the PL. These are to be expected as a result of irradiation-induced structural changes that may give rise to new recombination paths, either radiative or nonradiative. Changes in PL have been studied on samples after exposure to ions (Engemann *et al.*, 1975; Street *et al.*, 1979), electrons (Street *et al.*, 1979; Voget-Grote *et al.*, 1980; Schade and Pankove, 1981), and photons (Pankove and Berkeyheiser, 1980; Morigaki *et al.*, 1980). In all these cases, qualitatively the same observation was made, namely, a decrease of the luminescence intensity for the main transitions at 1.2–1.4 eV, and, furthermore, a change (either an increase or a decrease, depending on the type of radiation and experimental conditions) at about 0.8–0.9 eV.

In the following subsections, we shall review specific observations for the different irradiations. Moreover, we shall compare the effects of photon irradiation with those of electron irradiation side-by-side for the same a-Si: H film.

a. Light-Induced Effects

The application of a-Si: H to photovoltaic power generation renders the effects of prolonged exposure to light extremely important. Morigaki *et al.* (1980) reported the first observation of PL degradation (often called luminescence fatigue) in a-Si: H produced by glow discharge. They measured the spectral dependence of the PL at 4 and 77°K for various excitation levels. For both temperatures, their main finding was a substantial decrease (by about 40%) of the PL peak at about 1.2 eV after excitation with above-bandgap light (see Fig. 1). Isothermal annealing required temperatures up to 420°K to reach full recovery of the original PL intensity.

To interpret these results, the authors proposed as a possible model the creation of dangling bonds that give rise to nonradiative recombination paths and thus explain the decrease in PL. Since no PL recovery was

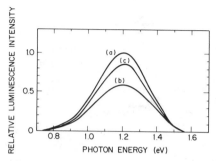

FIG. 1. Spectral dependence of photoluminescence at 4.2°K. (a) Before light exposure, (b) after prolonged illumination at 514.5 nm, and (c) after annealing at 270°K. [Reprinted with permission from *Solid State Communications* Vol. 33, K. Morigaki, I. Hirabayashi, M. Nakayama, S. Nitta, and K. Shimakawa, Fatigue effect in luminescence of glow-discharge amorphous silicon at low temperature. Copyright 1980, Pergamon Press, Ltd.]

obtained solely by infrared illumination, as is typically observed for chalcogenide glasses (Street, 1976), purely electronic mechanisms were ruled out. Instead, the nonradiative recombination at the dangling bonds was thought to supply energy for local structural changes that stabilize the atomic configuration around the dangling bonds and thus cause metastable states. These structural changes could be reversed only by annealing at elevated temperatures. It is interesting to note, however, that a so-called intermediate fatigued state could be totally reversed to the initial state by using infrared light instead of heat (Tomozane *et al.*, 1982).

Further evidence for the introduction of nonradiative centers by high optical excitation was found in PL decay measurements (Hirabayashi *et al.*, 1980). Prolonged light exposure results in a faster PL decay, particularly in the long delay-time region. The authors explained this observation by pointing out that for longer delay times, corresponding to recombination of more distant electron–hole pairs, the probability of nonradiative recombination at dangling bond centers becomes greater. Hence the decay times are shortened with the introduction of nonradiative recombination centers.

In order to positively identify the light-induced nonradiative centers with dangling bonds, additional information was derived from a correlation between PL and ESR results. A relationship between luminescence efficiency at 1.4 eV and equilibrium spin density was shown to exist, independent of the growth conditions (see Fig. 2). This correlation was explained by associating spins with a random distribution of nonradiative recombination centers. Therefore, the PL degradation can be expected to be accompanied by an increase in the spin density, and light-induced defects may thus be characterized by their ESR "signature." In fact, Hirabayashi *et al.* (1980)

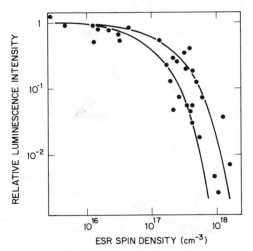

FIG. 2. Photoluminescence intensity as a function of the spin density at 300°K for samples prepared under a variety of deposition conditions. [After Street *et al.* (1978).]

ascribed the appearance of a line with $g = 2.005 \pm 0.001$ to optically created centers; the spin density of these centers was estimated to be 1×10^{17} cm^{-3}, consistent with the dangling-bond density required for the fatigue effect in PL.

Improved evidence for light-induced dangling bonds was given by Dersch *et al.* (1981). After white-light illumination, they found an over-two-fold increase in the ESR signal with $g = 2.0055$ to about 2×10^{16} spins cm^{-3}. This *g*-value is very close to the value found also after the effusion of hydrogen (Bachus *et al.*, 1979; Biegelsen *et al.*, 1980), which generates dangling bonds due to the breakage of Si–H bonds. The light-induced dangling bonds, however, cannot result directly from broken Si–H bonds, since the Si–H bonding states lie too deep in the valence band [about 5 eV (von Roedern *et al.*, 1979)] to be excited by nonradiative recombination of photoexcited carriers. Instead, Pankove and Berkeyheiser (1980) and Dersch *et al.* (1981) proposed models in which weak Si–Si bonds are broken by the photoexcitation of carriers. From the absence of exchange narrowing of the ESR line Dersch *et al.* concluded that the dangling bonds are well separated and thus cannot be represented just by pairs of dangling bonds arising from broken Si–Si bonds. This separation would occur by a relaxation of the network that surrounds broken Si–Si bonds and by a migration of hydrogen atoms from neighboring Si–H bonds to the freshly broken Si–Si bonds, whereby the dangling bonds would move far enough apart to form metastable states. Since this model involves the diffusion of hydrogen atoms, the increase of light-induced defects with temperature up to about 100°C (Pankove and Berkeyheiser, 1980) is conceivable. It is not clear, however, how the same diffusion process would also lead to the annealing of the defects at slightly higher temperatures (nor is it clear why atomic hydrogen would bind to a pair of adjacent dangling bonds rather than combine to form a volatile H_2 molecule that would make the change permanent). Note that Crandall (1981) has postulated an activation energy of about 1.0 eV for both creation and annealing of defects, although he attributed the metastable state to a different type of defect.

b. Optically Detected Magnetic Resonance

Further evidence for the nature of light-induced defect centers is provided by optically detected magnetic resonance (ODMR). This technique, described by Morigaki in Chapter 4 of Volume 21C, supplies information on spin-dependent radiative and nonradiative transitions and shows how they are affected by light-induced defects. Two of the three centers usually detected by ODMR are influenced on prolonged exposure to laser light (360 mW at 514.5 nm for 30 min at 2°K) (Morigaki *et al.,* 1982). These are the D_2 center with $g = 2.006$, attributed to dangling bonds, and the A center with a variable g-value that depends on the emitted photon energy and microwave power, which is attributed to trapped holes. These centers respond oppositely to light-irradiation (see Fig. 3). The relative PL intensity

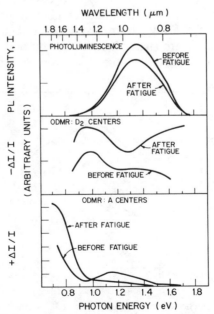

FIG. 3. Spectral dependences of photoluminescence and optically detected magnetic resonance of D_2 and A centers before and after light exposure. [After Morigaki *et al.* (1982).]

at the D_2 center resonance decreases throughout the spectral region. This decrease stems from an increase in nonradiative transitions due to the formation of D_2 centers, i.e., dangling bonds, during exposure to light.

However, PL around 0.8 eV, has been observed to increase on exposure to light (see Fig. 4), and this increase can be well correlated with the ODMR results for the A center. At photon energies below 0.9 eV, the relative change in PL intensity at the A center resonance rises, indicating radiative transitions between trapped electrons in the conduction-band tail and trapped holes at the A center, about 0.25 eV above the valence band (Morigaki *et al.*, 1982). On exposure to light the relative change in PL intensity becomes even larger (see Fig. 3), which is interpreted as an increased concentration of electron centers. The increased trapping of electrons may be due either to an increased number of dangling-bond centers (Street, 1980) or to additional irradiation-induced electron centers (Morigaki *et al.*, 1982).

c. Electron Irradiation-Induced Effects

The threshold energy for electron damage in a-Si:H is considerably smaller than in crystalline Si, namely, about 1 keV versus 100 keV (Schade, 1980). According to the energy dependence of the electron range (Gledhill, 1973), the electron penetration in a-Si:H films can thus be tailored from almost homogeneous absorption, by using electron energies in the mega-electron-volt range (Street *et al.*, 1979; Voget-Grote *et al.*, 1980), to a penetration depth comparable to that for photon absorption, by using electron energies in the kilo-electron-volt range (Schade and Pankove, 1981). In all cases of electron irradiation, the energy of the incident electrons is orders of magnitude higher than that of photons in visible light; therefore,

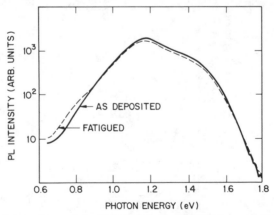

FIG. 4. Spectral dependence of photoluminescence at 78°K before and after light exposure for 10 min at about 5 W cm^{-2}. [After Pankove and Berkeyheiser (1980).]

electron irradiation may create additional types of defects for which the photon energy would not be sufficient. Besides the difference in energy, electrons may carry sufficient momentum to produce atomic displacements in elastic collisions. Thus electron energies in the kilo-electron-volt range would supply sufficient momentum for the displacement of hydrogen. However, since the damage threshold energy determined for both a-Si:H and a-Si:D is about equal (Schade, 1981), we can rule out direct atomic displacements of hydrogen as one possible type of electron damage. Instead, electronic excitations and processes of the Franck–Condon type must give rise to changes in certain bonding configurations, initiated by either electrons or photons.

The effect of electron irradiation on the main PL at about 1.2 eV is qualitatively the same as with light irradiation, namely, a reduction; however, the magnitude of the effect far exceeds that due to light. There is also a change in PL intensity around 0.8 eV, but both an increase (Schade and Pankove, 1981) similar to that due to light exposure and a decrease (Street *et al.*, 1979; Schade and Pankove, 1981) have been observed. Figure 5 shows the spectral dependence of PL before and after electron irradiation at 5 keV. Compared with the relatively small photon-induced decrease of about 15% (see Fig. 4), the main PL is now decreased by about 80%; the increase at 0.8 eV in this particular case is about equal for both photon and electron irradiation. Note that for the two cases shown in Figs. 4 and 5, the effects of electron and photon irradiation may be considered to be equivalent if one assumes the same energy of about 2 eV for defect creation in both cases. For the electron energy of about 5 keV, the electron penetration about equals the light penetration. Therefore, if one neglects the difference in the depth dependence of energy dissipation for electrons and photons, the applied doses of 1×10^{18} electrons cm^{-2} or 3×10^{21} photons cm^{-2} may give rise to

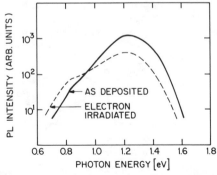

FIG. 5. Spectral dependence of photoluminescence at 78°K before and after 5 keV electron irradiation with a typical dose of 1×10^{18} electrons cm^{-2}. [After Schade and Pankove (1981).]

about equal concentrations of defects, since the electron carries about 2000 times more energy than the photon. Although this estimate may account for the comparable increase of the 0.8 eV PL on photon and electron irradiation, the appreciably different degradation for PL at 1.2 eV suggests that electrons generate more efficiently than photons the centers responsible for the decrease of the main PL peak.

Apart from the suspected difference in defect generation by photons and electrons, there are at least three observations that the types of defects, although not necessarily their individual concentrations are very similar (Schade and Pankove, 1981).

1. The annealing of irradiation-induced effects occurs at the same temperatures for photons and electrons. This is shown, e.g., in Fig. 6 for isochronal annealing of the 0.8 eV and 1.2 eV PL, after side-by-side photon and electron irradiation of an undoped sample. The PL is recovered at about the same annealing temperature of 240°C.

2. The 0.8 eV PL is a radiative transition that is observed on irradiation by both photons and electrons, and thus it is taken as evidence for a specific defect center, independent of its cause. Although the PL intensity at 0.8 eV initially may be decreased on electron irradiation as shown in Fig. 6, a moderate annealing treatment can lead to an enhancement, as after photon irradiation. Furthermore, note that the 0.8 eV PL intensity rises with the temperature during photon irradiation (Pankove and Berkeyheiser, 1980). It thus appears that the defect giving rise to enhanced radiative transitions at 0.8 eV is the result of a temperature-assisted rearrangement of irradiation-affected bonds that occurs either during the irradiation or thereafter at

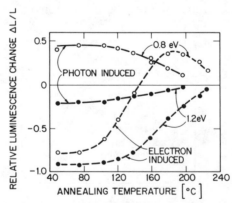

Fig. 6. Relative photoluminescence change at 1.2 and 0.8 eV as a function of isochronal annealing (5 min periods) of photon- and electron-induced irradiation damage. [After Schade and Pankove (1981).]

slightly elevated temperatures. A similar reversal from decrease to increase of the 0.8 eV PL on annealing of damage by electron bombardment has also been observed by Street *et al.* (1979).

3. As for light exposure, discussed in Subsection 1a, the spin density with the characteristic line at $g = 2.0055$ ascribed to dangling bonds is substantially increased by electron bombardment (Street *et al.*, 1979). In this context, Voget-Grote *et al.* (1980) suggest that both dangling-bond and vacancy-type defects are created by electron bombardment. Both types of defects would contribute to the quenching of the PL, as well as to an increase in the spin density, at about the same *g*-value.

It would be tempting to ascribe the large decrease in PL at 1.2 eV on electron bombardment compared with that due to light exposure (see Subsection 1a) to these additional vacancy-type defects. However, this difference in PL quenching has been observed even after electron irradiation in the kilo-electron-volt range, in which we have found no evidence for atomic displacements, as discussed above. Therefore, the generation of vacancy-type defects, at least with these relatively low electron energies, appears unlikely, although other additional types of defects may act as nonradiative centers and may thus be responsible for the stronger PL quenching at 1.2 eV on electron bombardment.

2. PHOTOCONDUCTIVITY

a. Conventional Photoconductivity

Staebler and Wronski (1977) observed that after exposure to light, a-Si : H films, particularly undoped or lightly doped *n*-type samples, show decreases in photoconductivity as well as in dark conductivity. These changes are reversible on annealing above 160°C, and are similar to the changes observed in PL (see previous section). For a detailed description of the Staebler–Wronski effect see Chapter 11 by Wronski in Volume 21C.

By now it has been well established that these light-induced changes in conductivity are bulk effects that can be separated from surface effects (Komuro *et al.*, 1983). These bulk effects are attributed to changes in the density of gap states and/or their occupation probability.

The photoconductivity depends strongly on the conditions of sample preparation and temperature treatment, and comparisons among different measurements are thus complicated. Generally, however, an increase in the total spin concentration entails a decrease of the photoconductivity (Voget-Grote *et al.*, 1980; Street, 1982a). Such a correlation is shown in Fig. 7 for a sample in which different spin densities were obtained by various deposition conditions. The photoconductivity also decreases after electron bombardment; in this case, the corresponding ESR is typical of the usual dangling

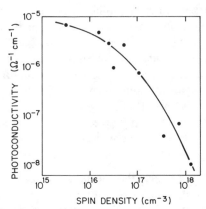

FIG. 7. Photoconductivity (300°K) as a function of the spin density for samples prepared under a variety of deposition conditions. [After Street (1982a).]

bond at $g = 2.0055$, and the loss in photoconductivity can thus be ascribed to an increase in the number of dangling bonds, as shown in Fig. 8. A similar conclusion can be reached also for light-soaked samples that show both an increase in the ESR intensity for $g = 2.0055$ and a decrease in photoconductivity (Street, 1982a). However, as we shall discuss in Subsection 4b, this interpretation is not unique. Vanier (1982) has related light-induced changes in photoconductivity measured at lower temperatures (100–300°K) to the generation of two types of centers, a donor and an acceptor center, both below midgap, rather than to an increase in the concentration of dangling bonds.

Although PL changes reflect both nonradiative and radiative changes in recombination, photoconductivity cannot distinguish between these recombination processes. Provided that the quantum efficiency (number of free carriers produced per absorbed photon) remains unaffected by previous irradiation, photoconductivity changes are due to changes in the product of carrier mobility and lifetime ($\mu\tau$ product).

There is strong evidence that the carrier mobility remains unaffected by the introduction of defect centers (Street, 1983). From changes in the intensity dependence of the photoconductivity, it was concluded that irradiation had increased the density or recombination cross section of deep gap states (Staebler and Wronski, 1977; Fuhs et al., 1978). On exposure to light, a transition from $\sigma_{ph} \propto I^{0.5}$ to $\sigma_{ph} \propto I^{0.9}$, even at high light intensities, indicates a transition from bimolecular to monomolecular recombination kinetics due to an increase in recombination via deep gap states. Dersch et al. (1983) also observed monomolecular recombination ($\sigma_{ph} \propto I^{1\pm0.05}$) over a wide temperature and concentration range.

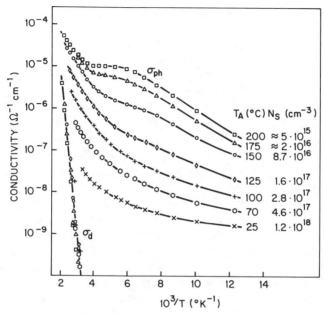

FIG. 8. Temperature dependence of the dark conductivity σ_d and photoconductivity σ_{ph} of electron-irradiated, undoped a-Si:H after different annealing temperatures T_A. [After Dersch et al. (1983).]

Even more direct evidence for the change in lifetime rather than mobility was reported by Komuro et al. (1983). A transient grating method allowed them simultaneously and separately to determine the diffusion coefficient D and the lifetime τ of photoexcited carriers (Aoyagi et al., 1982). From their experiment, the decrease of the photoconductivity on exposure to light (80 mW cm^{-2} for 10 hr) was attributed mainly to a decrease in the lifetime (by a factor of about 10), since the diffusion coefficient and thus the mobility remained practically unchanged. Decreases in lifetime are consistent with increases of the density of deep gap states, as has also been concluded from fatigue of PL, and increases in the spin density (see Section 1). With regard to the values of τ and D obtained by Komuro et al. (1983), we should note that they lead to an unusually high carrier diffusion length of over 4 μm, which significantly exceeds typical values for present state-of-the-art a-Si:H material.

b. Spin-Dependent Photoconductivity

Recombination transitions may be spin-dependent if the electronic states involved are paramagnetic. The spin selection rules then influence the transition probabilities. When the spins are brought into resonance, the change in recombination rate can give information about the type of the recombination center (from the observed g-value) and also about the dynamics of the recombination (from spin relaxation times). Similarly to ODMR (see Subsection 1b), spin-dependent photoconductivity (SDPC) is measured as a change $\Delta\sigma_{ph}$ in photoconductivity σ_{ph} when the sample is brought into microwave resonance by applying microwave power in a magnetic field (Solomon, *et al.* 1977). Since the recombination rate is enhanced by the microwave resonance, the SDPC is negative. The SDPC spectrum (i.e., photocurrent versus magnetic field) appears with a g-value of 2.0055, which is characteristic for dangling bonds (Solomon *et al.*, 1977; Schiff, 1981; Street, 1982a). Dersch *et al.* (1983) compared the SPDC line changes for samples with a range of defect concentrations (5×10^{15}–10^{18} cm^{-3}) that were obtained by 1 MeV electron bombardment and subsequent annealing steps. They were able to detect the resonances of both states involved in the recombination. Thus they could distinguish the following recombination processes: For high defect densities ($> 10^{17}$ cm^{-3}), spin-dependent tunneling takes place between localized band-tail electrons and singly occupied dangling bonds. For lower defect densities, there is additional spin-dependent diffusion of localized band-tail holes to doubly occupied dangling bonds.

The effect of the spin density on the photoconductivity σ_{ph} has already been shown in Figs. 7 and 8. The SDPC, however, is affected oppositely: $\Delta\sigma_{ph}/\sigma_{ph}$ increases with the spin density as shown in Fig. 9. The effect of prolonged light soaking on photoconductivity and SDPC is illustrated in Fig. 10. These results, in conjunction with the g-values derived from the SDPC spectra provide ample evidence that the irradiation-induced defects are dangling bonds that act as recombination centers. From the similarity between the spin density dependence of both the PL (Fig. 2) and photoconductivity (Fig. 9) one may conclude that the main recombination processes additionally introduced by irradiation are the same.

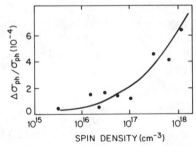

Fig. 9. Spin-dependent change in photoconductivity (300°K) as a function of the spin density. [After Street (1982a).]

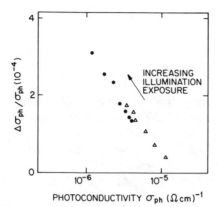

FIG. 10. Spin-dependent change in photoconductivity as a function of the photoconductivity ($300°K$) after increasing levels of exposure to white light. [After Street (1982a).]

3. INFRARED ABSORPTION

Light-induced effects on the optical absorption in a-Si:H are most pronounced in the energy range below the band-gap absorption. Skumanich *et al.* (1982) observed an increase in the tail absorption (absorption coefficients $\alpha < 10$ cm^{-1}) of about a factor of 3 due to the generation of metastable states. This absorption increase correlates well with the light-induced concentration of the dangling bonds (Yamasaka *et al.*, 1981).

More detailed information on the nature of defects introduced by photons, electrons, or other types of radiation can be obtained from the optical absorption further in the infrared (IR) region. Here characteristic structures in the absorption spectrum yield information on specific molecular bonding configurations and their modes of vibration. A detailed description of IR spectroscopy and its application to a-Si:H is given by Zanzucchi in Chapter 4. Here we shall limit the discussion to *changes* in the IR spectra that occur as a result of light soaking or electron irradiation.

The IR absorption spectra of a-Si:H show distinct features that are related to the types and local environment of Si-H$_x$ bonds. There are three groups of absorption bands:

1. *Bond Stretching Vibrations.* These have two components that are usually ascribed to Si-H single hydrogen bonds at 2000 cm^{-1} and to Si-H$_2$ and/or Si-H$_3$ multiple hydrogen bonds at 2100 cm^{-1}. The assignment of the modes around 2100 cm^{-1}, however, has not yet been fully resolved (Shanks *et al.*, 1980).

2. *Bond-Bending Vibrations.* These occur between 845 and 900 cm^{-1} and are ascribed to the presence of SiH$_2$ and SiH$_3$ radicals.

3. *Bond-Wagging* (*Rocking and Rolling*) *Vibrations.* These occur at 640 cm^{-1}

In terms of the relative intensity of these absorption bands, the IR spectra depend on the film preparation method, the hydrogen content, and the structure and morphology of the film, as well as on the type and concentration of defect sites in the film. All of these properties affect the local bonding configurations, and thus the relative concentrations of the various bonding types and/or the oscillator strengths.

The influence of defects on IR spectra has been studied by intentionally introducing defects into a given film with He$^+$ ion bombardment (Oguz *et al.*, 1980a), light exposure (Carlson *et al.*, 1982), and electron irradiation (Schade, 1982).

a. He$^+$ Ion Bombardment

Oguz *et al.* (1980a) reported significant changes in the IR spectra of rf-sputtered a-Si:H on bombardment with 100 and 200 keV He$^+$. They found that the intensities of the bands at 2000 and 850 cm^{-1} were strongly enhanced after the bombardment, whereas little change was observed at 2100, 890, and 650 cm^{-1}. It would be most interesting to correlate the observed intensity changes directly with concentration changes of the corresponding bonding types. It was argued, however, that the intensity change is caused by selective changes in the oscillator strengths. These were ascribed to bombardment-produced defects near IR-active H sites, whereby the local charge distribution was perturbed. These defects can also be considered to be metastable, since annealing for 60 min at 320°C reverses the effect of bombardment. Note that annealing treatments (up to 60 min at 225°C) comparable to the annealing of light- and electron-induced effects (discussed in Sections 1 and 2) were not sufficient to remove the bombardment damage. We conclude, therefore, that this damage consists of other types of defects that are absent or do not contribute after photon or electron irradiation.

b. Light Exposure

Carlson *et al.* (1982) were the first to report light-induced changes in IR spectra. Since the intensity changes after exposure to light are rather small, the measurements were performed on films deposited on Ge prisms to obtain multiple passes (typically about 20) of light through the film (for further experimental details see Chapter 4 by Zanzucchi). Figure 11 shows a sequence of absorbance differences as a function of wave number for an undoped a-Si:H film. The absorbance difference between two states of the film, e.g., irradiated compared with original or annealed compared with

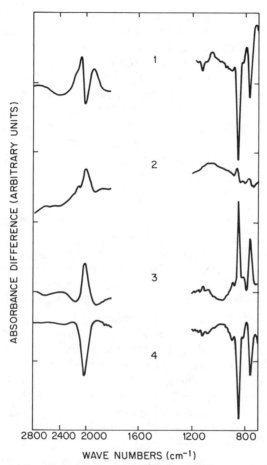

FIG. 11. Spectral dependence of absorbance differences between the following treatments:
(1) First light exposure minus original, (2) first anneal minus first light exposure, (3) second
light exposure minus first anneal, and (4) second anneal minus second light exposure.

irradiated, is directly proportional to the relative change in transmission
between such treatments, and provided that the film thickness is unaffected
by the film treatment, is also directly proportional to the change in the
absorption coefficient. The stretching modes between 2000 and 2100 cm^{-1}
are distinctly affected by the light exposure and annealing treatments.
Changes occur also between 1050 and 1200 cm^{-1} in a range that is ascribed
to modes related to oxygen (see Chapter 4 by Zanzucchi). In addition, the
modes at 850 and 2000 cm^{-1} change in the same direction; these hydrogen-
related intensity changes at 850 and 2000 cm^{-1} can be seen as reversible, at
least in part, after annealing at about 200°C for 30 min. Recall that

annealing also reverses the light-induced effects in photoluminescence and spin density (see Section 1). Although after the first light exposure Fig. 11 shows an increase of the mode at 2080 cm^{-1} and a decrease at 2000 cm^{-1}, other samples with different impurity content may exhibit different behavior; the relative contributions at 2080 and 2000 cm^{-1} and the direction of change after light exposure may vary. Thus, a sample deposited in the presence of water vapor by Carlson *et al.* (1982) showed decreases for both the 2080 and 2000 cm^{-1} modes after the first light exposure and subsequent recovery of the 2080 cm^{-1} but not the 2000 cm^{-1} mode; a second light exposure again decreased the 2080 cm^{-1} mode and further reduced the 2000 cm^{-1} mode, which was followed again by a recovery of the 2080 cm^{-1} but not of the 2000 cm^{-1} mode. Another sample deposited with an intentional air bleed (2.8% in SiH$_4$) by Carlson *et al.* (1982) showed an increase of the 2000 cm^{-1} mode after light exposure that, in contrast to the other sample, was reversible on annealing and repeatable after a second light exposure.

The incomplete recovery on annealing and the opposite changes after subsequent light exposures can be explained by concurrently observed irreversible changes in the oxygen-related vibrations that are due to a rearrangement of some Si–O bonds. The possible effects of oxygen on the hydrogen-related modes have been discussed by Freeman and Paul (1978) and Lucovsky (1982). Local structural changes involving oxygen may lead to energy shifts in the Si–H modes when oxygen is incorporated into the Si bonding cluster and/or may lead to intensity changes that result from changes in the oscillator strength.

The observed changes in the Si–H intensities, if completely interpreted in terms of concentration changes, amount to the order of 1% and thus would imply bonding changes on the order of 10^{19} cm^{-3}. The concentration of metastable states inferred from spin density measurements (Dersch *et al.,* 1981) or from deep level transient spectroscopy (Crandall, 1981) is typically only 10^{16} cm^{-3}. We are thus led to believe that bonding changes, such as light-induced movement of hydrogen between different bonding sites, do not necessarily result in the creation of gap states that determine many of the transport properties and the optical properties, such as PL. Nevertheless, the bonding changes observed in IR spectroscopy may represent a concentration of sites that are available for the formation of metastable gap states.

c. Electron Irradiation

We have also studied the effects of electron irradiation on IR absorption bands. In order to compare these electron-induced effects with those induced by light, films were grown onto identical Ge prisms side-by-side during the same deposition run. The electron irradiation was performed with a scanning electron beam covering the whole film area. The electron energy was chosen to obtain the maximum energy dissipation (Everhart and Hoff, 1971) at about one half of the film thickness (i.e., 5 keV for a film thickness of about 330 nm). The beam current at 50 μA was sufficiently small to avoid heating the film. The total dose typically amounted to 7×10^{17} electrons cm^{-2}.

Figure 12 shows the difference spectra for the companion film to the one shown in Fig. 11. Similar to the effects of light soaking, electron irradiation caused the most distinct changes in the Si–H modes around 2000 cm^{-1} and at 850 cm^{-1}. Changes in the oxygen-related modes between 1050 and 1200

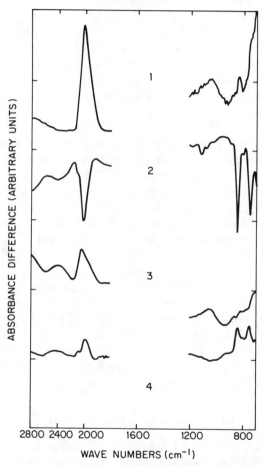

FIG. 12. Spectral dependence of absorbance differences between the following treatments: (1) First electron irradiation minus original, (2) first anneal minus first electron irradiation, (3) second electron irradiation minus first anneal, and (4) second anneal minus second electron irradiation.

cm^{-1} were also seen. Different samples were affected differently by alternate electron exposures and annealing treatments. As after light soaking, annealing may not lead to a complete reversal of the electron-induced changes, and subsequent electron irradiation may even cause changes opposite to the previous ones.

If we compare the effects of light exposure and electron irradiation on the IR spectra, we find roughly similar magnitudes for the intensity changes, even though a closer comparison may prove to be difficult due to the differences in energy dissipation of photons and electrons (see Subsection 1c). However, if we disregard the magnitudes of the intensity changes, the fact remains that alternate light exposures or electron irradiations and annealing treatments do not necessarily cause even the same direction of change. These differences appear to be inconsistent with the fairly parallel effects on PL and ESR after both light exposure and electron irradiation. The reasons for the sometimes apparently more diverging effects on IR spectra are indicated in the following subsection.

d. Interpretation of IR Absorption Changes

Although the concentration of a certain type of bond can be inferred from IR spectra, provided that the corresponding oscillator strength is known (Freeman and Paul, 1978; Shanks et al., 1980), it has become increasingly clear that the oscillator strengths depend critically on the local charge distribution. This local charge distribution will vary with the preparation conditions of the film, such as glow-discharge power and substrate temperature and, furthermore, with the structure and concentrations of the defects. Any observed change in IR absorption, particularly irradiation-induced changes, may thus consist of two possibly compensating contributions due to (1) a change in the concentration of bonds by an exchange of hydrogen between IR-active sites or between IR-active and -nonactive sites and (2) a change in the local field distribution produced by the movement of hydrogen or by defect generation near IR-active sites. The possibility of several concurrent processes leads to accidentally compensating effects (Oguz et al., 1980a) and may explain the various apparently incoherent results described in the previous sections.

III. The Nature of Metastable Defects

It has been well established that metastable defects are introduced into the bulk of a-Si:H by various types of radiation, such as by photons, electrons, and ions. The energy necessary to create these defects is supplied by the recombination energy of excess electron–hole pairs that may be generated by irradiation as well as by carrier injection, such as in forward-biased

$p-i-n$ devices (Staebler *et al.*, 1981). The similarity of various manifestations of these defects, such as the increase in the spin density or the same behavior on annealing, regardless of whether the defects result from photon or electron irradiation, is evidence for a common type of defect created by different excitations. However, defects created by ion bombardment have shown properties (Street *et al.*, 1979) and annealing behavior (see Subsection 3a) that indicate the presence of additional types of defects, possibly generated by direct momentum transfer.

In the preceding sections we have described the influence of defects on various optical properties and have begun to interpret the effects in terms of defect properties. Here we shall attempt to find a characterization for the metastable defects that is consistent with most of the observed effects. Due to the differences in a-Si : H material of various origins and the vast number of reported observations, it is inherently difficult to compare experimental results. The picture of the metastable defects presented here may thus be incomplete.

There are two aspects of defect characterization: (1) the atomic or molecular configuration of the defects and (2) the densities of states in the electronic energy-band diagram. The two aspects are, of course, related; ESR and IR absorption contain configurational information, whereas PL and photoconductivity and also transport phenomena (see Volume 21C, Chapters 5–8) refer mainly to the electronic density of states.

From the configurational point of view, one would like to find the types of molecular bonding, the formation energies and concentrations of defects, and possibly the relation of the defects to already existing configurations and sites that enhance the formation of metastable defects. From the electronic point of view, energies, densities of states, and carrier capture cross sections are of most interest. One must bear in mind, however, that the observed effects may not necessarily reflect the properties of the defects proper but instead may refer to properties of already existing bonding types or states that are modified by the presence of the metastable defects. In fact, these two possibilities are at the root of the discrepancies and controversies in the current interpretations.

4. CONFIGURATIONAL PROPERTIES

The bonding configurations in a-Si : H are influenced during the deposition of the film by temperature and glow-discharge conditions, such as surface bias (Street, 1982b). Consequently, depending on the specific configurations obtained, the irradiation-induced changes in bonding may differ.

Although the direct atomic displacement of hydrogen has been ruled out as the cause of metastable defects (see Subsection 1c), the suggestion was made (Staebler and Wronski, 1977) and evidence provided (Schade, 1981) that hydrogen is involved. The energy for Si–H bonds is 3.4 eV and thus is significantly larger than that of Si–Si bonds (Adler, 1978). It is unlikely, therefore, that Si–H bonds will be broken directly; instead, it has been proposed that Si–Si bonds are broken (Pankove and Berkeyheiser, 1980; Schade and Pankove, 1981; Dersch et al., 1981). The involvement of hydrogen is indirect in that Si–Si bonds are weaker when they are close to Si–H bonds (Elliott, 1979; Hirabayashi et al., 1980).

Based on this "softening" of the a-Si : H network, Elliot (1979) has explained many optical and electrical properties and particularly optically induced effects by the properties of "intimate charge transfer defects" (ICTD). These defects occur as a result of changes in the hybridization of sp^3 bonds that are favored around weaker bonds in the network. Thus dehybridization from sp^3-like to p-like bonding leads to spinless negatively charged defects (T_3^-), whereas hybridization from sp^3-like to sp^2-like bonding results in spinless positively charged defects (T_3^+). These two diamagnetic defects then occur as oppositely charged close pairs stabilized by Coulomb interaction, thus forming the ICTDs. The presence of ICTDs cannot be detected by ESR in the dark due to their spin pairing; however, light-induced ESR is observed (Knights et al., 1977) and is explained by the capture of electrons or holes that transform the diamagnetic defects into paramagnetic ones.

It has been postulated that the concentration of ICTDs is increased on light exposure (Elliot, 1979); the irradiation-induced effects would then be due to preferential trapping of electrons and holes by the T_3^- and T_3^+ centers. Due to this trapping, neutral centers with unpaired spins are formed that now give rise to the observed increase of the ESR, even in the dark after irradiation.

For more details on possible bonding configurations, we refer to Chapter 14 by Adler in Volume 21A. Here we shall discuss only further information on metastable defects deduced from certain configuration-related properties such as IR absorption and ESR.

a. IR Absorption

The IR absorption results (Section 3) clearly show that irradiation-induced configurational changes affect various Si–H vibrational modes. The difficulties in interpreting these changes via oscillator strengths and/or concentrations have been discussed in that section. Here we must stress that based on the IR spectra some of the configurational changes are not reversible on annealing. This behavior is not in disagreement with the generally observed reversal of the electrical and optical properties. In this context it is interesting to note that hydrogen in "unconventional," non-IR-active sites may still lead to electronic states that are responsible for electrical and optical properties (Singh *et al.*, 1977). Consistent herewith may be the observation of significant hydrogen evolution around 300°C, possibly from these unconventional sites, without concurrent changes in IR absorption (Oguz *et al.*, 1980b).

In attempting to correlate IR vibrational mode changes with electronic properties, Knights *et al.* (1979) found the mode at 845 cm^{-1} to increase with the spin density and thus with a decrease in PL and photoconductivity. This mode at 845 cm^{-1} is characteristic of coupled SiH$_2$ units, i.e., (SiH$_2$)$_n$. It was further shown that the 845 cm^{-1} mode is directly related to the film morphology, in particular, to the presence of microvoids that are thought to accommodate the (SiH$_2$)$_n$ groups (Knights *et al.*, 1979; Knights, 1980). These groups are less stable than the monohydride (John *et al.*, 1980) and are thus more easily affected by irradiation. Due to the local confinement of these groups to voids or internal surfaces, bonding changes can proceed only in poor communication with the a-Si:H matrix. This could explain, at least qualitatively, both the temperature-activated generation (Pankove and Berkeyheiser, 1980) and annealing of irradiation-induced defects.

The existence of voids has been further correlated to the chemical reactivity of the films (Knights *et al.*, 1979), i.e., films with a high portion of (SiH$_2$)$_n$ groups also contain Si–O absorption modes. These modes are attributed to oxygen adsorption on internal surfaces during exposure to air and differ from modes caused by the intentional incorporation of oxygen during film deposition. Note that metastable centers also appear to be associated with oxygen (Crandall, 1981; Carlson, 1982). Here we shall not discuss the various Si–O configurations that have been proposed as metastable defects (Pontuschka *et al.*, 1982). It is worth pointing out, however, that the poor communication between void surfaces and the a-Si:H matrix mentioned above might also apply to changes in oxygen bonding and thus account for the temperature-activated generation and annealing of metastable defects.

It must be assumed that the concentration of (SiH$_2$)$_n$ groups is generally sufficient to be responsible for irradiation-induced effects. At the same time, the concentration may still be too small to be followed by IR absorption measurements. If the presence of (SiH$_2$)$_n$ groups in microvoids were the main cause of degradation, one should expect larger effects in more porous films. The possible role of oxygen and its kind of incorporation, whether in voids or directly in the matrix, in causing metastable defects remains presently unresolved.

b. Electron Spin Resonance

The main feature of ESR upon irradiation is an increase in the line at $g = 2.0055$ that has been identified with unsatisfied tetrahedral Si bonds having an unpaired electron, i.e., neutral Si dangling bonds. The same resonance has also been observed in optically detected magnetic resonance (ODMR, see Subsection 1b) and in spin-dependent photoconductivity (SDPC, see Subsection 2b). Further evidence for this identification is derived from hydrogen-effusion measurements that also lead to an increase in the ESR signal with $g \cong 2.0055$. A slight difference between g-values obtained after irradiation or hydrogen effusion has been attributed (Voget-Grote et al., 1980) to vacancies in addition to dangling bonds, compared with dangling bonds only. From a comparison of equilibrium ESR and light-induced ESR in undoped and doped material it was concluded that the ESR resonance at $g = 2.0055$ is due to a neutral defect with an unpaired electron, which further corroborates its identification as a single dangling bond. Consistent with that identification is the observed lack of dependence of the g-value on preparation methods and sample composition, as well as on the kind of irradiation. This independence also indicates that the defect is fairly localized and thus, unlike vibrational modes, rather independent of the local environment.

There are two possibilities for the increase of the ESR intensity with $g = 2.0055$, namely:

(1) Irradiation causes, either directly or indirectly, an increase in the concentration of dangling bonds.

(2) Irradiation causes the generation of gap states and thus a shift in the Fermi level that leads to an increase in the ESR signal due to a change in occupancy of already existing dangling-bond states.

At present, both possibilities must still be considered valid in light of the following observations.

Street (1983) concluded from transient photoconductivity studies that light soaking increases the concentration of dangling bonds to $10^{16} - 10^{17}$ cm^{-3}. This conclusion is based on a quantitative relationship between the reduction of the $\mu\tau$ product of electrons and holes and the increase in the dangling-bond spin density. However, the difficulty remains in identifying the mechanism of bond breaking and subsequent annealing.

With regard to the alternate model of changed occupancy of dangling-bond states, deep level transient spectroscopy measurements have shown no change in the dangling-bond density but have shown increases in either hole (Cohen et al., 1981) or electron traps (Crandall, 1981) causing a shift of the Fermi level and thus a change in occupancy of the already existing dangling-bond states. A similar conclusion has also been reached on the basis of low temperature photoconductivity measurements by Vanier (1982), who deduced the presence of a large number of donorlike states and also a small number of acceptors below midgap.

So far, both models appear to be well founded within the framework of the individual studies. Note, however, that thorough ESR measurements on undoped and oxygen- and nitrogen-doped a-Si:H after x-ray irradiation (Pontuschka *et al.,* 1982) may indicate a reconciliation of both viewpoints. These measurements have uncovered, in addition to an increase in the Si dangling-bond line, three other centers, namely, an oxygen-associated hole center, an sp^3-hybridized dangling bond on Si bonded to three oxygens, and neutral atomic hydrogen, stabilized by the presence of oxygen and electrostatically trapped on internal oxygenated surfaces. Note that the oxygen content, even in unintentionally doped a-Si:H, is sufficiently high to account for at least some of the additional centers. Evidence has thus been provided in a single investigation that there are not only dangling-bond defects due to silicon and hydrogen bonding changes but also defects due to impurities, particularly those involving oxygen.

Based on this information, we further suggest here that irradiation-induced changes in the $(SiH_2)_n$ groups consist of dangling bonds and atomic hydrogen. As described above, the $(SiH_2)_n$ groups are predominantly located at internal surfaces that are easily oxidized and can thus provide the required oxygen atoms for the electrostatic trapping of hydrogen.

5. ENERGY BAND DIAGRAM

The identification of local bonding configurations is required in order to develop quantitative models for the electronic density of states and hence to deduce the electrical and optical properties. As we must conclude from the preceding section, there are presently still different configurational models, and consequently no unique electronic description can exist.

However, a majority of experiments leads us to the conclusion that dangling-bond defects are, in fact, generated by various kinds of irradiation, possibly in conjunction with other types of defects. The dangling-bond centers are generally accepted as lying at about midgap. In order to explain different recombination paths, as observed in PL and ODMR (see Section 1), either the dangling-bond defect has been assumed to act in different charge states (Dersch *et al.,* 1983) or additional states have been proposed, such as more electron tail states or trapped hole centers (Street *et al.,* 1979; Morigaki, 1981).

It is common to all of these models that both nonradiative and radiative transitions are introduced by the addition of dangling-bond states. Thus electrons from electron tail states are captured by neutral dangling-bond centers in nonradiative transitions, possibly by tunneling (Morigaki *et al.,* 1982; Dersch *et al.,* 1983), whereas radiative transitions are ascribed to the recombination of electrons in dangling-bond states and holes in hole tail states. Additional radiative transitions between electrons in tail states below the dangling-bond state and hole tail states have also been proposed on the

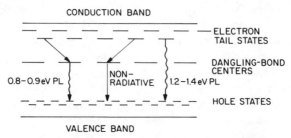

Fɪɢ. 13. Energy band diagram showing the main radiative and nonradiative recombination paths.

basis of ODMR results (Morigaki *et al.*, 1982). The energy band diagram illustrating these transitions is shown in Fig. 13, but no information on the transition rates of the various recombination paths is contained therein. Note particularly, that the increase in nonradiative transitions is not necessarily correlated with increases or decreases in radiative transitions. Generally, several competing recombination paths are simultaneously available, which may depend differently on various material parameters, such as doping, impurities, and morphology, as well as on excitation intensity and temperature. The positions of the equilibrium Fermi level and the quasi-Fermi levels are key quantities that determine the charge state of the impurity and defect centers and thus their capture cross sections. Further complexity arises from the fact that the densities and types of defect centers may change with temperature, as is observed during annealing (see Fig. 6) when one defect is transformed into another (Street *et al.*, 1979; Schade and Pankove, 1981).

With all these variables affecting the density of states and the recombination processes, it is obvious that more detailed analyses can be conducted only for specific cases.

IV. Summary

Metastable states changing the electrical and optical properties of a-Si : H can be generated by irradiation with visible light, x rays, electrons, and ions or by carrier injection. The generation of metastable defects relies on the recombination energy of electron–hole pairs. Annealing at temperatures above 160°C typically results in removal of the defects.

Most probably, several types of metastable defects occur as a consequence of bond breaking or bonding changes. The most prominent defects are dangling bonds; other types may involve complex formation. Although hydrogen plays an important role, most probably indirectly by providing softening of the a-Si : H network, impurities such as oxygen may also cause metastable defects.

Difficulties in finding a unique description of metastable defects stem from the dependence of defect formation on the original material properties. Furthermore, concurrent processes can result in compensating contributions to the observed effects.

ACKNOWLEDGMENTS

The author is grateful to D. E. Carlson and J. I. Pankove for valuable discussions. Thanks are also due to D. E. Carlson and R. W. Smith for providing the a-Si:H films for the infrared measurements, to E. F. Hockings for electron irradiations, to Z E. Smith for annealing treatments, and to P. J. Zanzucchi and W. R. Frenchu for the infrared measurements. The permission of the cited authors to use their figures is also gratefully acknowledged.

REFERENCES

Adler, D. (1978). *Phys. Rev. Lett.* **41**, 1755.
Aoyagi, Y., Segawa, S., and Namba, S. (1982). *Phys. Rev. B* **25**, 1453.
Bachus, R., Movaghar, B., Schweitzer, L., and Voget-Grote, U. (1979). *Philos. Mag. B* **39**, 27.
Biegelsen, D. K., Street, R. A., Tsai, C. C., and Knights, J. (1980). *J. Non-Cryst. Solids* **35/36**, 285.
Carlson, D. E. (1982). *J. Vac. Sci. Technol.* **20**, 290.
Carlson, D. E., Smith, R. W., Zanzucchi, P. J., and Frenchu, W. R. (1982). *In Proc. IEEE Photovoltaics Specialists Conf., 16th, San Diego, California*, pp. 1372–1375. IEEE, New York.
Cohen, J. D., Lang, D. V., Harbison, J. P., and Sergent, A. M. (1981). *J. Phys. Colloq. Orsay, Fr.* **42**, C4-371.
Crandall, R. S. (1981). *Phys. Rev. B* **24**, 7457.
Dersch, H., Stuke, J., and Beichler, J. (1981). *Appl. Phys. Lett.* **38**, 456.
Dersch, H., Schweitzer, L., and Stuke, J. (1983). *Phys. Rev. B* **28**, 4678.
Elliott, S. R. (1979). *Philos. Mag. B* **39**, 349.
Engemann, D., Fischer, R., Richter, F. W., and Wagner, H. (1975). *In Proc. Conf. Amorphous Liquid Semicond., 6th* (B. T. Kolomiets, ed.), p. 217. Nauka, Leningrad.
Everhart, T. E., and Hoff, P. H. (1971). *J. Appl. Phys.* **42**, 5837.
Freeman, E. C., and Paul, W. (1978). *Phys. Rev. B* **18**, 4288.
Fuhs, W., Milleville, M., and Stuke, J. (1978). *Phys. Status Solidi B* **89**, 495.
Gledhill, J. A. (1973). *J. Phys. A: Gen. Phys.* **6**, 1420.
Hirabayashi, I., Morigaki, K., and Nitta, S. (1980). *Jpn. J. Appl. Phys.* **19**, L357.
John, P., Odeh, I. M., Thomas, M. J. K., Tricker, M. J., Riddoch, F., and Wilson, J. I. B. (1980). *Philos. Mag. B* **42**, 671.
Knights, J. C. (1980). *J. Non-Cryst. Solids* **35/36**, 159.
Knights, J. C., Biegelsen, D. K., and Solomon, I. (1977). *Solid State Commun.* **22**, 133.
Knights, J. C., Lucovsky, G., and Nemanich, R. J. (1979). *J. Non-Cryst. Solids* **32**, 393.
Komuro, S., Aoyagi, Y., Segawa, Y., Namba, S., Masuyama, A., Okamoto, H., and Hamakawa, Y. (1983). *Appl. Phys. Lett.* **42**, 807.
Lucovsky, G. (1982). *Solar Energy Mat.* **8**, 165.
Morigaki, K. (1981). *J. Phys. Soc. Jpn.* **50**, 2279.
Morigaki, K., Hirabayashi, I., Nakayama, M., Nitta, S., and Shimakawa, K. (1980). *Solid State Commun.* **33**, 851.

Morigaki, K., Sano, Y., and Hirabayashi, I. (1982). *J. Phys. Soc. Jpn.* **51**, 147.

Oguz, S., Anderson, D. A., and Paul, W. (1980a). *Phys. Rev. B* **22**, 880.

Oguz, S., Collins, R. W., Paesler, M. A., and Paul, W. (1980b). *J. Non-Cryst. Solids* **35/36**, 231.

Pankove, J. I., and Berkeyheiser, J. E. (1980). *Appl. Phys. Lett.* **37**, 705.

Pontuschka, W. M., Carlos, W. W., Taylor, P. C., and Griffiths, R. W. (1982). *Phys. Rev. B* **25**, 4362.

Schade, H. (1980). *Vide, Couches Minces, Suppl.* **201**, 999.

Schade, H. (1981). *Int. Conf. Solid Films Surfaces, 2nd, Univ. of Maryland, College Park, Maryland.*

Schade, H. (1982). Unpublished data.

Schade, H., and Pankove, J. I. (1981). *J. Phys. Colloq. Orsay, Fr.* **42**, C4-327.

Schiff, E. A. (1981). *A.I.P. Conf. Proc.* **73**, 233.

Shanks, H., Fang, C. J., Ley, L., Cardona, M., Demond, F. J., and Kalbitzer, S. (1980). *Phys. Status Solidi B* **100**, 43.

Singh, V. A., Weigel, C., Corbett, J. W., and Roth, L. M. (1977). *Phys. Status Solidi B* **81**, 637.

Skumanich, A., Amer, N., and Jackson, W. (1982). *Bull. Am. Phys. Soc.* **27**, 146.

Solomon, I., Biegelsen, D. K., and Knights, J. C. (1977). *Solid State Commun.* **22**, 505.

Staebler, D. L., and Wronski, C. R. (1977). *Appl. Phys. Lett.* **31**, 292.

Staebler, D. L., Crandall, R. S., and Williams, R. (1981). *Appl. Phys. Lett.* **39**, 733.

Street, R. A. (1976). *Adv. Phys.* **25**, 397.

Street, R. A. (1980). *Phys. Rev. B* **21**, 5775.

Street, R. A. (1982a). *Philos. Mag. B* **46**, 273.

Street, R. A. (1982b). *Phys. Rev. Lett.* **49**, 1187.

Street, R. A. (1983). *Appl. Phys. Lett.* **42**, 507.

Street, R. A., Knights, J. C., and Biegelsen, D. K. (1978). *Phys. Rev. B* **18**, 1880.

Street, R., Biegelsen, D., and Stuke, J. (1979). *Philos. Mag. B* **40**, 451.

Tomozane, M., Hasegawa, F., Kawabe, M., and Nannichi, Y. (1982). *Jpn. J. Appl. Phys.* **21**, L497.

Vanier, P. E. (1982). *Appl. Phys. Lett.* **41**, 986.

Voget-Grote, U., Kümmerle, W., Fischer, R., and Stuke, J. (1980). *Philos. Mag. B* **41**, 127.

von Roedern, B., Ley, L., Cardona, M., and Smith, F. W. (1979). *Philos. Mag. B* **40**, 433.

Yamasaki, S., Hata, N., Yoshida, T., Okeda, H., Matsuda, A., Okushi, H., and Tanaka, K. (1981). *J. Phys. Colloq. Orsay, Fr.* **42**, C4-297.

SEMICONDUCTORS AND SEMIMETALS, VOL. 21, PART B

CHAPTER 12

Photoelectron Emission Studies

L. Ley

MAX-PLANCK-INSTITUT FÜR FESTKÖRPERFORSCHUNG
STUTTGART, FEDERAL REPUBLIC OF GERMANY

I. Introduction

Photoelectron spectroscopy (PES) provides direct information about the energy distribution of occupied states in solids, gases, and under certain circumstances, liquids (Cardona and Ley, 1978, Ley and Cardona, 1979; Brundle and Baker, 1977, 1978). Energy distribution curves (EDCs) of the valence electrons are, aside from cross-section effects, a true replica of the valence density of states (DOS) and thus reflect the changes in the electronic structure of silicon upon amorphization or incorporation of hydrogen, to mention only two examples of the topics to be discussed later. Valence-band spectra complement the information obtained from optical absorption measurements. Core-level spectra are characteristic for the elements present in the samples and are therefore employed for elemental analysis. The sensitivity is limited to a few percent in the concentration of the minority species but the extremely short sampling depth of ~20 Å makes PES particularly well suited for the analysis of surfaces.

The formation of heteropolar bonds such as Si–H or Si–F results in a charge transfer from Si to the more electronegative ion (H, F, etc.) that can be measured through the small energy shift it induces in the core levels of Si. We shall see how an analysis of these so-called chemical shifts aids in the unraveling of the frequency distribution of Si–F_n units in a-Si:F. Even the minute charge fluctuations due to topological disorder and bond-angle variations in a-Si have been measured through the chemical shifts they induce in the core levels of silicon.

A third aspect of PES that we shall discuss is the possibility of determining the position of the Fermi level (E_F) relative to the valence-band edge as a function of doping and hydrogen content. The use of PES to determine the position of E_F has been demonstrated in c-Si by Wagner and Spicer (1974). The relevance of this information to transport measurements is evident.

Two kinds of spectroscopies are closely related to PES: Auger and partial yield spectroscopy. The decay of a core hole that was created by electron or photon bombardment results in the emission of an Auger electron. The spectral distribution of Auger electrons contains information similar to that of photoelectrons. However, because the Auger emission is a two-electron process, the interpretation of Auger spectra is not as straightforward as that of PES (Allie *et al.,* 1980).

Partial yield, finally, is a way of measuring the density of empty conduction states. It requires a tunable photon source in the range 95 eV $\lesssim$ $\hbar\omega$ $\lesssim$ 150 eV to excite electrons from the Si 2p core state into the conduction band. Such a source is available in the form of synchrotron radiation emitted from an electron accelerator and we shall present some results obtained on a-Si and a-Si:H.

The scope of this chapter is tutorial rather than encyclopaedic. The basic concepts of PES are developed phenomenologically and are illustrated by examples from the field of amorphous silicon and its alloys. The reader interested in a rigorous approach to PES is referred to monographs such as the books by Cardona and Ley (1978), Ley and Cardona 1979), Brundle and Baker (1977, 1978), and Carlson (1975). A comprehensive review of all photoemission results on amorphous silicon up to 1982 has been given by the present author (Ley, 1983).

II. The Basics of Photoelectron Spectroscopy

Photoelectron spectroscopy is based on the external photoelectric effect. The sample is irradiated with monochromatic light of energy $h\nu$ and the emitted photoelectrons are analyzed with respect to their energy in an electrostatic energy analyzer equipped with electron detector and counting electronics (Fig. 1a). The measured quantity is the energy spectrum $I(E_k)$ of the photoelectrons, where E_k denotes their kinetic energy. It is advantageous and customary to introduce binding energies E_B through

$$E_B = h\nu - E_k - \phi_A,\tag{1}$$

where ϕ_A is the work function of the analyzer used to determine E_k. The binding energies E_B of electronic states so defined are characteristic for a given material and do not depend on the photon energy $h\nu$ as does E_k. The photoelectron spectrum of silicon in Fig. 1b excited with Al K_α x rays ($h\nu = 1486.6$ eV) shows two sharp lines at 99.0 and 144 eV binding energy due to the excitation of electrons from the atomic-like Si 2p and 2s core levels. The Si 3s and 3p electrons form the valence bands which give rise to the structures between 0 and ~ 20 eV binding energy in Fig. 1b.

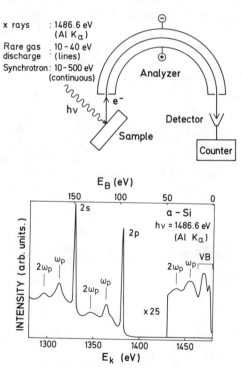

FIG. 1. (a) Schematic representation of a photoemission experiment. The sources of monochromatic light commonly employed are indicated. (b) The electron energy spectrum $I(E_k)$ of amorphous silicon as obtained with Al K_α x rays. The Si 2s and 2p core levels and the valence bands (VB) are identified. Single and double plasmon losses are labeled by ω_p and $2\omega_p$, respectively.

Valence bands and core levels are superimposed on a background of inelastically scattered electrons. On their way to the surface, photoelectrons interact strongly with the elementary electronic excitations of the solid, namely, electron–hole excitations and plasmons, and thereby lose energy. The probability for these energy losses defines an inelastic mean free path λ_e for the electrons, which limits the information of PES that is contained in the unscattered photoelectrons to a depth of roughly $3\lambda_e$ (Ibach, 1977). Figure 2 summarizes the energy dependence of λ_e in silicon according to a recent compilation of experimental and theoretical data (Ley, 1983). At low energies λ_e is determined by electron–hole excitations, which have a threshold equal to two times the energy gap (2.2 eV in c-Si). The minimum in λ_e above ~ 25 eV is due to the onset of plasmon excitations, which require an energy $\hbar\omega_p \simeq 16.4$ eV in silicon.

In the spectrum of Fig. 1b each core level is accompanied by satellites labeled ω_p, $2\omega_p$ which correspond to photoelectrons which have excited one or two plasmons on their way through the sample. The plasmon energy depends on the electron density, which may thus be extracted from PES spectra (Ley, 1983). An analysis of plasmon energies in terms of bonding configurations in a-Si_xC_{1-x}:H has been given by Katayama *et al.* (1981a). The mean free path rises again for electron energies above ~ 100 eV (Fig. 2) but stays below ~ 25 Å even for the highest energies encountered in PES ($E_k \simeq 1500$ eV). The λ_e versus E_k curve of Fig. 2 applies within a factor of approximately two in λ_e for all solids. Photoelectron spectroscopy is thus a rather surface-sensitive method.

This has two consequences. Since each spectrum contains only information about the last 20 to 100 Å of the sample, one has to assess in each case whether this information is representative of the bulk or the surface of the

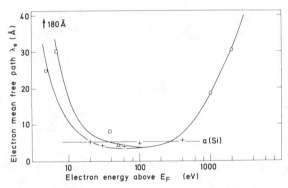

FIG. 2. The electron mean free path λ_e in silicon as a function of electron energy. The two curves below 100 eV reflect the scatter in measured values of λ_e. The line labeled a(Si) marks the lattice constant of c-Si. [After Ley (1983).]

specimen. The electronic structure of the surface differs from that of the bulk due to the loss of a half-space of neighbors. However, in crystalline silicon, and for that matter in all elemental solids, these differences extend no further than two atomic layers into the bulk, and the surface contribution to PES is therefore small except for electron energies near the minimum in λ_e (Appelbaum and Hamann, 1976).

On the other hand, in a-Si:H or a-Si:F we shall present evidence that the hydrogen or fluorine concentration measured with PES differs from that in the bulk, a result that sheds some light on the growth process of a-Si:H and a-Si:F (See Section 3 and Part V). Similar gradients in stoichiometry are in principle possible in all binary or ternary systems such a-Si$_{1-x}$C$_x$:H, a-Si$_{1-x}$N$_x$:H and they should be considered in the interpretation of photoemission data of these materials.

The short sampling depth requires that photoemission measurements be performed on atomically clean surfaces. This is in our experience best achieved if samples are prepared in a chamber attached to the spectrometer. The sample is transferred after deposition into the spectrometer without breaking the vacuum. The spectrometer vacuum is in the 10^{-10}-Torr range, which ensures that surfaces will be free of contamination for periods of at least 24 hr in the case of a-Si. Hydrogenated silicon is much less susceptible to contamination with oxygen or water vapor and longer periods with clean surfaces are obtained.

A sample of a-Si that has been exposed to air shows oxide contamination that obscures the valence-band spectra even if the contamination amounts to less than $\frac{1}{30}$ of a monolayer (Kärcher and Ley 1982). The oxide layer can be removed by argon-ion bombardment, but that might also leave the freshly exposed surface with a composition different from that of a virgin specimen.

The spectrum of binding, or ionization, energies obtained in PES according to Eq. (1) is usually interpreted in the framework of the single electron approximation to the photoemission process as indicated schematically in Fig. 3 . Optical transitions between occupied states below the Fermi level E_F (energy eigenvalues $\varepsilon_j < 0$) and empty conduction states above E_F ($\varepsilon_j > 0$) (Fig. 3a) generate the spectrum of high-energy photoelectrons as sketched in Fig. 3b. In this approximation the binding energy $E_B(j)$ of a particular electron state j is simply related to its orbital energy ε_j according to Koopmans' theorem (Koopmans, 1933) as

$$E_B(j) = -\varepsilon_j. \qquad (2)$$

The photoemission spectrum may thus be interpreted directly in terms of the ground state properties of the material under investigation: the valence-band spectrum will be compared with the density of states as obtained in a band structure calculation (see Part IV) and shifts in core-level binding

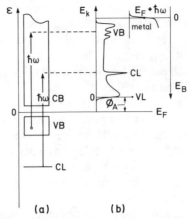

FIG. 3. One-electron scheme for the interpretation of photoelectron energy distributions in terms of initial state energies. The abbreviations refer to core level (CL), valence band (VB), conduction band (CB), Fermi level (E_F), vacuum level (VL), and analyzer work function ϕ_A.

energies in terms of changes in the electrostatic potential at the atomic core (Section 2).

Koopmans' theorem [Eq. (2)] holds rigorously provided that the ε_j are calculated in a self-consistent-field Hartree–Fock scheme (SCF–HF) and the relaxation of the passive orbitals in response to the suddenly created photohole is neglected. In reality, however, the "passive" electrons tend to screen the core hole and thereby lower the binding energy with respect to the value given in Eq. (2) by an amount E_R, the relaxation energy.

The relaxation energies are by no means negligible; they amount typically to a few percent of the binding energy for core levels (Fadley, 1978) and can reach values as high as 10 eV for the valence bands of solids (Ley et al., 1973). Part of the relaxation energy in solids is provided by the electrons on neighboring atoms (extra-atomic relaxation) and this contribution may change in going from Si to SiO_2, for example (Hollinger et al., 1975).

Equation (2) is nevertheless used in the interpretation of PES data because E_R varies monotonically with E_B and the *changes* in E_R are small over limited energy ranges such as the width of the valence bands (10–20 eV) or the range of binding energies encountered in the chemical shifts of core levels ($\lesssim 4$ eV).

Finally, a word about the reference energy in photoemission. For conducting samples in ohmic contact with the electron spectrometer, the common Fermi level of sample and spectrometer is the natural reference energy, i.e., $E_B(E_F) \equiv 0$. The position of E_F in terms of its kinetic energy in the I versus E_k spectrum is easily identified by the sharp drop in emission intensity observed for a metal as shown schematically in Fig. 3b. The position of E_F so determined remains valid also for a semiconductor fulfilling the requirements stated above as long as the work function of the electron analyzer does not change. It is thus possible to measure directly the separation between the top of the valence bands (TOVB) and E_F (see Part V). In doing so one has to take the possibility of band-bending at the surface into account since PES measures E_F − TOVB in the first 20–50 Å only.

III. Core-Level Spectra

1. ELEMENTAL ANALYSIS

The binding energies of core levels are characteristic for each element, and the core-level spectrum may therefore be used for a qualitative and quantitative (via the core-level intensities) chemical analysis of the sample within the sampling depth of PES. The sensitivity is typically a few atomic percent —a limit that is nevertheless sufficient to detect submonolayer quantities of impurities such as oxygen or carbon adsorbed at the surface of specimens (Brundle, 1974; Menzel, 1978). An up-to-date list of core-level binding energies of all elements is given in Cardona and Ley (1978).

The relative concentrations of two elements $C(Z_1)/C(Z_2)$ may be obtained from the measured intensities of two core levels $I(j_1,Z_1)$ and $I(j_2,Z_2)$ according to Powell (1978) as

$$\frac{C(Z_1)}{C(Z_2)} = \frac{I(j_1, Z_1)\sigma(j_2, Z_2, \omega)\lambda_e(j_2, Z_2, \omega)A(j_2, Z_2, \omega_2)}{I(j_2, Z_2)\sigma(j_1, Z_1, \omega)\lambda_e(j_1, Z_1, \omega)A(j_1, Z_1, \omega)}, \tag{3}$$

where $\sigma(j_i,Z_i,\omega)$ is the photoelectric cross section of level j_i in element Z_i at a photon energy of $\hbar\omega$. The cross sections have either been calculated (Scofield, 1976; Band et al., 1979) or they have been determined experimentally from measurements on samples with known composition (Wagner, 1972; Leckey, 1976; Berthou and Jørgensen, 1975). Here λ_e is the electron mean free path discussed earlier, and it is assumed that the photons penetrate much further into the sample than λ_e, an assumption that holds for the x-ray energies [$\hbar\omega = 1486.6$ eV (Al K$_\alpha$) or 1253.6 eV (Mg K$_\alpha$)] employed in core-level spectroscopy. The factor A takes the energy-dependent transmission function of the electron analyzer into account. For the commonly used analyzers with fixed pass energy and preretardation of the photoelectrons, $A(E_k) \propto E_k^{-1}$ (Klemperer, 1971). Tables with $\sigma \times \lambda \times A$ values appropriate for all elements using Al K$_\alpha$ radiation, an electron energy analyzer of the retarding type, and theoretical cross sections have been given by Elliot et al. (1983). A selection of similar values for elements likely to be encountered in work on a-Si and its alloys has been compiled by Ley (1983). The accuracy of concentrations obtainable using Eq. (3) and theoretical σ values is estimated to be ~20%. Better results are achieved if a sample with known composition is used as a standard to determine the factors on the right-hand side of Eq. (3).

The intensity of phosphorus and boron core levels relative to those of Si has been used by von Roedern et al. (1979a) to determine the incorporation rate of P and B in doped a-Si:H. They found that the incorporation rate was 80% for P and 70% for B relative to the [P]/[Si] or [B]/[Si] ratio in the gas phase, respectively. These values are in agreement with those obtained from Auger electron spectroscopy (Thomas, 1980).

As an example of the determination of surface concentrations with PES, consider the oxidation of silicon shown in Fig. 4. Here the intensity ratio of the O 1s and the Si 2p line is taken as a measure of the oxygen coverage of a silicon surface as a function of oxygen exposure (Ley et al., 1981a). The oxygen signal on a surface of c-Si increases initially with oxygen exposure and levels off when a coverage of one monolayer of oxygen is reached and the sticking coefficient is reduced. At this coverage, a ratio of 0.6 is reached for $I(O\ 1s)/I(Si\ 2p)$, i.e., fractions of a monolayer may easily be detected (Kärcher and Ley, 1982).

The oxygen uptake of a-Si follows that of c-Si closely (Müller et al., 1981), whereas molecular oxygen does not stick to a-Si:H, as is shown in Fig. 4. The conclusion is that the dangling bonds at the surface, which are mainly responsible for the initial uptake of oxygen at the Si surface, are saturated with hydrogen in a-Si:H, thereby lowering its reactivity.

2. CHEMICAL SHIFTS IN a-Si:F AND a-Si:H

The second kind of information extracted from core-level spectra is the so-called chemical shift. Consider as an example the Si 2p core-level spectra

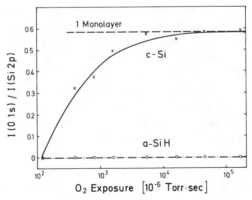

FIG. 4. O 1s to Si 2p core-level intensity ratio as a function of O_2 exposure in c-Si and a-Si:H. [After Ley et al. (1981a).]

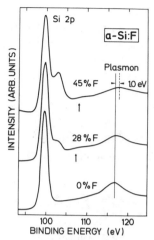

FIG. 5. Si 2p core-level spectra for three films of a-Si:F containing increasing amounts of fluorine. [After Gruntz *et al.* (1981).]

of fluorinated a-Si presented in Fig. 5 (Gruntz *et al.*, 1981). The samples were prepared by sputtering a c-Si target in an Ar–SiF$_4$ mixture. The pure a-Si sample shows a single Si 2p level at 99.6 ± 0.1 eV binding energy. With increasing F content, a second peak shifted by ~3.5 eV toward higher binding energy develops. The intensity of this satellite scales with fluorine content, and it has therefore been associated with 2p electrons emitted from Si atoms bonded to one or more fluorine atoms. The 3.5 eV difference in binding energy induced by the formation of a Si–F bond is referred to as the chemical shift.

The chemical shift ΔE_B of a core level reflects the increase in binding energy E_B due to the reduction of valence charge that is transferred from Si to the more electronegative F. Experimentally, it is found that the chemical shift is approximately the same for all inner core levels (Fadley *et al.*, 1968) and scales with the charge transfer Δq (Hagström *et al.*, 1964) according to

$$\Delta E_B = \beta \, \Delta q. \tag{4}$$

The proportionality constant β takes a value of approximately -2.2 eV per electron for the 2p level of silicon, as derived from chemical shift data on SiO$_x$ (Grunthaner *et al.*, 1979). The numerical value assigned to β depends obviously on the way the valence electrons are divided among the partners participating in an ionic or partly ionic bond. As a result, one arrives at different values for β, depending on the choice of the particular scheme used to estimate the Δq's from the electronegativities of the elements. This point has been comprehensively discussed by Gray *et al.* (1976), based on a large number of silicon compounds. An analysis of ΔE_B values in terms of absolute charges should therefore be treated with caution. An estimate of chemical shifts in a-Si based on electronegativity arguments has been given by Katayama *et al.* (1981b).

There is, however, one aspect of chemical shifts that is of considerable analytical value: their additivity. The point is illustrated in Fig. 6, which was taken from the work of Kelfve *et al.* (1980) on chemical shifts in substituted silanes SiX_i ($i = 1, 2, 3, 4$). The plot illustrates the excellent correlation between the chemical shifts of the Si $2p_{3/2}$ level measured in 18 compounds and those calculated on the basis of the group shift model (Gelius *et al.*, 1970)

$$\Delta E_B = \sum_i n_i \, \Delta E_B^{gr}(i). \tag{5}$$

The shift $\Delta E_B^{gr}(i)$ is associated with one of the groups or substituents listed in Fig. 6 and n_i is the number of times a particular group X_i is attached to Si. The concept of a group chemical shift thus expresses the experimental fact that the chemical shift induced by the formation of a particular Si–X bond is *independent* of the charge transfer associated with any other bonds already formed with the Si atom under consideration. The additivity of chemical shifts is, of course, not limited to silicon.

For the case at hand — a-Si:F — we expect an equidistant series of chemically shifted components for each $Si–F_m$ ($m = 0, 1, \ldots, 4$) configuration separated by $\Delta E(F)$, the chemical shift per attached fluorine atom. Two fits made to Si 2p spectra of a-Si:F samples containing different amounts of fluorine confirm this hypothesis as shown in Fig. 7. From similar fits to ~ 45

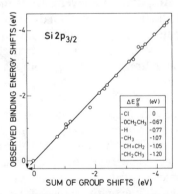

FIG. 6. Calculated relative (to $SiCl_4$) group shifts from the Si $2p_{3/2}$ binding energy data of 18 gaseous SiX_4 compounds. The diagram correlates the sum of the group shifts with the observed binding energy chemical shifts. [After Kelfve *et al.* (1980).]

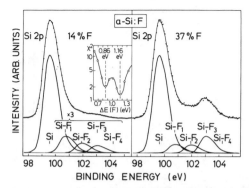

FIG. 7. Si 2p core-level spectra of two a-Si : F specimens (dots). The spectra were obtained with monochromatized Al K_α radiation. The solid line is the result of a fit to five components corresponding to Si atoms bonded to $n = 0, 1, \ldots, 4$ fluorine atoms. The inset represents the χ^2 test for a particular fit with the chemical shift per fluorine atom $\Delta E(F)$ as the parameter. [After Gruntz et al. (1981).]

Si 2p spectra Gruntz et al. (1981) obtain a value of 1.15 ± 0.02 eV for $\Delta E(F)$ in a-Si.

With a calibration factor of 2.2 eV per electron, this shift corresponds to a charge transfer of about half an electron from Si to F per Si–F bond. For less electronegative elements, smaller charge transfers and therefore smaller chemical shifts are expected. In particular, hydrogen has an electronegativity (on Pauling's scale) of 2.1 that is only slightly higher than the 1.8 of Si. By comparison, fluorine, oxygen, and nitrogen have electronegativities of 4.0, 3.5, and 3.0, respectively. The hydrogen-induced chemical shift was thus only recently determined as $\Delta E(H) = 0.34 \pm 0.01$ eV, using high-resolution [0.14 eV full-width-at-half-maximum (FWHM)] photoemission spectra employing synchrotron radiation (Ley et al., 1982). Even with the improved resolution, separate Si 2p components can no longer be resolved, but the comparison of Si 2p spectra for a-Si and a-Si : H in Fig. 8a shows extra emission at higher binding energies in a-Si : H that has been fitted successfully to a total of four lines, as illustrated in Fig. 8b. The four components correspond to emission from silicon atoms where $n = 0, 1, 2, 3$ Si neighbors have been replaced by hydrogen. Unlike a-Si : F, for which one finds evidence of trapped SiF_4 molecules in photoemission and IR absorption spectroscopy (Fang et al., 1980), no SiH_4 is found in a-Si : H. The charge transfer per attached hydrogen atom is $\Delta q(H) = 0.15$ electrons which should be compared to $\Delta q(H) = 0.12$ electrons obtained by Kelfve et al. (1980) from core-level studies of gaseous Si compounds and the 0.27 electrons deduced by Kramer et al. (1983) from model calculations on a-Si : H clusters.

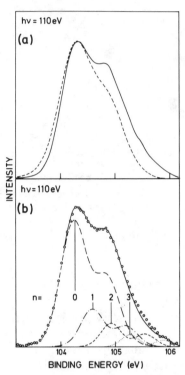

FIG. 8. (a) Comparison of the Si 2p spectra of a-Si and a-Si : H. The a-Si spectrum is shifted by 0.2 eV to higher binding energies. Binding energies are referenced to the vacuum level. ——, a-Si : H; –––––, a-Si. (b) Least squares fit of the a-Si : H spectrum to four spin-orbit doublets ($2p_{3/2}$ and $2p_{1/2}$) corresponding to Si atoms bonded to $n = 0, 1, 2,$ or 3 hydrogen atoms, respectively. [After Ley et al. (1982).]

3. FLUORINE AND HYDROGEN CONCENTRATIONS FROM Si 2p SPECTRA

Once the positions and the intensities of the shifted components have been determined from a least squares computer fit this information can be used to determine the concentration of fluorine or hydrogen atoms and, more important, their distribution among the different bonding configurations. Let us consider again a-Si : F as an example. If we denote the intensity of the unshifted component by I_0 and those of the shifted components by $I(Si-F_m)$ ($m = 1, 2, 3, 4$) we have for the ratio of F to Si atoms [according to Gruntz *et al.* (1981)]

$$\frac{N_F}{N_{Si}} = \frac{\Sigma_m N_{F_m}}{N_{Si}} = \frac{\Sigma_{m=1}^4 mI(Si-F_m)}{\Sigma_{m=1}^4 I(Si-F_m) + I_0}. \tag{6}$$

In the numerator each shifted component $I(\text{Si}-\text{F}_m)$ is weighted by the number m of fluorine atoms that give rise to that particular line. The denominator equals the total area under the Si 2p spectrum, a quantity that is obviously a measure of the Si concentration sampled. Total $C_F(\text{Si }2p)$ and partial $C_F(\text{Si}-\text{F}_m)$ fluorine concentrations are then defined according to

$$C_F(\text{Si }2p) = N_F/(N_F + N_{\text{Si}}) \tag{7}$$

and

$$C(\text{Si}-\text{F}_m) = N_{\text{F}_m}/(N_F + N_{\text{Si}}). \tag{8}$$

The validity of these expressions has been confirmed by comparing $C_F(\text{Si }2p)$ with fluorine concentrations obtained from the F 1s/Si 2p intensity ratios, using Eq. (3) (Gruntz *et al.*, 1981).

The advantage over other methods of using Eq. (7) to measure the fluorine, or by the same token the hydrogen (Ley *et al.*, 1982) concentration lies in the fact that it does not depend on cross sections and thus allows an absolute determination of fluorine concentrations.

In Fig. 9a the distribution of the fluorine atoms among the different $\text{Si}-\text{F}_n$ configurations is evaluated by plotting $C(\text{Si}-\text{F}_n)$ versus $C_F(\text{Si }2p)$ for sputtered a-Si : F samples with fluorine concentrations between 0 and 45 at. %. The same data are replotted in Fig. 9b so as to indicate the *relative* contribution each configuration makes to the total fluorine content. The distribution of the $\text{Si}-\text{F}_n$ units derived from a simple statistical model for the incorporation of fluorine in a-Si is also shown for comparison in Fig. 9c. The model considers a $\text{Si}-\text{X}_4$ cluster and asks for the probability of finding a $\text{Si}-\text{F}_n$ configuration under the assumption that the four X places are statistically occupied by F or Si atoms. The boundary condition is that the F/Si ratio corresponds to a given fluorine concentration C_F.

For low C_F the $\text{Si}-\text{F}_1$ units dominate as expected. With increasing fluorine concentration the relative contribution of $\text{Si}-\text{F}_1$ decreases, and $\text{Si}-\text{F}_2$ and $\text{Si}-\text{F}_3$ configurations become more important. Clusters of SiF_4 are formed only according to the model for fluorine concentrations higher than ~ 40 at. %. The concentration dependent distribution of the F atoms among the possible configurations found experimentally is in basic agreement with the model. A notable exception is the excess in $\text{Si}-\text{F}_3$ configurations compared to their statistical probability. Units of $\text{Si}-\text{F}_3$ take their intensity initially from $\text{Si}-\text{F}_1$ units and, for C_F above about 20 at. %, from the then statistically dominant $\text{Si}-\text{F}_2$ units. Below ~ 12 at. %, an enrichment of $\text{Si}-\text{F}_2$ is also evident.

The overemphasis of $\text{Si}-\text{F}_3$ (and $\text{Si}-\text{F}_2$) configurations has been attributed to a surface enrichment in fluorine (Gruntz *et al.*, 1981; Ley *et al.*, 1981b). An excess fluorine concentration of ~ 9 at. % is concentrated in a

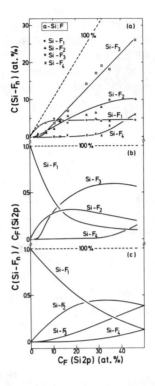

FIG. 9. (a) The absolute and (b) relative fluorine concentrations in each Si–F_n configuration as a function of the total fluorine content C_F. (c) The relative contribution of each Si–F_n unit to C_F calculated for a statistical distribution of F atoms in a-Si:F. [After Gruntz *et al.* (1981).]

surface layer no thicker than 9 Å. This corresponds to an excess surface density of F atoms of 3.4×10^{15} cm^{-3} or 2.5 fluorine atoms per surface Si atom. This concentration is reached for example if 80% of the available surface sites are saturated with Si–F_3 units. The excess fluorine concentration builds up for $0 \lesssim C_F$ (Si 2p) $\lesssim 12$ at. % and remains constant thereafter.

These results impose certain requirements on any growth model of a-Si:F films. (i) The growth has to take place in such a way as to ensure a surface rich in Si–F_3 units (and possibly Si–F_2 for $C_F \lesssim 12$ at. %). (ii) The incorporation of F into the growing network is governed primarily by the laws of statistics and not by chemical effects. Chemical effects would, for instance, favor the bonding of a F^- ion with an already existing Si^+–F^- unit because of the Coulomb interaction between Si^+ and F^-. It is also expected that the saturation of the a-Si:F film surface with fluorine makes these films particularly resistant to oxidation (see Part VI).

A similar analysis of the Si 2p spectra of a-Si:H has been performed by

Ley *et al.* (1982), and Reichardt *et al.* (1983) have used the energy dependence of the electron mean free path to determine a depth profile of the hydrogen content from the analysis of Si 2p core-level spectra excited with photon energies between 107 and 140 eV employing synchrotron radiation (see also Part VI).

4. CHARGE FLUCTUATIONS IN AMORPHOUS SILICON

In addition to the charges just discussed that are due to the formation of partially ionic bonds, there are other possible sources of charges on the Si atoms that give rise to chemical shifts, namely, charged defects and charge fluctuations on the Si atoms as a result of bond-length and bond-angle variations. According to the model calculations of Guttman *et al.* (1980), the last mechanism leads to a Gaussian distribution of charges on the Si atoms with an rms width of 0.2 elementary charges. A detailed analysis reveals that bond-angle variations and the concomitant changes in the second-nearest-neighbor distance are mostly responsible for these charges. Nearest-neighbor distances are virtually rigid in a-Si and the orbital overlap between neighbors more than one removed is too small to induce a sizable rehybridization upon distance changes.

The most likely charged defect in a-Si, on the other hand, is a dangling bond either unoccupied or occupied with two electrons. In the terminology of Adler (1978) this corresponds to states T_3^+ and T_3^-, respectively. The analog of these defects is the dangling-bond state on a reconstructed c-Si (111) surface. The reconstruction is driven by a Jahn–Teller distortion in which the originally singly occupied dangling bonds acquire alternating positive and negative charges via a charge transfer between neighboring rows of dangling-bond states (Haneman, 1975). This corresponds to the reaction $T_3^0 \rightarrow T_3^+ + T_3^-$, and it is accompanied by the loss of the spin on the T_3^0 dangling-bond state. A similar situation is expected to hold for the inner surfaces of voids in a-Si. It would explain the disparity between the spin density in unhydrogenated a-Si and the hydrogen concentration necessary to quench the spin signal (Brodsky and Kaplan, 1979).

The charge transfer connected with the reconstruction of the Si (111) surface has been measured through the chemical shift of the Si 2p core level of the surface atoms (Brennan *et al.*, 1980; Himpsel *et al.*, 1980; Ley, 1983). In addition to emission from bulk Si atoms, two doublets (S1 and S2) that correspond to Si atoms in the reconstructed top surface layer have been observed. The energy separation of S1 and S2 from the bulk lines is -0.39 and $+0.33$ eV, respectively. This corresponds to an average charge transfer between the surface atoms, $\Delta q_{surf} \approx 0.16$ electrons, using the calibration factor of 2.2 eV per electron mentioned earlier. The charge transfer is less than one because of the formation of bands of surface states. Nevertheless this example gives an indication of the magnitude of chemical shifts expected for reconstructed inner void surfaces in a-Si.

Upon hydrogenation, the reconstruction is removed and a uniform chemical shift toward *higher* binding energy is expected as a result of the charge transfer to the more electronegative hydrogen atom. Measurements on a-Si and a-Si:H (Ley *et al.*, 1982) indicate, however, that the charged dangling-bond states are of secondary importance in determining the charges on Si atoms in a-Si. The other two mechanisms dominate. In Fig. 10 we compare the 2p spectra of c-Si and a-Si obtained with $hv = 107.5$ and 110 eV, respectively. The spectra represent bulk Si atoms because of the large (~ 40 Å) electron mean free path of electrons with only 5 eV kinetic energy. The Si 2p doublet of the Si (111) surface (Fig. 10a) is well described by the convolution of a Lorentzian ($1/[(\omega - \omega_0)^2 + \Gamma^2]$) that takes into account the lifetime Γ of the core hole and a Gaussian ($\exp -[(\omega - \omega_0)^2/2\sigma^2]$) representing the combined resolution of monochromator and electron analyzer (Ley *et al.*, 1982). The parameters obtained from a least squares fit to the data points are listed in Table I.

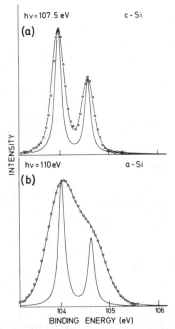

FIG. 10. Si 2p core-level spectra of a-Si and c-Si. The photon energy chosen ensures that mainly bulk atoms contribute to the spectra. Binding energies are referenced to the vacuum level. The two lines correspond to emission from the Si $2p_{3/2}$ and $2p_{1/2}$ levels, respectively. They are not resolved in the lower resolution spectra of Fig. 7. [After Ley *et al.* (1982).]

TABLE I

PARAMETERS OBTAINED IN THE LEAST SQUARES FITS OF Si 2p Core Levels of Figs. 8 and 10

Sample		Photon energy (eV)	Binding energy relative to vacuum level Si $2p_{3/2}$ (eV)	2Γ (meV)	σ_{exp} (meV)	σ_{amorph} (meV)	Δq(rms) ($|e|$)
c-Si		107.5	103.99(05)[a]	210(30)	60(5)	—	—
a-Si		110.0	104.01(05)	106(40)	80(5)	256(10)	0.11
a-Si:H	Si_0[b]	110.0	104.21(05)	106(40)	60(5)	192(10)	0.09
a-Si:H	Si_1	110.0	104.55	106(40)	60(5)	128	0.06
a-Si:H	Si_2	110.0	104.89	106(40)	60(5)	64	0.03
a-Si:H	Si_3	110.0	105.23	106(40)	60(5)	0	0

[a] Errors in the last digits are given in parentheses.
[b] Subscripts of Si refer to the number of H atoms.

The corresponding spectrum (Fig. 10b) taken with $hv = 110$ eV from an a-Si sample prepared in situ by sputter deposition is considerably broader, so that the spin–orbit components are no longer resolved. The data points are again fitted by the convolution of a Lorentzian and a Gaussian with parameters as displayed in Table I. The width of the Gaussian is partitioned into a contribution σ_{exp} due to the experimental resolution obtained from a spectrum of c-Si measured under identical conditions and a contribution σ_{amorph} that takes the homogeneous broadening into account such that $\sigma_{exp}^2 + \sigma_{amorph}^2 = \sigma^2$.

The Si 2p core-level spectrum ($hv = 110$ eV) of a-Si:H (prepared by adding 50 vol % H_2 to the sputter gas) has already been discussed in connection with the hydrogen–induced chemical shift (Fig. 8).

Notice, first, that the leading (low-binding-energy) edge of the spectrum of a-Si:H is somewhat steeper than that of a-Si. The charges on silicon atoms not bonded directly to hydrogen are thus *slightly* reduced. This excludes the possibility that all broadening in a-Si is due solely to surface chemical shifts of atoms surrounding voids, because in this case an *unbroadened* spectrum is to be expected.

The rms deviation from charge neutrality is estimated to be 0.11 electrons in a-Si, about half as much as that obtained by Guttman *et al.* (1980). In a-Si:H the incorporation of hydrogen leads apparently to an overall reduction in the bond-length fluctuations as manifested in the reduced σ_{amorph}, corresponding to charge fluctuations (rms) of only 0.09 electrons. This is in agreement with the 20% attenuation of the TO bonds in the IR spectrum of a-Si upon hydrogenation (Shen *et al.*, 1980).

Connected with these charges are, of course, potential fluctuations that can be determined once the local distribution of the charges is known in addition to their magnitude. Recalling what we said about the origin of the chemical shift in Section 2, it is clear that the core-level shifts reflect these potential fluctuations directly. Thus the potential fluctuations have a Gaussian distribution with an rms magnitude of 256 meV in a-Si and 192 meV in nonannealed a-Si : H. These potential fluctuations may contribute to the exponential valence-band and absorption tail in a-Si : H that will be discussed in Section 9.

IV. Valence and Conduction Bands

5. VALENCE-BAND SPECTRA OF a-Si

The valence-band spectra of a-Si and a-Ge have been obtained by low-energy photoemission (Pierce and Spicer, 1972; Eastman and Grobman, 1972; von Roedern *et al.*, 1979b), by x-ray photoelectron spectroscopy (XPS) (Ley *et al.*, 1972) and by x-ray emission spectra involving transitions from the valence bands of a-Si (Wiech and Zöpf, 1973).

Figure 11 gives a comparison of the densities of states of c-Si and of a-Si as obtained with XPS after correction for differences in photoemission cross section across the valence bands.

The four valence electrons of Si are distributed among three peaks in the valence–band density of states (VB-DOS) in such a way that the area under peak I corresponds to 2 electrons, and the areas under peaks II and III to one electron per Si atom each. States at the top of the valence bands are

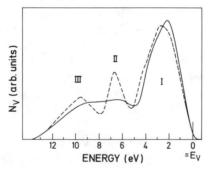

FIG. 11. Experimental valence densities of states of c-Si and a-Si derived from x-ray induced photoemission spectra (Ley *et al.*, 1972) after correction for inelastically scattered electrons and variations in photoelectric cross sections. ————, a-Si; — — —, c-Si.

predominantly 3p-like and those at the bottom (peaks II and III) are mainly of 3s character.

A comparison of the densities of states of c-Si and a-Si may be summarized as follows:

(1) The overall appearance of the DOS is similar for the amorphous and crystalline materials. We find in particular that the total width of the valence bands remains unchanged as well as the valley between the region of predominantly s-derived states (peaks II and III) and the p-like peak at the top of the DOS.

(2) The two lower lying peaks (corresponding to the L_1 and L'_2 critical points in the crystals) merge into one hump. Since the edges of this hump do not broaden, one has to conclude that the one-hump structure occurs as a result of states that fill the dip between L_1 and L'_2 rather than through a mere broadening of the two peaks.

(3) The centroid of peak I shifts toward the top of the valence bands by 0.4 eV. This results in a distinct steepening of the leading edge in the DOS of the amorphous samples.

The shift in peak I is responsible for the red shift in the maximum of $\varepsilon_2(\omega)$ in a-Si that cannot be accounted for by the loss of k-conservation alone (Pierce and Spicer, 1972). The first point emphasizes the fact that the principal features of the DOS are to a large extent determined by the tetrahedral environment of a Si atom, the number of nearest neighbors, and the interatomic distance, all of which remain unchanged in the amorphous modifications (Weaire and Thorpe, 1971a). The considerable density deficit of up to 20% in amorphous group IV semiconductors has consequently no influence on the DOS, as was first assumed by Herman and van Dyke (1968), because it is due to voids (Moss and Graczyk, 1969; Ohdomari *et al.*, 1977; Knights, 1980) and not to a dilatation of the lattice.

The loss of long-range order is obviously of secondary importance for the VB-DOS. It introduces some broadening (~ 0.5 eV in peak I) into the VB-DOS as was shown by the calculations of Kramer (1971) and of Brust (1969), hardly sufficient to wipe out critical-point structure in the density of occupied states completely, let alone to reproduce the characteristic blue shift of peak I and the filling of the valley between peaks II and III.

The extended conduction states are much more susceptible to the absence of long-range order. The structure in the density of unoccupied states is almost completely lost in a-Si according to these calculations (Kramer, 1971), a result that is in good agreement with the experimental results to be discussed later (Section 7).

Peaks II and III in the DOSs of c-Si and c-Ge are related to the presence of sixfold rings in the diamond structure (Weaire and Thorpe, 1971b; Joannopoulos and Cohen, 1976). In models of the amorphous network the topological freedom of the atoms allows for five-, seven-, and perhaps eightfold rings while preserving the basic tetrahedral units. The most common fivefold ring, for example, can be formed with only a slight adjustment of the tetrahedral bond angle. It has been suggested by Weaire and Thorpe (1971b) that fivefold rings should introduce states between those of peaks II and III and thereby fill the gap. This conjecture has been confirmed by a series of calculations involving either simple models (Weaire and Thorpe, 1971b; Joannopoulos and Cohen, 1976), or complex crystalline polytypes of Si and Ge containing five-, six-, and sevenfold rings (Alben *et al.,* 1972; Ortenburger *et al.,* 1972; Joannopoulos and Cohen, 1973), and by the building of actual cluster models of the amorphous structure with as many as 519 atoms (Ching *et al.,* 1976, 1977; Kelly and Bullet, 1976). In summary, we attribute the featureless appearance of the lower portion of the a-Si DOSs to an averaging of a variety of local bonding topologies, i.e., a distribution of rings of bonds.

As a result of the same calculations (Joannopoulos and Cohen, 1976), it was concluded that the blue shift of peak I is due largely to bond-angle variations. The electrons in this part of the valence bands are mainly localized in the bond region between atoms. Deviations from the equilibrium tetrahedral bond angle lead, on the average, to an increase in the energies of these states as a result of increasing Coulomb repulsion between neighboring bond charges. Such an increase in the energies of bonding states corresponds, of course, to a weakening of the bonds with a concomitant reduction of the Penn gap, i.e., the difference between the average energies of bonding and antibonding states (Penn, 1962).

A different point of view is held by Yonezawa and Cohen (1981) and by Singh (1981). They showed that the states at the top of the valence bands are strongly affected by variations in the dihedral angle and thus cause at least some of the observed changes in the shape and position of peak I.

The results of the VB-DOS and their interpretation, sketched here for a-Si, apply virtually unchanged to a-Ge as well (Ley *et al.,* 1979).

6. VALENCE-BAND SPECTRA OF a-Si : H

The addition of hydrogen to a-Si introduces Si–H bonding states between approximately 5 and 11 eV below the top of the valence bands. Figure 12 shows a series of EDCs obtained with photon energies 21.2 eV (HeI) and 40.8 eV (HeII) (von Roedern *et al.,* 1977). The spectra are for reactively sputtered films of a-Si prepared at room temperature with increasing amounts of hydrogen in the sputter gas (argon) as indicated in the figure. In the absence of hydrogen the spectrum is dominated by emission from Si 3p states at the top of the valence bands that correspond to peak I in Fig. 11

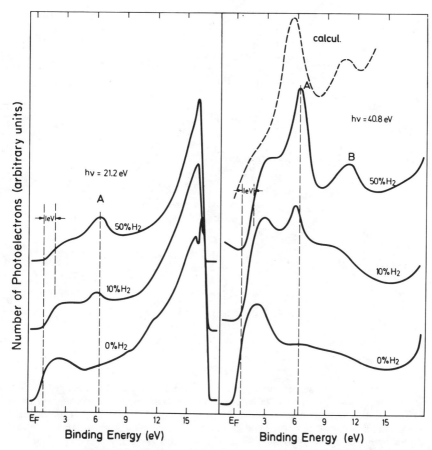

FIG. 12. (a) HeI and (b) HeII valence-band spectra of amorphous silicon films prepared by reactive sputtering in an argon–hydrogen mixture with increasing amounts of hydrogen. The hydrogen concentrations are those of the sputter gas. The dashed curve reproduces the spectrum for hydrogen adsorbed on a Si (111) surface in the form of SiH_3. [After von Roedern *et al.* (1977).]

(notice that the energy scale in the UV photoelectron spectroscopy (UPS) spectra runs in a direction opposite to that of Fig. 11). The contribution from the Si 3s states shows up as the structureless hump between 5 and 14 eV binding energy in the HeII spectrum, but it is lost in the steeply rising background of secondary electrons in the HeI spectrum. The cross-section ratio $\sigma(3s)/\sigma(3p)$ is noticeably smaller in the UPS regime than it is for $h\nu = 1486.6$ eV. The cross section for H 1s states is negligible compared to $\sigma(3s)$ and $\sigma(3p)$ at x-ray energies, and hydrogen-induced states are visible only around $h\nu \simeq 20-100$ eV (Ley, 1983).

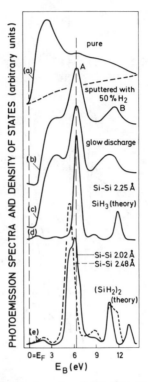

PHOTOEMISSION SPECTRA AND DENSITY OF STATES (arbitrary units)

E_B (eV)

FIG. 13. (a)–(c): HeII valence-band spectra of a-Si and a-Si:H samples prepared at room temperature by reactive sputtering (a), (b), and by the glow discharge decomposition of silane (c). (d) and (e): Calculated local densities of hydrogen-derived states in a-Si:H_x for different bonding geometries by Ching *et al.* (1980). The calculated spectra have been shifted by 1.8 eV toward lower binding energy so that peaks A line up. [After von Roedern *et al.* (1979b).]

With the addition of hydrogen, two new peaks labeled A and B appear with an intensity that is roughly proportional to the hydrogen concentration in the sputter gas. Peak B is hidden in the background of the HeI spectrum. The binding energies (relative to E_F) are 5.9 eV (10% H_2) and 6.2 eV (50% H_2) for peak A and 10.2 eV (10% H_2) and 11.2 eV (50% H_2) for peak B. The FWHM of peak A increases from 1.2 eV to 1.9 eV. The width of peak B is 2.2 eV in the spectrum with 50% H_2. The intensity ratio $I(A):I(B)$ is approximately 2:1 for $hv = 40.8$ eV. The A–B structure—ascribed to the polyhydride configuration (see below)—is the one always observed for specimens deposited below ~ 300°C. In Fig. 13 we compare the HeII EDCs of sputtered a-Si:H with that of glow-discharge (GD) sample, both prepared at room temperature (von Roedern *et al.*, 1979b). Peak A lines up almost perfectly for both specimens. The width of peak A in the glow-discharge sample is slightly (~ 0.2 eV) narrower. Peak B on the other hand is noticeably wider, somewhat skewed towards the low-binding-energy side, and has a binding energy of 10.8 eV, which is 0.5 eV lower than that in the sputtered film.

Specimens of a-Si:H sputtered onto substrates held at 350°C exhibit a distinctly different photoemission spectrum (Fig. 14). Two small peaks at 5.3 eV (peak C) and 7.3 eV (peak D) take the place of peak A. Instead of peak B a similar broad peak E at ~ 10.3 eV below E_F is observed. This form of the EDC for $T_D = 350°C$ is independent of the hydrogen concentration for the sputter gas as long as it exceeds 10 vol %. The spectra for GD a-Si:H films prepared at 350°C are again virtually identical with the C–D–E structure, which is somewhat less well-defined than that in the sputtered sample (see Fig. 14). On annealing, the A–B structure transforms into the C–D–E structure around $T_A = 300°C - 350°$. Further annealing to $\sim 600°C$ removes all hydrogen from the film and the spectrum of unhydrogenated a-Si is recovered. It has been argued (von Roedern *et al.*, 1979b) that the transition A–B $\rightarrow$ C–D is not merely due to a loss of hydrogen from polyhydride configurations until only Si–H configurations (structure C–D)

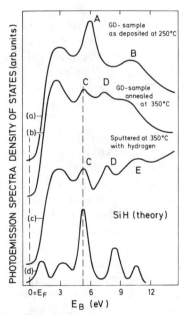

FIG. 14. (a)–(c): HeII valence-band spectra of a-Si:H annealed or deposited at elevated temperatures. (d) Calculated local density of states for Si–H configuration, obtained by Ching *et al.* (1980). The theoretical spectrum has been shifted by 1.8 eV toward lower binding energy to line up peaks C. [After von Roedern *et al.* (1979b).]

remain, but that some of the polyhydride configurations actually transform into the monohydrides [see Fig. 3 of von Roedern *et al.* (1977)].

The first identification (von Roedern *et al.*, 1977) of the hydrogen-induced features in terms of specific Si–H bonding configurations was based on the photoemission spectra of hydrogenated c-Si surfaces and their interpretation in terms of tight binding or pseudopotential calculations (Pandey *et al.*, 1975; Pandey, 1976; Ho *et al.*, 1977). The spectrum of a c-Si (111) surface saturated with hydrogen in the Si–H_3 configuration shows indeed a remarkable similarity to that of the low-temperature phase of a-Si:H (Fig. 12, the dashed line) (Pandey *et al.*, 1975; Fujiwara, 1981). The energy separation between A and B is, however, reduced at the surface compared to that of a-Si:H.

Based on these calculations, the origin of peaks A and B as Si 3p–H 1s and Si 3s–H 1s bonding states has been established. Gruntz *et al.* (1980) measured the intensity of peak A in a-Ge:H as a function of $\hbar\omega$ and, using known photoemission cross sections for $\sigma(\text{H 1s})$ and $\sigma(\text{Ge 4p})$, derived a contribution of 64% H 1s and 36% Ge 4p to peak A. This ratio should not be too different in a-Si:H, and it is in fact in agreement with the 60% H 1s contribution to peak A in a-Si:H calculated by Ching *et al.* (private communication).

The local (i.e., H 1s derived) densities of states (LDOSs) have been calculated by Ching *et al.* (1979, 1980) for several possible types of Si–H bonding configurations, each of them embedded in an amorphous silicon cluster. These are Si–H_x ($x = 1, 2, 3$) units, a $(\text{SiH}_2)_2$ chain fragment, and Si–H H–Si (a broken Si–Si bond with two H atoms taking the place of the broken bond). Some of these results are compared, in Figs. 13 and 14, with the UPS spectra.

It is apparent from this comparison that in addition to Si–H_3 (Fig. 13d), Si–H_2 (not shown) and $(\text{SiH}_2)_2$ also have a similar two-peak structure with a splitting of about 5 eV that could correspond to peaks A and B. Furthermore, the two examples—given for the silane fragment $(\text{SiH}_2)_2$ in Fig. 13—emphasize how sensitively the exact splitting and the shape of the LDOS depend on the model parameters. Under these circumstances it appears at present impossible to assign a particular polyhydride configuration to a given spectrum. We may at best comment on the tendency of the A–B splitting to increase with increasing hydrogen content that we mentioned before. The calculations of Ching *et al.* (1979, 1980) indicate a splitting that is ~0.3 eV larger for Si–H_3 than for SiH$_2$ or $(\text{SiH}_2)_2$. That would indicate that the increase in the A–B splitting is associated with a shift from $(\text{Si}-\text{H}_2)_x$ ($x = 1, 2$) to Si–H_3 with increasing hydrogen content.

There is ample evidence both from hydrogen adsorbed on c-Si (Sakurai and Hagstrum, 1976; Fujiwara, 1981) and from a number of calculations (Ching *et al.,* 1979, 1980; Economou and Papaconstantopoulos, 1981; Picket, 1981; Lemaire and Gaspard, 1981; Kramer *et al.,* 1983) that the C–D–E structure signals hydrogen bonded as monohydride Si–H. The theoretical result of Ching *et al.* (1979), obtained for an isolated Si–H unit in a Si cluster, is shown in Fig. 14. The agreement with the measured spectra is seen to be good if we assume that the two leading peaks in the LDOS are masked by the Si 3p emission between 0 and 5 eV. The LDOS calculated for the broken-bond model (Si–H H–Si) can similarly be made to agree well with experiment, although the peak positions depend on the parameters used for the bond lengths (Ching *et al.,* 1980). For this configuration a strong peak in the DOS appears at the bottom of the valence bands at ~ 14 eV. Such a resonance appears to be characteristic of strongly interacting Si–H units, as they are present also in the LDOS calculated for hydrogenated vacancies (Economou and Papaconstantopoulos, 1981). The corresponding peak is not observed in the photoemission spectra.

Allan *et al.* (1982) have investigated the conditions under which Si–H configurations yield the C, D, E structure using the cluster Bethe-lattice approach. They conclude that the signature of an isolated monohydride is a single peak C, whereas structures D and E appear only under two circumstances:

(1) The monohydrides are present as strongly interacting clusters either in the form of a hydrogenated vacancy or on the surfaces of inner voids. In the latter case the situation would be similar to the hydrogenated surface of c-Si.

(2) An isolated monohydride is part of a network with mainly sixfold rings in the immediate neighborhood as in c-Si. Peaks D and E reflect then the DOS of the remainder of the network corresponding to peaks II and III in the DOS of c-Si (Fig. 11). The idea is (Phillips, 1979) that hydrogenation of a-Si leads *locally* to the formation of crystalline regions with hydrogenated boundaries. Should this picture apply to a-Si : H, XPS measurements on a-Si : H samples, which are sensitive to the Si LDOS only, would resemble those of c-Si and exhibit peaks II and III. This is, however, not the case, as is illustrated in Fig. 15, and this second possibility appears not to be the main reason for structures C, D, and E. A number of other calculations on monohydride configurations give two peaks in reasonable agreement with peaks C and D but fail to reproduce peak E (Economou and Papaconstantopoulos, 1981; Lemaire and Gaspard, 1981; DiVincenzo *et al.,* 1981a,b).

Therefore, it seems premature to try to understand the subtle differences observed in the monohydride spectra that depend on preparation condi-

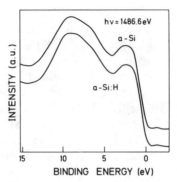

FiG. 15. XPS valence-band spectra of a-Si (upper curve) and a-Si:H (lower curve). [H. Richter and L. Ley (private communication).]

tions. The possibility of hydrogen clusters is, of course, a very real one. Their presence has been postulated independently by Shanks *et al.* (1980), based on IR spectroscopy, and by Reimer *et al.* (1981), based on NMR studies.

An estimate of the hydrogen concentration within the sampling depth of photoemission is possible, based on a comparison of the intensities of peak A for a-Si:H and for Si–H_3 units formed upon chemisorption of H on c-Si (111) surfaces (see Fig. 12) (von Roedern *et al.*, 1977). A Si (111) surface saturated with SiH_3 has a surface density of hydrogen of 8×10^{14} cm^{-2}. This corresponds to an effective bulk concentration ratio, [H]/[Si], of ~35 to 50 at. % assuming an average escape depth of 10 ± 2 Å (cf. Fig. 2). Inspection of Fig. 12 indicates a comparable hydrogen concentration in a-Si:H films prepared at room temperature. We shall see in Part V, however, that most of the hydrogen is concentrated in one or two surface layers. The high concentration of hydrogen and the Si–H bonding configurations determined from photoemission spectra are thus not always representative of the bulk of a-Si:H.

Valence-band spectra of a-Ge:H and a-Si:F have also been published (Gruntz *et al.*, 1980, 1981), as well as those of a-Si–C alloys (Katayama *et al.*, 1981b) and a-Si–N alloys (Kärcher *et al.*, 1983).

7. Partial Yield Spectroscopy and the Density of Conduction States in Amorphous Silicon

As was mentioned in the introduction, optical absorption from sharp core levels into the unoccupied conduction states gives the most direct information about their energy distributions. Since the Si 2p core levels have a binding energy of ~99 eV, such experiments require synchrotron radiation. The absorption coefficient of a-Si:H was, however, not determined from conventional transmission measurements but was found indirectly via the yield of secondary electrons that follow the decay of the core hole left behind. This has the advantage that this so-called partial yield can be measured with the same experimental setup that is used for photoemission measurements (Kunz, 1976). As long as the escape depth of the electrons does not exceed the penetration depth of the light, the secondary electron yield is proportional to the absorption coefficient (Gudat and Kunz, 1972).

The Si 2p → conduction-band (CB) *optical* absorption spectra for c-Si and a-Si have been measured by Brown and Rustgi (1972) and the corresponding yield spectra by Gudat and Kunz (1972). In Fig. 16 we present the L_{III} (Si $2p_{3/2}$ → CB) partial yield spectra of c-Si, a-Si, and a-Si : H (50 vol % H_2 in sputter gas, T_D = RT), all three taken under identical conditions (Reichardt *et al.*, 1983). The spectra in Fig. 16 have been corrected for contributions from the overlapping L_{II} (Si $2p_{1/2}$ → CB) transitions. The silicon $2p_{3/2}$–$2p_{1/2}$ spin–orbit splitting is 0.60 ± 0.05 eV (Ley *et al.*, 1982).

The threshold of the L_{III} ($2p_{3/2}$) absorption (point of maximum slope, E_T) is 99.90 ± 0.05 eV for c-Si and 99.85 ± 0.05 eV for sputtered a-Si. In a-Si : H it is shifted by 0.2 eV to 100.10 eV. The values obtained by Brown and Rustgi (1972) from direct absorption measurements are 99.84 ± 0.06 eV for c-Si and a-Si. The structures between 100 and 103 eV in the spectrum of c-Si

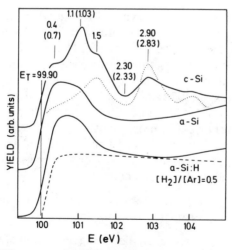

FIG. 16. The L_{III} (Si $2p_{3/2}$ → conduction band) partial yield spectra of c-Si, a-Si, and a-Si : H. The amorphous films were prepared by reactive sputtering without and with 50 vol % H_2, respectively. Energies above threshold are given for characteristic features in the spectrum of c-Si and the corresponding numbers obtained by Brown and Rustgi (1972) are added in parentheses. The dotted line is the density of conduction states calculated for c-Si by Kane (1966), and the dashed line indicates the one-electron density of conduction states appropriate for a-Si.

are in reasonable agreement with those obtained by Brown and Rustgi who identify them with transitions to maxima in the conduction DOS (Kane, 1966) (dotted line in Fig. 16). These critical points are, of course, absent in the spectra of the noncrystalline modifications. The complete loss of structure in these spectra — aside from the 1.3 eV-wide hump at the onset — is in keeping with the complex band-structure results of Kramer (1971): the loss of long-range order is sufficient to wipe out all structure in the conduction DOS.

A remarkable result is the complete lack of discernible hydrogen-derived antibonding states in the yield spectrum of a-Si : H. The antibonding states have predominantly Si sp^3 character since the bonding states are mainly H 1s derived. They are expected to lie at the bottom of the conduction bands (Ching et al., 1980; Economou and Papaconstantopoulos, 1981). According to the calculation of Ching et al. (1980), the hybridization of the antibonding states is such that the Si 3p partial DOSs exceed the Si 3s ones by a factor of about 4 to 5. The Si–H antibonding states are therefore expected to be weak in the 2p yield spectrum according to the dipole selection rule. It is therefore not unlikely that the weak Si–H antibonding states are hidden under the initial hump in the yield spectrum of a-Si : H. An altogether satisfying explanation of the partial yield spectrum of a-Si : H is nevertheless still lacking.

It has been pointed out by Brown and Rustgi (1972) that the initial rise and the region up to about 1 eV above threshold is greatly enhanced over the density of states in the $L_{II,III}$ spectrum of c-Si. They ascribe this enhancement to the strong Coulomb interaction between the core hole and conduction electron — a view that has been subsequently confirmed by the calculations of Altarelli and Dexter (1972).

Thus the spectra of Fig. 16 do not represent the one-electron density of conduction states. For a-Si a more realistic shape would be a simple step function as indicated by the dashed line in Fig. 16. A similar DOS was suggested by Spicer and Donovan (1970) for a-Ge, based on photoemission spectra taken at different photon energies.

8. The Influence of Hydrogen on the Energies of Valence- and Conduction-Band Edges

The spectra of Fig. 12 reveal an increasing separation of the valence-band edge from the Fermi level with the addition of hydrogen to a-Si. When measured at the point of maximum slope, this shift amounts to 1 eV, as indicated in Fig. 12, for the sample with ~ 50 at. % hydrogen. Von Roedern et al. (1977) proved that the shift is the result of a recession of the valence-band edge and not a shift of E_F within the gap because no comparable change in the binding energy of the Si 2p core levels was observed. They also showed that the correlation between optical gap and hydrogen concentration could be explained by the recession of the valence band alone (von Roedern *et al.*, 1978). Recently the position of the valence-band maximum (VBM), of E_F and the L_{III} threshold E_T have been simultaneously measured by using synchrotron radiation (Reichardt *et al.*, 1983). The results of these measurements are summarized, in the form of energy diagrams, in Fig. 17. All energies are given relative to the unshifted component (see Section 2) of the Si $2p_{3/2}$ core level. In this way, mere Fermi-level shifts within the gap can be distinguished from changes in the VBM and E_T, brought about by a redistribution of states near the band edges. The VBM is defined in the usual way through the extrapolation of the steepest descent of the leading edge of the valence-band spectrum, and the definition of E_T has been explained in Section 7. Also given in Fig. 17 is the value of the gap E_g, which is defined as

$$E_g = E_T - \text{VBM} + E_{xc}, \qquad (9)$$

where we used the value of 150 meV for the core-level exciton binding energy E_{xc} derived from partial yield spectra of c-Si (Reichardt *et al.*, 1983).

The energy levels so defined are plotted in Fig. 17 for four amorphous silicon films and for the (111) surface of c-Si. The first three amorphous samples are a hydrogen-free film sputtered at room temperature (sample A) and two films sputtered at RT with different amounts of H_2 in the sputter gas (samples B and C). Finally, film D is sample C after it has been annealed at

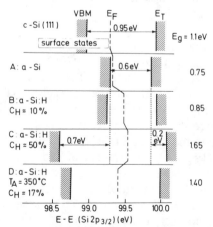

FIG. 17. The energies of the valence-band maximum (VBM), the Fermi level (E_F) and the photoemission threshold (E_T) plotted for four samples relative to the binding energy of the Si $2p_{3/2}$ core level. The value of E_g was obtained from $E_T - \text{VBM} + E_{xc}$. The hydrogen concentrations quoted are deduced from the intensities of the chemically shifted Si 2p lines as explained in the text. [After Reichardt *et al.* (1983). Copyright North-Holland Publ. Co., Amsterdam, 1983.]

350°C for ~20 min. The hydrogen concentrations C_H in Fig. 17 were obtained from the intensities of the chemically shifted Si 2p lines as explained in Section 3. Sample D has H only in monohydride configurations. The results of Fig. 17 may be summarized as follows:

(1) The top of the valence bands recedes by as much as 0.7 eV with increasing hydrogen content. This is 0.3 eV less than the recession measured at a point halfway up the leading edge (compare Fig. 12), because the slope of the leading edge decreases also with hydrogenation. In a-Ge:H and a-Si:F (Gruntz et al., 1980; Gruntz et al., 1981) a similar narrowing of the valence bands by 0.3 and 0.8 eV, respectively, and a concomitant decrease in the slope of the valence-band edge, has been observed.

(2) The conduction-band edge (E_T) is by comparison little affected. The maximum recession is 0.2 eV compared to its position in unhydrogenated a-Si. In fact, the position of E_T is within that same margin equal in a-Si and c-Si.

(3) The Fermi level is pinned near VBM in a-Si films. The distance between E_F and VBM may vary between 0 and 0.2 eV for undoped specimens. The smallest amount of hydrogen added to the film frees E_F to move toward the middle of the gap. Further addition of hydrogen results only in minor movements of E_F in such a way that the difference $E_T - E_F$ remains virtually constant at 0.52 ± 0.07 eV, corresponding to an energy of 0.67 eV below the conduction-band edge after correction for the exciton binding energy.

The recession of the VBM with hydrogenation is in agreement with the results of most calculations (Ching et al., 1979, 1980; Picket, 1981; Allan et al., 1982; DiVincenzo et al., 1981a,b). This recession is ascribed to the replacement of the Si–Si bond with the stronger Si–H or Si–F bond that moves states from the top of the valence bands to a position deep inside the valence bands, where they are observed as peaks A–E. The depletion of states is not limited to the Si atoms bonded directly to H or F. According to DiVincenzo et al. (1981a,b), the presence of hydrogen on a dangling bond reduces the bonding charge associated with valence states in the top 0.25 eV of the valence bands out to the third nearest neighbor. That is the reason why we observe an actual *recession* of VBM even for small hydrogen concentrations and not just an attenuation in the emission near the top of the valence bands that is due to the few atoms bonded directly to H or F. The bottom of the conduction bands is little affected by the addition of H, which is in agreement with the calculations (Economou and Papaconstantopoulos, 1981; DiVincenzo, 1981a,b), and the shift in VBM accounts for almost all the change in the optical gap E_G with hydrogen content. This result suggests that fluctuations in hydrogen content as they occur apparently in most samples (Knights and Lujan, 1979; Reimer *et al.*, 1980; Knights, 1980) will lead to spatial fluctuations of the valence-band edge and as a consequence to localized states near the valence-band edge. The localized states exist in regions with low hydrogen concentration and it has been suggested (Allan and Joannopoulos, 1980) that they are responsible for the E_y peak in the density of gap states derived from field-effect measurements (Madan *et al.*, 1976). Moreover, these states could act as deep traps for holes and thus provide a natural explanation for the large disparity between hole and electron mobilities (LeComber *et al.*, 1972; Moore, 1977). Brodsky (1980) has taken these considerations further and discusses tail states localized in quantum wells that are formed by regions with low H content surrounded by those with high H content.

The pinning of E_F near the VBM in hydrogen-free a-Si is in agreement with the observation of Beyer *et al.* (1975) that these samples are *p*-type. The activation energy of 0.77 eV measured by Beyer *et al.* (1975) at high temperatures requires, however, that the mobility edge E_v lies about 0.5 to 0.6 eV below the VBM and that the states between E_v and VBM are localized. A region of localized states below VBM that is at least 0.3 eV wide is necessary in sample B to ensure the *n*-type conductivity that is generally observed for a-Si : H. It is not possible to place E_v in the spectra of samples C and D.

The position of E_F in sample A could be due to the high bulk defect density in the lower half of the gap; it may, however, equally well reflect the surface position of E_F. In c-Si the Fermi level is pinned 0.35 ± 0.15 eV above the valence-band edge by a band of surface states (Rowe *et al.*, 1975), as indicated in Fig. 17. It is possible that a similar band of surface states pins E_F in unhydrogenated a-Si and that E_F moves toward midgap as these surface states are passivated by hydrogen. The surface charge Q/e necessary to induce a band-bending V_s is (Solomon *et al.*, 1978)

$$Q/e = -V_s \sqrt{\varepsilon \varepsilon_0 g}, \tag{10}$$

where g, the density of gap states, is assumed to be constant. With $g = 10^{20}$ eV^{-1} cm^{-3} and $V_s = 0.19$ eV (the shift in E_F between samples A and B), we have $Q/e = 6 \times 10^{12}$ cm^{-2}. This is a reasonable number corresponding to 1 electron per 100 surface atoms, and it would reduce the width of localized states below VBM in the bulk of sample A to 0.3–0.4 eV—a value comparable to that in sample B. Alternatively, one could argue that the states at the VBM in sample A correspond to inner surface states associated with voids. These states are then primarily saturated when a-Si is hydrogenated. The question of surface states in a-Si : H has been discussed by Ley (1983) and by Miller *et al.* (1981).

9. VALENCE-BAND TAILS AND THE DISTRIBUTION OF OCCUPIED GAP STATES IN a-Si: H

Photoelectron spectroscopy is — in principle — suited to determining the energy distribution of occupied tail and gap states in a-Si: H. The dynamical range of PES measurements is, however, limited to about two orders of magnitude in the density of valence states $N_v(E)$ due to a background of inelastically scattered electrons. These electrons originate from higher energy photons that cannot be sufficiently suppressed in the photon sources used in conventional PES. As a result, tail and gap states in a-Si: H can only be explored down to about 10^{20} eV^{-1} cm^{-3} as demonstrated by von Roedern et al. (1979b). Griep and Ley (1983) succeeded, however, in extending this range down to 2×10^{16} eV^{-1} cm^{-3} using a precursor of PES known as photoelectric yield measurements. This technique, originally developed to determine the work function of metals (Fowler, 1931), was later adapted to study surface states on semiconductors (Sébenne et al., 1975). The measured quantity is the total photoelectric current emitted by the sample as the energy of a highly monochromatized photon beam is scanned through the threshold for photoemission ($\equiv$VL-E_F in Fig. 3). It turns out that for amorphous samples the differential of the yield with respect to the photon energy is proportional to the DOS. Applied to a-Si: H the measurements reveal an exponential valence-band tail that extends over two to four orders of magnitude in $N_v(E)$ depending on preparation condition and a band of defect states centered about 0.4 above VBM (as determined in conventional PES). The logarithmic slope of the tail E_0 has a value of 75 meV for samples prepared at RT, reaches a minimum of 50 meV for $T_D = 220°C$ and rises again for higher deposition temperatures. The density of defect states drops similarly from 10^{19} ($T_D =$ RT) to 1.5×10^{17} eV^{-1} cm^{-3} ($T_D = 220°C$) and increases again steeply with $T_D > 220°C$. The incorporation of phosphorus, finally, raises the defect density dramatically, and a value of 3×10^{20} eV^{-1} cm^{-3} is reached for a [PH$_3$]/[SiH$_4$] ratio of 0.06.

These results resemble the optical absorption edge data (Cody et al., 1980; Jackson and Amer, 1981, 1982) so closely that there remains little doubt that the latter represent mainly the state distribution at the valence-band edge. This implies also that the conduction-band edge exhibits considerably less tailing as suggested on the basis of other evidence by Tiedje et al. (1981). The quantitative agreement between the logarithmic slope of the Urbach edge in the optical absorption data (Cody et al., 1981) and the E_0 values of the valence-band tails (Griep and Ley, 1983) poses rather stringent restrictions on any theory of the Urbach edge in a-Si: H. This point has been discussed at length elsewhere (Ley, 1983).

V. The Effect of Doping on the Position of E_F

The position of E_F within the gap varies with doping and with temperature. Changes in the activation energy E_σ obtained from temperature-dependent dark conductivities are interpreted in terms of changes in the position of E_F relative to a mobility edge. An implicit assumption usually made is that E_F shifts linearly with T as

$$E_F(T) = E_F(0) - \beta T. \tag{11}$$

In that case the term $\beta T/kT$ is temperature independent and $E_\sigma = (E_F - E_v)_0$ or $(E_c - E_F)_0$ where E_c and E_v are the mobility edges, and the subscript indicates that the energies are appropriate for zero temperature.

The position of E_F as a function of doping has been investigated by Williams *et al.* (1979) and by von Roedern *et al.* (1979a), following the method explained in connection with Fig. 3. Williams *et al.* (1979) determined the change in the binding energy of the Si 2p core levels relative to E_F for three samples: one undoped, one heavily P doped, and one heavily B doped. The samples were prepared by glow discharge outside the spectrometer and their surfaces were oxidized when measured. The binding energy difference was 1.2 eV between the two doped samples, which is in agreement with the 1.2 eV difference in E_σ obtained from conductivity measurements performed on the same samples. They concluded therefore that changes in E_σ were indeed due to corresponding shifts in E_F.

Von Roedern *et al.* (1979a) came essentially to the same conclusion from measurements on a series of GD a-Si:H ($T_D = 250°C$, $T_A = 350°C$) films prepared *in situ* and therefore unoxidized. This enabled von Roedern *et al.* to follow the separation between E_F and the top of the valence bands (E_v in Fig. 18) in addition to $E_F - E_B$(Si 2p). The two quantities track each other well as a function of doping, and the total variation in E_F of 0.75 eV agrees with the change in E_σ observed for the same samples. By adding (or subtracting for *p*-type samples) the activation energies to (or from) E_F, von Roedern *et al.* obtained the average energies of the conduction paths as indicated by the hatched regions in Fig. 18. For *n*-type samples this region coincides with E_c. For *p*-type samples a region 0.3 eV above, the VBM appears to carry the hole current. This is a direct spectroscopic confirmation of the analysis of transport data by Persans (1980), who used the formalism proposed by Döhler (1980). The average energy $\bar{\varepsilon}$ of the conductance path defined through the Kubo–Greenwood formula for $\sigma(T)$ is (Persans, 1980)

$$\bar{\varepsilon} = \int \varepsilon e^{-\varepsilon/kT} g(\varepsilon)\mu(\varepsilon)\, d\varepsilon \bigg/ \int e^{-\varepsilon/kT} g(\varepsilon)\mu(\varepsilon)\, d\varepsilon, \tag{12}$$

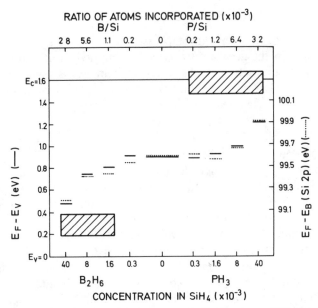

FIG. 18. Position of the Fermi level E_F in the gap of a-Si : H as determined spectroscopically for a range of P- and B-doped samples. The hatched regions indicate the energies at which charge transport takes place as deduced by combining the activation energies with E_F. The conduction-band mobility edge E_c is derived by adding an optical gap of 1.6 eV to the valence-band maximum E_v. The number of B or P atoms incorporated (upper abscissa) has been derived as described in the text. [Adapted with permission of *Solid State Communications*, Vol. 29, by B. von Roedern, L. Ley, and M. Cardona, Spectroscopic determination of the position of the Fermi-level in doped amorphous hydrogenated silicon. Copyright 1979, Pergamon Press, Ltd.]

where all energies are measured from E_F and $\mu(\varepsilon)$ is the energy-dependent carrier mobility. A large Boltzmann factor (small ε) in conjunction with a large $g(\varepsilon)$ may thus shift $\bar{\varepsilon}$ into a region of low mobility, i.e., into the tails. A high density of gap states $g(\varepsilon)$ near VBM has been observed in undoped samples by von Roedern *et al.* (1979b) and by Griep and Ley (1983), and there is evidence that B doping increases this density further (Overhof and Beyer, 1980). The exact character of the tail states and the transport mechanisms in these states is still a matter of controversy (LeComber *et al.*, 1972; Persans, 1980). There appears to be no comparable tailing of the conduction-band edge (cf. Section 9).

Another aspect of Fig. 18 worth noting is the very small shifts of E_F obtained for low dopant concentrations. Significant movement sets in only for $[PH_4]/[SiH_4] \geq 8 \times 10^{-3}$ and $[B_2H_6]/[SiH_4] \gtrsim 1.6 \times 10^{-3}$. This is in contrast to the variation in E_σ with doping: rapid changes for small concentrations and saturation for concentrations approaching 1×10^{-3} in the gas phase (Spear and LeComber, 1976). It is possible that this difference is due to a near-surface density of gap states $g_s(E)$ that differs from that of the bulk. The photoemission measurements are, of course, sensitive to the position of E_F within the first 20 Å or so, whereas E_σ measures the shift of E_F in the bulk unless we are dealing with an accumulation layer. The maximum difference of 0.2 eV (the width of the hatched regions) allowed between ΔE_F and ΔE_σ according to Fig. 18 corresponds to a difference of 2×10^{19} eV^{-1} cm^{-3} between $g_s(E)$ and $g(E)$. For a Debye length of 100 Å this difference may be accounted for by one true surface state for every 50 surface atoms. It seems worthwhile to pursue this question further using more accurate data in view of the known fact that the surfaces of a-Si:H films exhibit electrical properties that differ from those of the bulk (Ast and Brodsky, 1980; see also Chapter 9, by Fritzsche, in Volume 21C.). Williams *et al.* (1979) conclude from a similar analysis for their films that the density of states at the a-Si:H–SiO$_2$ interface was less than 10^{11} cm^{-2}.

That doping shifts the Fermi level in a-Ge:H in much the same way as in a-Si:H has been demonstrated by Gruntz *et al.* (1980), using the Ge 3d$_{5/2}$ and the Ar 3p core levels. Their results indicate a maximum shift of 0.5 eV between heavily B- and P-doped samples. This is much smaller than the band gap of a-Ge:H of about 1.0 eV.

We have already mentioned that the actual concentration of dopant atoms has also been determined by von Roedern *et al.*, who used the relative intensities of the B 1s, P 2p, and Si 2p core levels of the most heavily doped films (Section 1). The result is indicated at the top of Fig. 18.

The temperature dependence of E_F — the so-called statistical shift — has also been determined using photoemission (Ley, 1983). The result indicates a deep minimum in the density of gap states at approximately $E_v + 1.2$ eV in agreement with the analysis of Overhof and Beyer (1980) and recent field effect and deep level transient spectroscopy (DLTS) measurements (Grünewald *et al.*, 1981; Lang *et al.*, 1977).

VI. Hydrogen Enrichment of the a-Si:H Surface

Since photoemission is a rather surface-sensitive method, it is ideally suited to study the growth interface of a film or the oxidation of a-Si and a-Si:H in comparison with c-Si. The latter has been comprehensively discussed (Ley *et al.*, 1981a), with the main result that a-Si:H has a sticking coefficient for O$_2$ that is orders of magnitude lower than that of c-Si, whereas the rate of oxidation of unhydrogenated a-Si is the same as that of c-Si (cf. Section 1 and Fig. 4). Valence-band spectra of oxidized a-Si and a-Si:H have been published by Kärcher and Ley (1982) and by Miller *et al.* (1981).

Here we should like to discuss briefly the hydrogen enrichment of the growth surface of a-Si : H, since this is expected to have far-reaching consequences for the properties of any device based on a-Si : H.

Broadly speaking, the photoelectron spectrum of clean a-Si : H surfaces distinguishes only between a polyhydride phase characterized by peaks A – B and a monohydride phase with structures C – E (cf. Section 6). The polyhydride phase is always observed for specimens prepared at temperatures below ~ 300°C and is independent of the deposition method used (glow discharge, dc or rf sputtering). It has therefore been conjectured that the polyhydride phase is indigenous to the growing a-Si : H surface and is independent of the concentration and prevailing bonding configurations of H in the bulk of the specimen (Ley *et al.*, 1981a). The conjecture has been substantiated by a series of experiments in which photoemission provided the surface-sensitive method and IR spectroscopy was the means of determining the concentration and bonding geometries of hydrogen in the bulk of the specimens. Figure 19 shows the IR and HeII valence-band spectra of three a-Si : H specimens. The photoemission spectra were taken during the growth of the specimens and the IR spectra after the films had grown to a thickness d of between 0.7 and 1.4 μm. All films were prepared by rf glow

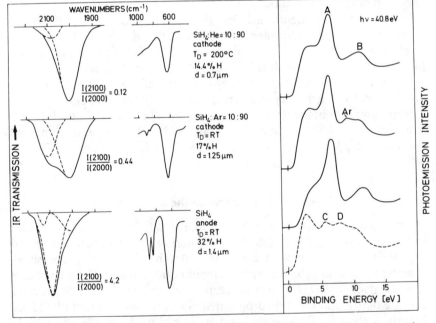

FIG. 19. Infrared and valence-band spectra of three a-Si : H films prepared in such a way that the bulk Si – H bonding configurations are dominated by monohydride (a), polyhydrides (c), and a mixture of both (b). The photoelectron spectrum containing only monohydrides at the surface ($T_D = 350°C$) is shown for comparison by the dashed line.

discharge in situ. The deposition conditions are also specified in Fig. 19. They are varied in such a way that for sample (a) the hydrogen is almost exclusively bonded in the form of monohydride as shown by the weak absorption at 2100 cm^{-1} and the absence of bending modes around 850 cm^{-1} (Lucovsky et al., 1979). In sample (c) the polyhydride configurations Si–H$_2$, Si–H$_3$, and possibly (SiH$_2$)$_n$, are dominant, and sample (b) finally contains a mixture of poly- and monohydrides as indicated by the comparable strengths of the 2100 and 2000 cm^{-1} absorption bands. The hydrogen content of these films was determined from the strength of the 640 cm^{-1} wagging band according to the description of Shanks et al. (1980).

Despite the considerable variations in hydrogen bonding and hydrogen concentration in the bulk of these samples, the valence-band spectra of all three samples are dominated by the signature of the polyhydride configurations. The hydrogen concentration furthermore lies between about 40 and 50 at. % within the sampling depth of ~ 10 Å, using the estimate described in Section 6. This exceeds the bulk concentration by factors of 2.5 and 3 for samples (b) and (a), respectively. It cannot be ruled out from the data that monohydrides are also present at the surface, but they are clearly of secondary importance. A fivefold increase in the hydrogen content at the surface of a-Si:H compared with subsurface layers has been directly measured by Reichardt et al. (1983) using the analysis of the Si 2p core-level spectra as mentioned at the end of Section 3. The correlation between VBM and the hydrogen content leads to a 0.2 eV increase in E_F-VBM at the surface over its bulk value as measured by Reichardt et al. (1983). This change in the band structure at the surface of a-Si:H has certainly far reaching consequences for many of its applications.

These results are not in contradiction to the hydrogen-profiling data of Müller et al. (1980).These authors found a subsurface hydrogen depletion layer ~ 1000 Å thick but dismissed a two- to threefold increase in hydrogen concentration right at the surface (with a 30-Å resolution) as being due to adsorbed water.

The observations just described are consistent with the growth model for a-Si:H proposed by Kampas and Griffith (1980, 1981) and by Scott et al. (1980). In this model (Fig. 20) the active species in the gas phase is the SiH$_2$ radical. It "inserts" itself into a Si–H bond at the surface and forms a SiH$_3$ surface species. The cross linking of the SiH$_{2,3}$ units at the surface occurs through hydrogen elimination reactions as indicated in Fig. 20. In this model a surface rich in Si–H$_2$ and Si–H$_3$ units is thus pushed ahead of the growing film. The fluorine enrichment of the a-Si:F films discussed in Section 3 suggests that a similar growth mechanism, with SiF$_2$ taking the place of SiH$_2$, might apply in this case.

At deposition temperatures above ~ 300°C, the a-Si:H surfaces exhibit only monohydride species. Whether this is due to an annealing process that

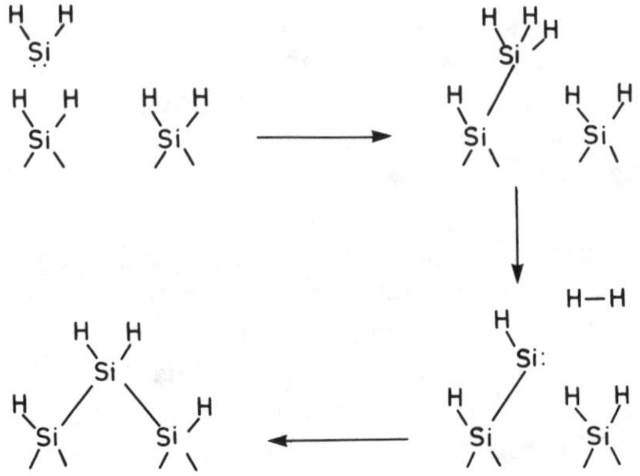

FIG. 20. The a-Si:H growth model due to Kampas and Griffith (1980).

occurs faster at the surface than it takes to transfer the sample from the deposition chamber to the measuring chamber (~ 1 min) or is due to a different growth mechanism remains to be seen (Kampas and Griffith, 1981).

ACKNOWLEDGMENTS

It is a pleasure to acknowledge the many and substantial contributions to the material presented here made by my former and present collaborators M. Cardona, K. J. Gruntz, R. L. Johnson, R. Kärcher, J. Reichardt, H. Richter, and B. von Roedern.

REFERENCES

Adler, D. (1978). *Phys. Rev. Lett.* **41**, 1755.
Alben, R., Goldstein, S., Thorpe, M. F., and Weaire, D. (1972). *Phys. Status Solidi B* **53**, 545.
Allan, D. C., and Joannopoulos, J. D. (1980). *Phys. Rev. Lett.* **44**, 43.
Allan, D. C., Joannopoulos, J. D., and Pollard, W. B. (1982). *Phys. Rev. B* **25**, 1065.
Allie, G., Lauroz, C., and Chenevas-Paule, A. (1980). *J.Non-Cryst. Solids* **35/36**, 267.
Altarelli, M., and Dexter, D. L. (1972). *Phys. Rev. Lett.* **29**, 1100.
Appelbaum, J. A., and Hamann, D. R. (1976). *Rev. Mod. Phys.* **48**, 479.
Ast, D. G., and Brodsky, M. H. (1980). *J. Non-Cryst. Solids* **35/36**, 611.
Band, I. M., Kharitonov, Y. I., and Trzhaskovkaya, M.B. (1979). *At. Data Nucl. Data Tables* **23**, 443.
Berthou, H., and Jørgensen, C. K. (1975). *Anal. Chem.* **47**, 482.
Beyer, W., Stuke, J., and Wagner, H. (1975). *Phys. Status Solidi A* **30**, 231.
Brennan, S., Stöhr, J., Jaeger, R., and Rowe, J. E. (1980). *Phys. Rev. Lett.* **45**,1414.
Brodsky, M. H. (1980). *Solid State Commun.* **36**, 55.
Brodsky, M. H., and Kaplan, D. (1979). *J. Non-Cryst. Solids* **32**, 431.
Brown, F. C., and Rustgi, O. P. (1972). *Phys. Rev. Lett.* **28**, 497.

Brundle, C. R. (1974). *J. Electron Spectrosc. Relat. Phenom.* **5,** 291.
Brundle, C. R., and Baker, A. D., eds. (1977, 1978). "Electron Spectroscopy," Vols. 1 and 2. Academic Press, London.
Brust, D. (1969). *Phys. Rev.* **186,** 768.
Cardona, M., and Ley, L., eds. (1978). "Photoemission in Solids I." Springer-Verlag, Berlin, Heidelberg, and New York.
Carlson, T. A. (1975). "Photoelectron and Auger Spectroscopy." Plenum, New York.
Ching, W. Y., Ling, C. C., and Huber, D. L. (1976). *Phys. Rev. B* **14,** 620.
Ching, W. Y., Ling, C. C., and Guttmann, L. (1977). *Phys. Rev. B* **16,** 5488.
Ching. W. Y., Lam, D. J., and Lin, C. C. (1979). *Phys. Rev. Lett.* **42,** 805.
Ching, W. Y., Lam, D. J., and Lin, C. C. (1980). *Phys. Rev. B* **21,** 2378.
Cody, G.D., Abeles, B., Wronski, C. R., Stephens, R. B., and Brooks, B. (1980). *Solar Cells* **2,** 277.
Cody, G. D., Tiedje, T., Abeles, B., Brooks, B., and Goldstein, Y. (1981). *Phys. Rev. Lett.* **47,** 1480.
DiVincenzo, D. P., Bernholc, J., Brodsky, M. H., Lipari, N. O., and Pantelides, S. T. (1981a). *AIP Conf. Proc.* **73,** 156.
DiVincenzo, D. P., Bernholc, J., and Brodsky, M. H. (1981b). *J. Phys. Colloq. Orsay, Fr.* **42,** C4-137.
Döhler, G. H. (1980). *J. Non-Cryst. Solids* **35/36,** 363.
Eastman, D. E., and Grobman, W. D. (1972). *Proc. Int. Conf. Phys. Semicond.,* p. 889. PWN Polish Scientific Publ., Warsaw.
Economou, E. N., and Papaconstantopoulos, D. A. (1981). *Phys. Rev. B* **23,** 2042.
Elliot, I., Doyle, C., and Andrade, J. D. (1983). *J. Electron Spectrosc. Relat. Phenom.* **28,** 303.
Fadley, C. S., Hagström, S. B. M., Klein, M. P., and Shirley, D. A. (1968). *J. Chem. Phys.* **48,** 3779.
Fadley, C. S. (1978). *In* "Electron Spectroscopy" (C. R. Brundle, A. D. Baker, eds.) Vol. 2, Chap. 1. Academic Press, London.
Fang, C. J., Ley, L., Shanks, H. R., Gruntz, K. J., and Cardona, M. (1980). *Phys. Rev. B* **22,** 6140.
Fowler, R. H. (1931). *Phys. Rev.* **38,** 45.
Fujiwara, K. (1981). *Phys. Rev. B* **24,** 2240.
Gelius, U., Heden, P.-F., Hedman, J. Lindberg, B. J., Manne, R., Nordberg, R., Nordling, C., and Siegbahn, K. (1970). *Phys. Scr.* **2,** 70.
Gray, R. C., Carver, J. C., and Hercules, D. M. (1976). *J. Electron Spectrosc. Relat. Phenom.* **8,** 343.
Griep, S., and Ley, L. (1983). *J. Non-Cryst. Solids* **59/60,** 253.
Grünewald, M., Weber, K. Fuhs, W., and Thomas, P. (1981). *J. Phys. Colloq. Orsay, Fr.* **42,** C4-523.
Grunthaner, F. J., Grunthaner, P. J., Vasquez, R. P., Lewis, B. F., Maserjian, J., and Madhukar, A. (1979). *Phys. Rev. Lett.* **43,** 1683.
Gruntz, K. J., Ley, L., Cardona, M., Johnson, R., Harbeke, G., and von Roedern, B. (1980). *J. Non-Cryst. Solids* **35/36,** 453.
Gruntz, K. J., Ley, L., and Johnson, R. L. (1981). *Phys. Rev. B* **24,** 2069.
Gudat, W., and Kunz, C. (1972). *Phys. Rev. Lett.* **29,** 169.
Guttman, L., Ching, W. Y., and Rath, J. (1980). *Phys. Rev. Lett.* **44,** 1513.
Hagström, S. Nordling, C., and Siegbahn, K. (1964). *Z. Phys.* **178,** 439.
Haneman, D. (1975). *In* "Surface Physics of Phosphors and Semiconductors" (C. G. Scott, C. E. Reed, eds.), p. 2. Academic Press, London.
Herman, F., and van Dyke, J. P. (1968). *Phys. Rev. Lett.* **21,** 1575.

424　　　L. LEY

Himpsel, F. J., Heimann, P., Chiang, T. C., and Eastman, D. E. (1980). *Phys. Rev. Lett.* **45**, 1112.

Ho, K. M., Cohen, M. L., and Schlüter, M. (1977). *Phys. Rev. B* **15**, 3888.

Hollinger, G., Jugnet, Y., Pertosa, P., and Tran Minh Duc (1975). *Chem. Phys. Lett.* **36**, 441.

Ibach, H. (1977). *In* "Electron Spectroscopy for Surface Science," (H. Ibach, ed.) p. 1. Springer-Verlag, Berlin and New York.

Jackson, W. B., and Amer. N. M. (1981). *J. Phys. Colloq. Orsay, Fr.* C4-293.

Jackson, W. B., and Amer, N. M. (1982). *Phys. Rev. B* **25**, 5559.

Joannopoulos, J., and Cohen, M. H. (1973). *Phys. Rev. B* **8**, 2733.

Joannopoulos, J., and Cohen, M. H. (1976). *Solid State Phys.* **31**, 71. Academic Press, New York.

Kampas, F. J., and Griffith, R. W. (1980). *Solar Cells* **2**, 385.

Kampas, F. J., and Griffith, R. W. (1981). *In* "Tetrahedrally Bonded Amorphous Semiconductors" (R. A. Street, D. K. Biegelsen, and J. C. Knights, eds.), p. 1. Amer. Inst. of Phys., New York.

Kane, E. O. (1966). *Phys. Rev.* **146**, 558.

Kärcher, R., and Ley, L. (1982). *Solid State Commun.* **43**, 415.

Kärcher, R., Johnson, R. L., and Ley, L. (1983). *J. Non-Cryst. Solids* **59/60**, 593.

Katayama, Y., Shimada, I., and Usami, K. (1981a). *Phys. Rev. Lett.* **46**, 1146.

Katayama, Y., Usami, K., and Shimada, T. (1981b). *Philos. Mag.* **43**, 283.

Kelfve, P., Blomster, B., Siegbahn, H., Siegbahn, K., Sanhueza, E., and Gosczinski, O. (1980). *Phys. Scr.* **21**, 75.

Kelly, M. J., and Bullet, D. W. (1976). *J. Non-Cryst. Solids* **21**, 155.

Klemperer, O. (1971). "Electron Optics." Cambridge Univ. Press, Cambridge.

Knights, J. C., and Lujan, R. A. (1979). *Appl. Phys. Lett.* **35**, 244.

Knights, J. C. (1980). *J. Non-Cryst. Solids* **35/36**, 159.

Koopmans, T. (1933). *Physica Amsterdam* **1**, 104.

Kramer, B. (1971). *Phys. Status Solidi B* **47**, 501.

Kramer, B., King, H., and MacKinnon, A. (1983). *Physica Amsterdam* **117B/118B**, 944.

Kunz, C. (1976). *In* "Optical Properties of Solids, New Developments" (B. O. Seraphin, ed.), p. 473. North-Holland, Amsterdam.

Lang, D. V., Cohen, J. D., and Harbison, J. P. (1977). *Phys. Rev. Lett.* **31**, 292.

Leckey, R. C. G. (1976). *Phys. Rev. A* **13**, 1043.

LeComber, P. G., Madan, A., and Spear, W. E. (1972). *J. Non-Cryst. Solids* **11**, 219.

Lemaire, P., and Gaspard, J. P. (1981). *J. Phys. Colloq. Orsay, Fr.* **42**, C4-765.

Ley, L., Kowalczyk, S. P., Pollak, R., and Shirley, D. A. (1972). *Phys. Rev. Lett.* **29**, 1088.

Ley, L., Kowalczyk, S. P., McFeely, F. R., Pollak, R. A., and Shirley, D. A. (1973). *Phys. Rev. B* **8**, 2392.

Ley, L., and Cardona, M. (eds.) (1979). "Photoemission in Solids II." Springer-Verlag, Berlin, Heidelberg, and New York.

Ley, L., Cardona, M., and Pollak, R. A. (1979). *In* "Photoemission in Solids II" (L. Ley and M. Cardona, eds.), Chap. 2. Springer-Verlag, Berlin, Heidelberg, and New York.

Ley, L., Richter, H., Kärcher, R., Johnson, R. L., and Reichardt, J. (1981a). *J. Phys. Colloq. Orsay, Fr.* **42**, C4-753.

Ley, L., Gruntz, K. J., and Johnson, R. L. (1981b). *In* "Tetrahedrally Bonded Amorphous Semiconductors" (R. A. Street, D. K. Biegelsen, J. C. Knights, eds.), p. 161. Amer. Inst. of Phys., New York.

Ley, L., Reichardt, J., and Johnson, R. L. (1982). *Phys. Rev. Lett.* **22**, 1664.

Ley, L. (1984). *In* "The Physics of Amorphous Silicon and its Applications" (J. Joannopoulos, G. Lucovsky, eds.). Springer-Verlag, Berlin and Heidelberg.

Lucovsky, G., Nemanich, R. J., and Knights, J. C. (1979). *Phys. Rev. B* **19**, 2064.

Madan, A., LeComber, P. G., and Spear, W. E. (1976). *J. Non-Cryst. Solids* **20**, 239.

Menzel, D. (1978). *In* "Photoemission and the Electronic Properties of Surfaces" (B. Feuerbacher, B. Fitton, R. F. Willis, eds.), Chap. 13. John Wiley & Sons, Chichester.

Miller, J. N., Lindau, I., and Spicer, W. E. (1981). *Philos. Mag. B* **43**, 273.

Moore, A. R. (1977). *Appl. Phys. Lett.* **31**, 766.

Moss, S. C., and Graczyk, J. F. (1969). *Phys. Rev. Lett.* **23**, 1167.

Müller, G., Demond, F., Kalbitzer, S., Damjantschitsch, H., Mannsperger, H., Spear, W. E., LeComber, P. G., and Gibson, R. A. (1980). *Philos. Mag. B* **41**, 571.

Ohdomari, I., Ikeda, M., and Yoshimoto, H. (1977). *Phys. Lett. A* **64**, 253.

Ortenburger, I. B., Rudge, W. E., and Herman, F. (1972). *J. Non-Cryst. Solids* **8–10**, 653.

Overhof, H., and Beyer, W. (1980). *J. Non-Cryst. Solids* **35/36**, 375.

Pandey, K. C. (1976). *Phys. Rev. B* **14**, 1593.

Pandey, K.C., Sakurai, T., and Hagstrum, H. D. (1975). *Phys. Rev. Lett.* **35**, 1728.

Penn, D. (1962). *Phys. Rev.* **128**, 2093.

Persans, P. D. (1980). *J. Non-Cryst. Solids* **35/36**, 369.

Phillips, J. C. (1979). *Phys. Rev. Lett.* **42**, 1151.

Picket, W. E. (1981). *Phys. Rev. B* **23**, 6603.

Pierce, D. T., and Spicer, W. E. (1972). *Phys. Rev. B* **5**, 3017.

Powell, C. J. (1978). *In* "Quantitative Surface Analysis of Materials" (N. S. McIntyre, ed.), p. 5. ASTM STP-643.

Reichardt, J., Ley, L., and Johnson, R. L. (1983). *J. Non-Cryst. Solids* **59/60**, 329.

Reimer, J. A., Vaughn, R. W., and Knights, J. C. (1980). *Phys. Rev. Lett.* **44**, 1936.

Reimer, J. A., Vaughn, R. W., and Knights, J. C. (1981). *Solid State Commun.* **37**, 161.

Rowe, J. E., Ibach, H., and Froitzheim, H. (1975). *Surf. Sci.* **48**, 44.

Sakurai, T., and Hagstrum, H. D. (1976). *Phys. Rev. B* **14**, 1593.

Scofield, J. H. (1976). *J. Electron Spectroscopy* **8**, 129.

Scott, B. A., Brodsky, M. H., Green, D. C., Kirley, P. B., Plecenik, R. M., and Simonyi, E. E. (1980). *Appl. Phys. Lett.* **37**, 727.

Sébenne, C., Belmont, D., Guichar, G., and Balkanski, M. (1975). *Phys. Rev. B* **12**, 3280.

Shanks, H. R., Fang, C. J., Ley, L., Cardona, M., Demond, F. J., and Kalbitzer, S. (1980). *Phys. Status Solidi B* **100**, 43.

Shen, S. C., Fang, C. J., Cardona, M., and Genzel, L. (1980). *Phys. Rev. B* **22**, 2913.

Singh, J. (1981). *Phys. Rev. B* **23**, 4156.

Solomon, I., Dietl, T., and Kaplan, D. (1978). *J. Phys. Paris* **39**, 1241.

Spear, W. E., and LeComber, P. G. (1976). *Philos. Mag.* **33**, 935.

Spicer, W. E., and Donovan, T. M. (1970). *J. Non-Cryst. Solids* **2**, 66.

Thomas, J. H. (1980). *J. Vac. Sci. Technol.* **17**, 1306.

Tiedje, T., Cebulka, J. M., Morel, D. L., and Abeles, B. (1981). *Phys. Rev. Lett.* **46**, 1425.

von Roedern, B., Ley, L., and Cardona, M. (1977). *Phys. Rev. Lett.* **39**, 1576.

von Roedern, B., Ley, L., and Smith, F. W. (1978). *In* "The Physics of Semiconductors" (L. H. Wilson, ed.), p. 701. Inst. of Phys., London.

von Roedern, B., Ley, L., and Cardona, M. (1979a). *Solid State Commun.* **29**, 415.

von Roedern, B., Ley, L., Cardona, M., and Smith, F. W. (1979b). *Philos. Mag. B* **40**, 433.

Wagner, C. D. (1972). *Anal. Chem.* **44**, 1050.

Wagner, L. F., and Spicer, W. E. (1974). *Phys. Rev. B* **9**, 1512.

Weaire, D., and Thorpe, M. F. (1971a). *Phys. Rev. Lett.* **27,** 1581.

Weaire, D., and Thorpe, M. F. (1971b). *Phys. Rev. B* **4,** 2548, 3517.

Wiech, G., and Zöpf, E. (1973). *In* "Band Structure Spectroscopy of Metals and Alloys" (D. J. Fabian and L. M. Watson, eds.), p. 637. Academic Press, New York.

Williams, R. H., Varma, R. R., Spear, W. E., and LeComber, P. G. (1979). *J. Phys. C* **12,** L209.

Yonezawa, F., and Cohen, M. H. (1981). *In* "Fundamental Physics of Amorphous Semiconductors" (F. Yonezawa, ed.), p. 11. Springer-Verlag, Berlin, Heidelberg, and New York.

Index

Contents of Previous Volumes